Sustainable
Steel Buildings

Sustainable Steel Buildings

A Practical Guide for Structures and Envelopes

Edited by

Bernhard Hauke
bauforumstahl, Düsseldorf, Germany

Markus Kuhnhenne
RWTH Aachen University, Aachen, Germany

Mark Lawson
The Steel Construction Institute, Ascot
and The University of Surrey, Guildford, UK

Milan Veljkovic
Delft University of Technology, Delft, The Netherlands

In collaboration with Raban Siebers
bauforumstahl, Düsseldorf and University of Duisburg-Essen, Essen, Germany

WILEY Blackwell

This edition first published 2016

Registered Office
John Wiley & Sons, Ltd, The Atrium, Southern Gate, Chichester, West Sussex, PO19 8SQ, United Kingdom.

Editorial Offices
9600 Garsington Road, Oxford, OX4 2DQ, United Kingdom.
The Atrium, Southern Gate, Chichester, West Sussex, PO19 8SQ, United Kingdom.

For details of our global editorial offices, for customer services and for information about how to apply for permission to reuse the copyright material in this book please see our website at www.wiley.com/wiley-blackwell.

Library of Congress Cataloging-in-Publication Data

Names: Hauke, Bernhard, editor. | Kuhnhenne, Markus, editor. | Lawson, R. M., editor. | Veljkovic, Milan, editor.
Title: Sustainable steel buildings : a practical guide for structures and envelopes / edited by Bernhard Hauke, Markus Kuhnhenne, Mark Lawson, Milan Veljkovic.
Description: Chichester, UK ; Hoboken, NJ : John Wiley & Sons, 2016. | Includes bibliographical references and index.
Identifiers: LCCN 2016012477 (print) | LCCN 2016016530 (ebook) | ISBN 9781118741115 (cloth) | ISBN 9781118740798 (pdf) | ISBN 9781118740811 (epub)
Subjects: LCSH: Building, Iron and steel. | Sustainable buildings–Materials.
Classification: LCC TH1611 .S87 2016 (print) | LCC TH1611 (ebook) | DDC 693/.71–dc23
LC record available at https://lccn.loc.gov/2016012477

A catalogue record for this book is available from the British Library.

Wiley also publishes its books in a variety of electronic formats. Some content that appears in print may not be available in electronic books.

Cover image: *"Diamond" Tower - Milan, Italy*
Client: HINES Italia SGR spa on behalf of Fondo Porta Nuova Varesine
Architect: Kohn Pedersen Fox Associates Pc
Structural design: ARUP
Contractors: CMB, Unieco
Steel Builder: Stahlbau Pichler srl
Picture: Oskar Da Riz – Stahlbau Pichler srl

Set in 10/12.5pt Minion by SPi Global, Pondicherry, India
Printed and bound in Malaysia by Vivar Printing Sdn Bhd

1 2016

Contents

List of contributors

Joana Andrade
University of Minho, Civil Engineering Department
Guimarães, Portugal

Anne Bechtel
Leibniz University Hannover, Institute for Steel Construction
Hannover, Germany

Jörn Berg
University of Duisburg-Essen, Institute for Metal and Lightweight Structures
Essen, Germany

Henning Bloech
Henning Bloech Sustainability Consulting
Göttingen, Germany

Jan Bollen
ArcelorMittal, Environment & Global CO2 Strategy
Brussels, Belgium

Luis Braganca
University of Minho, Civil Engineering Department
Guimarães, Portugal

Matthias Brieden
RWTH Aachen University, Institute of Steel Construction
Aachen, Germany

Tanja Brockmann
Federal Institute for Research on Building, Urban Affairs and Spatial Developement (BBSR)
Berlin, Germany

Martin Classen
RWTH Aachen University, Institute of Structural Concrete
Aachen, Germany

Murray Cook
European General Galvanizers Association
Birmingham, UK

Laure Delaporte
ConstruirAcier
Puteaux, France

Jan-Pieter den Hollander
Bouwen met Staal
Zoetermeer, The Netherlands

Bernd Döring
FH Aachen University of Applied Sciences, Faculty of Civil Engineering, subject area Building Services
Aachen, Germany

Christian Donath
RKDS & Partners
Essen, Germany

Johann Eisele
Technical University Darmstadt, Department of Architecture
Darmstadt, Germany

Lorenz Erfurth
Trimble
Eschborn, Germany

Markus Feldmann
RWTH Aachen University, Institute of Steel Construction
Aachen, Germany

Diana Fischer
Ingenieurbüro Fischer
Krefeld, Germany

Patric Fischli-Boson
SZS Stahlbau Zentrum Schweiz
Zürich, Switzerland

Fondazione Promozione Acciaio
Milan, Italy

Rolf Frischknecht
treeze Ltd.
Uster, Switzerland

Helena Gervasio
University of Coimbra, Departamento de Engenharia Civil – FCTUC
Coimbra, Portugal

Holger Glinde
Institut Feuerverzinken GmbH
Düsseldorf, Germany

Christopher Hagmann
University of Stuttgart, Institute for Construction Economics
Stuttgart, Germany

Bernhard Hauke
bauforumstahl
Düsseldorf, Germany

Josef Hegger
RWTH Aachen University, Institute of Structural Concrete
Aachen, Germany

Annette Hillebrandt
University of Wuppertal, Department of Building Construction | Design | Material Science
Wuppertal, Germany

Li Huang
Technical University of Munich, Chair of Metal Structures
Munich, Germany

Michael Huhn
Huhn IT Consulting
Karlsruhe, Germany

Kai Kahles
International Association for Metal Building Envelopes
Krefeld, Germany

Billie Kaufman
Trimble
Eschborn, Germany

Jyrki Kesti
Ruukki Construction Oy, Technology Center
Hämeenlinna, Finland

Thomas Kleist
GREYDOT engineering consultants
Düsseldorf, Germany

Ronald Kocker
bauforumstahl
Düsseldorf, Germany

Johannes Kreißig
German Sustainable Building Council - DGNB GmbH
Stuttgart, Germany

Markus Kuhnhenne
RWTH Aachen University, Institute of Steel Construction
Aachen, Germany

Simone Lakenbrink
DIFNI GmbH & Co. KG
Frankfurt, Germany

Mark Lawson
The Steel Construction Institute
Ascot, UK
University of Surrey
Guildford, UK

Burkhart Lehmann
Institut Bauen und Umwelt e.V. (IBU)
Berlin, Germany

Johan Löw
The Swedish Institute of Steel Construction
Stockholm, Sweden

Ricardo Mateus
University of Minho, Civil Engineering Department
Guimarães, Portugal

Marc May
ArcelorMittal Europe – Long Products
Esch-sur-Alzette, Luxembourg

Martin Mensinger
Technical University of Munich, Chair of Metal Structures
Munich, Germany

Lamia Messari-Becker
University of Siegen, Institute of Building Technology and Building Physics
Siegen, Germany

Quentin Olbrechts
Philippe Samyn and Partners
Brussels, Belgium

Gerry O'Sullivan
Mulcahy McDonagh & Partners Ltd.
Dublin, Ireland

Chris Oudshoorn
Slimline Buildings B.V.
Rotterdam, the Netherlands

Michael Overs
AkzoNobel – International Paint
Hamburg, Germany

Selcuk Ozdil
ArcelorMittal
Istanbul, Turkey

Dominik Pyschny
RWTH Aachen University, Institute of Steel Construction
Aachen, Germany

Vitali Reger
RWTH Aachen University, Institute of Steel Construction
Aachen, Germany

Peter Schaumann
Leibniz University Hannover, Institute of Steel Construction
Hannover, Germany

Eva Schmincke
Thinkstep AG
Tübingen, Germany

Ingo Schrader
Ingo Schrader Architekt BDA
Berlin, Germany

Raban Siebers
bauforumstahl
Düsseldorf, Germany

Christian Stoy
University of Stuttgart, Institute for Construction Economics
Stuttgart, Germany

Natalie Stranghöner
University of Duisburg-Essen, Institute for Metal and Lightweight Structures
Essen, Germany

Richard Stroetmann
Technical University Dresden, Institute for Steel and Timber Construction
Dresden, Germany

Benjamin Trautmann
Technical University Darmstadt, Department of Architecture
Darmstadt, Germany

Thomas Ummenhofer
Karlsruhe Institute of Technology (KIT), Research Center for Steel,
Timber and Masonry
Karlsruhe, Germany

Jo van den Borre
Infosteel
Brussels, Belgium

Ger van der Zanden
Slimline Buildings B.V.
Rotterdam, The Netherlands

Milan Veljkovic
Delft University of Technology
Delft, The Netherlands

Franziska Wyss
treeze Ltd.
Uster, Switzerland

Tim Zinke
Karlsruhe Institute of Technology (KIT), Research Center for Steel,
Timber and Masonry
Karlsruhe, Germany

Preface

Sustainability has been established for over one decade in building construction. It has become a high priority and is well established amongst professionals and authorities involved in design and construction. So why would one need another book explaining the same again? This is required because detailed information on the sustainability credentials of steel as a construction material is scattered over a wide range of publications, reports or company data. Hence, this book focuses on design and construction of sustainable steel buildings, looking at steel as a construction material, on steel structures, and steel envelopes and illustrates all this with many practical examples. For this purpose, we have brought together European experts and professionals from various fields.

The book starts with general introductory information on the background of sustainable construction, followed by highlights of the legal and normative frame. A discussion of basic concepts of sustainability assessment, such as life-cycle thinking and environmental product information in general and for steel construction products specifically, is the next focus, followed by the methods and design tools to deliver sustainable steel buildings and construction in general. Topics and structural elements that are crucial for sustainable steel buildings are addressed at. This comprises, for example, topics such as flexibility, benefits of high strength steel or design for deconstruction, and hot-dip galvanising and fire-protective coatings. In addition, the efficient design of structures and elements is discussed, such as multistorey buildings, bridges, and renewable energy structures, as well as columns, beams, floor systems and envelopes. Various sustainability certification labels have been established. The major labels DGNB, LEED and BREEAM are introduced and the performance of steel in these certification schemes is explained. Finally, several examples of sustainable steel buildings and contemporary case studies provide further guidance to the practitioner.

This book would not have been possible without the numerous contributions of authors from all over Europe with widespread backgrounds such as architectural or structural engineering practice, sustainability, steel production, steel construction, academia or various kinds of associations. Further thanks go to my coeditors Markus Kuhnhenne, Mark Lawson and Milan Veljkovic for the fruitful discussions and good cooperation in preparing this book. I thank particularly Raban Siebers, who has done most of the actual work of bringing together all the different authors and contributions.

Sustainable Steel Buildings shows specifiers, contractors, building authorities, lecturers and students how steel can be used to deliver buildings and structures with a high level of inherent sustainability.

Bernhard Hauke
February 2016

Chapter 1

What does 'sustainable construction' mean? An overview

Sustainable construction is a relatively new subject with which many of those involved in planning and construction are not familiar. It has been covered in numerous technical papers, but few of them present specific measures for implementing sustainability in the building and construction industry. This publication aims to improve the information available to those working in the construction sector using examples and guidance on steel construction in particular. The background and basic principles of how to achieve sustainable construction are presented and dealt with in a clearly structured manner. This publication also aims to convey a comprehensive understanding of sustainability and identifies the opportunities and essentials that can result from sensible implementation of sustainable steel construction strategies. The latest developments in steel construction provide a means to measure the success of the building and construction industry.

1.1 INTRODUCTION

Diana Fischer, Bernhard Hauke, Luis Braganca, Joana Andrade and Ricardo Mateus

The term 'sustainable' was first used in forestry to convey the idea that only as many trees could be felled in a given time period as were capable of growing again during the same period. A definition of the term 'sustainability' that is common today in the context of society can be found in the Brundtland report of the United Nations, which was published in 1987: 'Sustainable development is development that meets the needs of the present without compromising the ability of future generations to meet their own needs' [1]. These needs can be of an ecological, economic or social nature. A development or action is only sustainable if a minimum level of satisfaction is achieved in all areas and can be maintained in the future.

Sustainable Steel Buildings: A Practical Guide for Structures and Envelopes, First Edition.
Edited by Bernhard Hauke, Markus Kuhnhenne, Mark Lawson and Milan Veljkovic.

In 1992, the Earth Summit was held in Rio de Janeiro. It was an unprecedented event and attempted to establish sustainable development policies at a global scale. Among other documents, Agenda 21 was born during this conference [2]. It sought to move the interpretation of the sustainable development concept from just environmental protection to improvement of life quality and well-being, generation equity, ethics and healthy conditions [3].

Twenty years later, a new summit took place in Rio – Rio +20 Conference. The two main themes discussed were (1) a green economy in the context of sustainable development and poverty eradication; and (2) the institutional framework for sustainable development. Although still concerned with environmental and economic issues, this summit concluded that eradicating poverty is the greatest global challenge nowadays.

A shift in how sustainable development is seen is apparent. It started only as an environmental concern, and currently the social aspects of sustainability are highlighted. This shows the importance of going beyond environmental protection and considering also both the economic and social aspects. It implies that environmental protection is linked to maintaining and improving equity of the present and future generations, as follows: Sustainable development should be promoted by '*sustained, inclusive and equitable economic growth, creating greater opportunities for all, reducing inequalities, raising basic standards of living, fostering equitable social development and inclusion, and promoting integrated and sustainable management of natural resources and ecosystems that supports, inter alia, economic, social and human development while facilitating ecosystem conservation, regeneration and restoration and resilience in the face of new and emerging challenges*' [4].

Thus, the sustainability concept is based on the interrelation of three fields: environment, society and economy. A sustainable model should stimulate and pursue agreement and equality among the three (Figure 1.1).

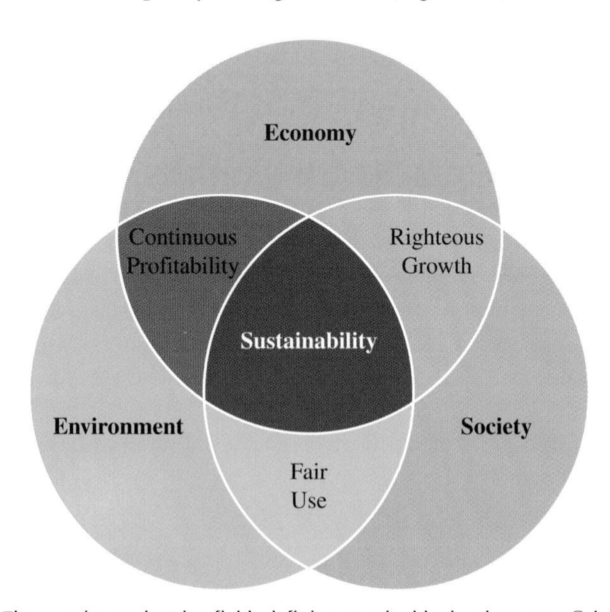

Figure 1.1 Three main overlapping fields defining sustainable development. © bauforumstahl.

1.1.1 The influence of the building sector

The building sector's influence on the above-mentioned problems is often underestimated. In 2013, €1162 billion was invested in construction in the countries of the European Union (EU-28). At the same time, the building sector was responsible for 8.8% of the EU-28 gross domestic product (GDP), providing 29% of the industrial employment and representing 6.4% of the total employment in Europe [5]. From the environmental side, the construction sector is responsible for 34.2% of the total waste produced in EU-28 in 2010 (851.6 million tonnes) [6]. In 2012, it was responsible for 11.7% of the greenhouse gases emission in EU-28 and accounts for approximately 47% of raw materials extraction. Besides economic and environmental impacts, the construction industry plays a major role in society. The employment of millions of world citizens depends directly and indirectly on construction. Buildings, roads, bridges and even water and energy infrastructures are all products from this industry. Buildings have a major influence on people's lives and well-being. In the past 60 years the world population has doubled, and most of our lives are spent inside buildings of all types.

Taking a closer look at buildings, their impact on people's lives is considerable. Data from the World Health Organization confirms that 90% of a person's lifetime is spent inside buildings [7]. With the current patterns, the expansion of the built environment will affect the natural habitats on more than 70% of earth's land by 2032 [8]. The economic influence of the property sector has also increased. Properties are now closely linked to the global finance markets via funds and credit guarantees. The last financial crisis showed the macroeconomic impacts that property can have. This clearly demonstrates that acting responsibly in the building sector can also result in an important contribution to preservation of the environment and conservation of resources as well as to economic efficiency.

This background data shows the influence of construction on the three pillars of sustainability. Charles Kibert defended this importance during the first international conference on sustainable construction in Tampa in 1994. He introduced the concept of 'sustainable construction' as being '*the creation and the responsible management of a healthy built environment based on resource efficient and ecological principles*' [9]. He highlighted the need for a life-cycle approach considering the impacts from the raw materials' extraction to the building's demolition [10]. With this holistic view, the following principles to achieve construction sustainability can be defined:

- efficiently use resources to avoid depletion of raw materials (energy, water and soil);
- protect ecosystems (waste, emissions, pollutants, land use);
- recycle materials in their end of life and use recyclable resources;
- eliminate hazardous products;
- minimize costs over the entire building's life-cycle;
- promote health, safety and well-being conditions for the inhabitants, neighbours and workers.

The growing shortage of resources and the high levels of emissions and waste production were the motivation for promoting sustainable construction more strongly. The building fabric plays a key role with regard to the primary energy consumption and global warming potential of buildings because it strongly affects the energy consumption that occurs during the building's life. Statutory requirements in the form of energy-saving ordinances and thermal efficiency of the building fabric and building services lead to lower energy use during the lifetime of a building. The choice of construction materials used is more important because they are increasingly also impacting the ecological quality of construction during its life cycle. Three questions particularly relevant here are: (1) *What environmental impacts occur during production?* As a matter of principle, materials and products with a small ecological footprint should be used as much as possible; specifically, construction materials should consume little energy and water during their production and should not contribute to emissions over their life and when dismantled at the end of their life (see below). (2) *How much 'construction' can be achieved using a unit of the product, in other words how efficiently can the product be used?* It is not sufficient to merely compare eco-relevant indicators in order to determine the ecological significance of a construction product.

It is also necessary to check how much functionality the use of a construction product offers, for example, how much useful area can be achieved using a kilogram of a construction material or what energy savings thermal insulation brings during the course of the life cycle. Here, comparisons are mostly only possible at building level; for example, lighter construction allows a certain method of construction and reduced foundation sizes. Environmental performance indicators can only be used to compare construction materials directly if the choice is between products of a similar type from different manufacturers. (3) *What happens to the construction material if the building is dismantled? Reuse, recycling or disposal site?* The question of whether a construction product can be reused or has to be disposed of after dismantling plays a decisive role in a sustainable – in other words, a future viable – approach. If products can be reused or recycled without any loss in quality they are available to be used by future generations. If they are disposed of as waste, they have to be replaced by primary resources. Reuse or recycling of materials and components also reduces the quantity of waste requiring disposal.

The building sector is therefore an important player in pursuing the sustainability goals. As one of the broad stakeholders, the building sector comprises the professionals involved in the building design, construction, maintenance and demolition (such as designers, engineers, urban planners, contractors, suppliers, manufacturers, etc.), decision makers (regulatory agencies at a local, national and international level, project developers and owners, etc.), and finally users and neighbours. In order to achieve the desired objectives, an integrated cooperation between all stakeholders is necessary. Only with having the will of all involved is it possible to change the way buildings are built and used to move towards a sustainable environment. The decisions made at building level are important but not sufficient, while the commitment of higher levels, such as regulatory agencies, is fundamental to achieving sustainability at building level. This imposes new challenges to each of the stakeholders, as presented in Table 1.1.

Table 1.1 Construction challenges for stakeholders to achieve sustainability in the built environment [11].

Stakeholder	Action
Authorities	Financial incentives
	Regulation and standards
	Effective labelling for consumers
	Research in improved construction practice
Clients, owners, developers and investors	Set appropriate and achievable environmental, social and economic targets
Users	Perform their own activities in a sustainable manner
	Operate the building and make environmentally friendly choices
Designers	Adopt an integrated design approach
	Make environmental improvements of the building fabric in the building design
	Adopt a 'life-cycle thinking'
Industry	Promote use of products with lower environmental impacts
	Promote recycling of materials and reuse, if possible
	Develop more efficient and less environmentally harmful products
Contractors and maintenance organizations	Reduce environmental impacts through better procedures
	Take environmental consciousness as a competitive factor
	Select partners based on their sustainable practices and standards

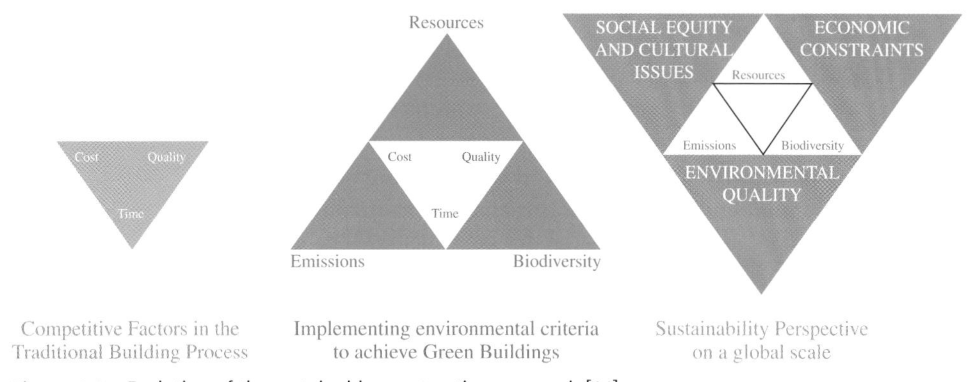

Figure 1.2 Evolution of the sustainable construction approach [14].

To reach this goal, a new approach to the way construction is thought of has recently been implemented (Figure 1.2). In the past, the planning of buildings concentrated primarily on construction costs and on the construction time as a measure of the return on investment. With greater environmental awareness, concerns about resources exploitation, pollution (emissions to air, soil and water) and the degradation of ecosystems biodiversity are increasingly considered in the construction approach. With a sustainability perspective, social and economic concerns on a global scale need to be added to the construction pyramid, as illustrated in Figure 1.2.

1.1.2 Can we afford sustainability?

When planned and designed well, projects can achieve a basic level of sustainability with little to no additional cost. However, society in general does not recognize the benefits of sustainable construction and does not understand the potential higher capital cost implications, thinking only of the initial cost.

A wider vision of the problem is required, as in construction there are two main types of costs: the initial cost required to build and the operational cost of the building. The life-cycle cost is the key parameter, together with labelling the resulting products as sustainable. Cost strategies, programme management and environmental strategies should be integrated into the design process right from the start.

Furthermore, a developer should keep in mind that sustainable construction may have positive economic effects by adding value to the building, as a sustainable image sells well. Furthermore, implementing criteria for sustainability at an early stage can result in short and medium term cost savings.

1.1.3 How can we achieve sustainability in the building sector?

In order to achieve a sustainable built environment, there is the need to work to achieve a balance between the three fundamental dimensions of sustainability – environment, society and economy. Unfortunately, there is no global formula to do that, in particular because societal concerns and economic status vary widely across the world. Having the same building under different conditions and trying to consider it as sustainable would not be sensible. Therefore, specific circumstances have to be considered when planning has to achieve sustainable design in the building sector, as good design is fundamental to sustainable construction. Decisions made at the initial design stage have the greatest effect on the overall sustainability impact of the construction project as well as over the lifetime of the building.

It is important to recognize that sustainability should be achieved in the context of existing buildings and cities, and it is hardly possible to build a sustainable city from zero. Nor would it make sense to demolish existing buildings, most of which are not considered sustainable, and build up a whole new set of sustainable buildings and localities. In this sense, rehabilitation of existent building stock is a cornerstone of achieving sustainability. Keeping existing buildings avoids unnecessary material use, saves on land use and preserves cultural identity and heritage. Rehabilitation projects should follow the principles of a sustainable design, as if a new building is being created. For instance, if structural components are in good condition and can be retained, this will reduce the building's life-cycle environmental impact, as the use of new materials is avoided.

There are some basics that can help during the planning process of sustainable buildings, such as early determination of the basic goals, integral planning that includes the whole life cycle of a building, and good-quality management. What this means and how the theoretical foundations can be applied to steel construction will be explained in Chapter 4.

1.2 AIMS OF SUSTAINABLE CONSTRUCTION

Diana Fischer, Bernhard Hauke, Luis Braganca, Joana Andrade and Ricardo Mateus

Sustainable construction is the process of creating buildings of high quality from ecological, economic and social points of view. Various measures are presented in Figure 1.3.

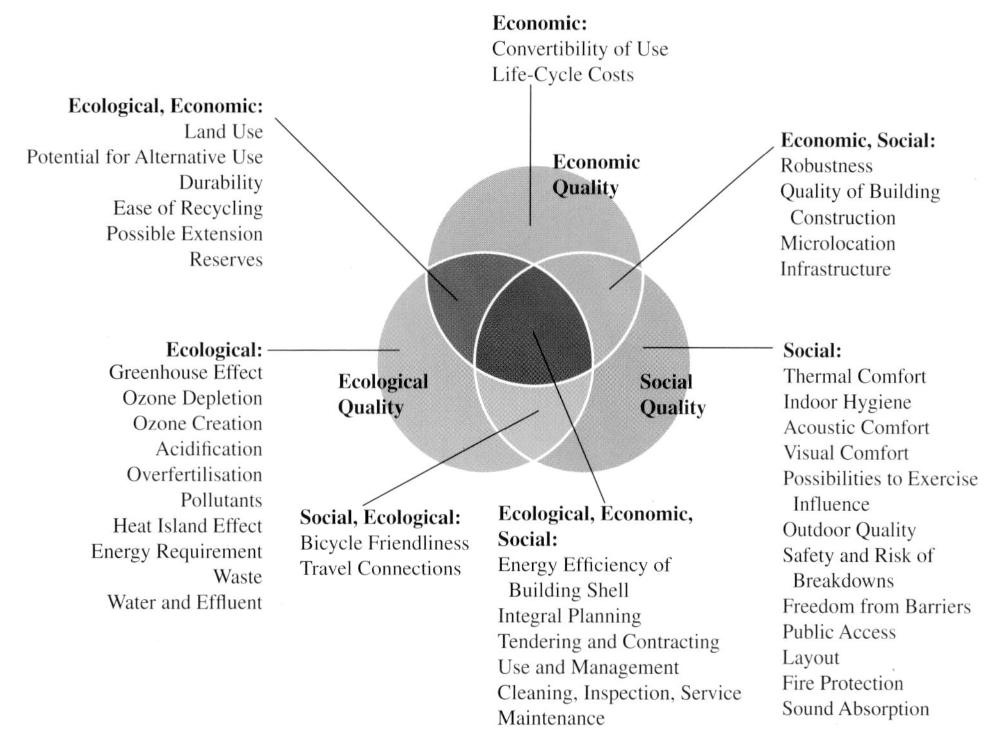

Figure 1.3 Illustration of sustainable buildings and their demands with respect to their ecological, social and economic quality. © bauforumstahl.

1.2.1 Ecological aims

1.2.1.1 Energy efficiency

Planning buildings that have high levels of moisture protection and thermal insulation has been standard practice in many European countries for years. Following the introduction of thermal insulation ordinances in the 1970s, planners' knowledge of building physics has improved and so has the quality of the thermal insulation used. However, given the improvements in products, higher energy prices and stricter statutory requirements, it is increasingly important to optimise the whole building

Docks Malraux

Location:	Strasbourg, France
Architect:	Heintz–Kehr Architectes
Building description:	Mixed-use development, culture, dwellings. Refurbishment/renovation of an old warehouse.
Steel details:	Replacement of the old tile roof by a three-storeys-high superstructure. A steel exo-skeleton of 800 t now shelters 67 prestige apartments.
Sustainability:	A high thermic performance was achieved – 20% lower than the 'BBC' standards (Batiment Basse Consommation) = 52 kWh/m^2/y
Awards:	Docks Malraux is selected in the 100 buildings of the year 2014 in AMC-Magazin. The project is the winner of 2015 steel architecture Eiffel Trophy (Learning category).

Figure 1 On the northern facade, fire escapes are externalized, connecting the different uses with some structural acrobatics. © Heintz–Kehr Architectes.

Figure 2 The old Seegmuller warehouse, built in 1932, was refurbished and enlarged by a three-storey steel structure. © ConstruirAcier.

Example provided by ConstruirAcier.

concept. Practical experience has shown that use of intelligent concepts makes it possible to achieve energy savings of up to 60% compared with conventional buildings.

In addition to the energy requirement during the use of a building, consideration is also increasingly given to the 'grey' energy tied up in the construction materials – in other words, the energy that is used in the production of the materials themselves. From an energy point of view, a well-planned building is characterised by the fact that it fulfils the demands made with respect to economy, comfort and the health of the user with the lowest possible total energy requirement over the whole life cycle, which includes production, use, deconstruction and eventual disposal.

1.2.1.2 *Resource efficiency*

Using resources carefully also implies that consideration is given to use of recycled materials in manufacturing of construction products as well as to the recovery and handling of materials in the post-use phase of buildings. Reuse and recycling are important aspects of resource efficiency because they contribute to reducing the use of primary raw materials. Reusable and recyclable materials are also available for future generations. At the same time, they also contribute to reducing the amount of waste produced, avoiding inefficient burning and dumping of nonrecoverable materials. Sensible life-cycle management therefore makes a double contribution to reducing emissions: the volume of waste is reduced and the effort expended in the mining of new raw materials and making products again can be avoided.

1.2.1.3 *Reduction of emissions*

In addition to the efficient use of resources, noxious emissions have an important influence on the ecological quality of a construction. Here, too, it is a case of considering the whole life cycle of a particular construction, including the production and disposal of the materials used. The focus of politics is in particular on reducing environmentally damaging greenhouse gases, of which carbon dioxide accounts for the major share of about 75%. Other emissions such as sulphur dioxide, which causes acid rain, or fluorinated hydrocarbons (FCKW), which damage the ozone layer, should also be avoided.

1.2.2 Social aims

In the building sector, social quality covers very different aspects. According to EN 15643 [12], the social concerns applicable to sustainable buildings inter alia

- accessibility;
- adaptability;
- maintenance;
- health and comfort;
- impact on the neighbourhood;
- safety/security;
- stakeholder involvement.

Accessibility is the ability of a space to be entered with ease, including provisions to facilitate access to and use of its facilities such as building services particularly for the physically disabled, elderly and parents with small children.

Adaptability is the ability of the building or its parts to be changed or modified to make it suitable for a particular use. Together with adaptability is the *robustness* of the building's structure, which is the capacity to resist disproportionate or progressive collapse from a natural or manmade hazard. Robustness is somehow considered across codes and standards, but adaptability is not. Also, *space efficiency* is a key aspect of social impacts of buildings, concerning the utilization of floor space inside buildings and the suitability to the function it was designed for.

The way *maintenance* operations are managed and performed is also a topic for social concerns as part of sustainability. The consequences for users and *neighbourhood* should be accounted for, and their importance to maintain the building's technical performance has to be considered. It is an expression of the quality of the building design, its construction, the maintainability of its structure, surfaces and services, and the quality of the maintenance plan.

Health and comfort accounts for (1) acoustic comfort, (2) visual comfort, (3) indoor air quality, and (4) thermal comfort. A building should provide healthy and acoustically acceptable comfort conditions to its inhabitants. An acoustically comfortable environment improves productivity and well-being. *Visual comfort* regards the indoor lighting, which should provide the right amount of light in the right place. This allows building users to perform their tasks efficiently without strain or fatigue. Good indoor lighting enhances the appearance of a space and provides a pleasant working environment or attractive leisure area.

Indoor air quality (IAQ) is one of the most important factors in a building's performance influencing directly the health of the building users; otherwise the building would not satisfy its occupants. IAQ affects their comfort and the ability to conduct activities. There are many sources of indoor air pollution in a building, such as (1) microbial contaminants (mold, bacteria), (2) gases (including carbon monoxide, radon and volatile organic compounds), (3) particulates, or (4) any mass or energy stressor that can induce adverse health conditions. There are immediate effects that may occur after a single exposure or repeated exposures – such as irritation of the eyes, nose, and throat, headaches, dizziness, and fatigue, and long-term effects that may show up either years after exposure has occurred or only after long or repeated periods of exposure, such as respiratory diseases, heart disease or cancer [13]. Using ventilation to dilute contaminants, filtration and source control are the primary methods for improving indoor air quality in most buildings.

Thermal comfort aims to provide a comfortable thermal environment inside the building both in summer and winter conditions. A pleasant temperature inside buildings promotes productivity and well-being of occupants. As is well known each person has his or her own thermal sensations, and so it is the designer's role to provide average conditions for comfort within which occupants will adapt.

Safety and security concerns the capacity of a building to resist projected current and future loadings from, for example, rain, heavy wind, snow, flooding, fire, earthquake, explosion and landslides, as well as to provide security from external sources of disruption of utility supply. It is a measure of the building's ability to provide safe and secure shelter during exceptional events that have a potential impact on the safety of its users and occupants, and the building's ability to maintain its function and appearance and to minimise any disruption as a result of these exceptional events.

1.2.3 Economic aims

The integration of economics into a holistic approach to sustainability is an important aspect. Social and ecological goals should be achieved at as small of a cost as possible. To evaluate this, it is necessary to consider both the expenditure and the possible income due to improved sustainability. On the expenditure side, the total costs that a building incurs over its whole life cycle, in other words from the fabrication of the building through to its end of life with recovery or disposal, are taken into account. Income is usually more difficult to estimate. The basic prerequisite for future income is the market value of the building. In a way, the extent to which a building holds its value is an indication of the sustainability of an investment. This sustainability depends to some extent on both the building itself, such as the durability of the materials and facilities, and external changes, such as user demand and the unit cost of energy. Considering and assessing buildings over the whole life cycle provides a useful tool for risk management. The aim of a sustainable construction method is to minimise total cost and for as-built construction to hold its value and to meet social and ecological goals. A building with a high degree of flexibility and convertibility of use can be adapted to meet changing social demands. This leads to the property having a prolonged life cycle, which is beneficial economically and ecologically.

REFERENCES

[1] United Nations. (1987) Our Common Future - Brundtland Report. Oxford: Oxford University Press.

[2] UN. (1992) *Rio Declaration on Environment and Development.* Rio de Janeiro: United Nations.

[3] UN. (1992) *Agenda 21: Earth Summit – The United Nations Programme of Action from Rio.* Rio de Janeiro: United Nations.

[4] UN. (2012) The future we want. In: *Report of the United Nations Conference on Sustainable Development.* New York: United Nations.

[5] FIEC. (2014) *Annual Report 2014.* Brussels: European Construction Industry Federation.

[6] European Union. (2014) *Generation of Waste by Economic Activity 1995–2014.* Brussels: Eurostat.

[7] Burgan, B.A., Sansom, M.R. (2006) Sustainable steel construction. *Journal of Constructional Steel Research* 62(11), pp. 1178–1183.

[8] UNEP. (2003) Sustainable building and construction: Facts and figures. In: *UNEP Industry and Environment.* United Nations Environment Programme Division of Technology, Industry and Economics, pp. 5–8.

[9] Kibert, C.J. (1994) Establishing principles and a model for sustainable construction. Proceedings of the First International Conference of CIB TG 16 on Sustainable Construction, Tampa, Florida.

[10] Bragança, L., Mateus, R., Koukkari, H. (2007) Perspectives of building sustainability assessment. In: *Portugal SB07 – Sustainable Construction, Materials and Practices: Challenge of the Industry for the New Millenium 2007.* Amsterdam: IOS Press, pp. 356–365.

[11] Gervásio, H. (2010) Sustainable design and integral life-cycle analysis of bridges. PhD diss., Institute for Sustainability and Innovation in Structural Engineering, Departamento de Engenharia Civil Faculdade de Ciências e Tecnologia da Universidade de Coimbra.

[12] EN 15643-3. (2012) *Sustainability of Construction Works – Assessment of Buildings – Part 3: Framework for the Assessment of Social Performance.* Brussels: European Committee for Standardization.

[13] Perfection Partners. (2011) *PERFECTION – Performance Indicators for Health, Comfort and Safety of the Indoor Environment.* FP7 EU Project Grant number 212998.

[14] Mateus R., Bragança L. (2015) Tecnologias Construtivas para a Sustentabilidade da Construção - eBook. Publindustria: Porto.

Chapter 2

Legal background and codes in Europe

Eva Schmincke and Jan Bollen

In December 2007, the European Commission officially declared that sustainable construction and the closely related recycling industries should be lead markets in Europe. The six lead markets also include eHealth, protective textiles, bio-based products and renewable energies. The aim of the lead market initiative is to purposefully develop innovation-friendly markets and to facilitate the market entrance of innovations.

According to the European Commission, the importance of the sustainable construction market is due to the fact that 42% of energy consumption in the EU is attributable to buildings. Furthermore, buildings are also responsible for about 35% of all greenhouse gas emissions. Sustainable construction represents a broad market sector in which environmental aspects (e.g. efficient electrical equipment and heating systems), health issues (e.g. air quality in buildings) and user comfort (e.g. accessibility for older people) all play a role. In addition, regulations for the building and construction industry were previously not well coordinated at the European and national level, which led to a fragmentation of the market for sustainable construction.

In the field of sustainable construction, the European Commission wants to extend the area of application of the overall efficiency of buildings by introducing EU-wide energy efficiency targets for new and refurbished buildings and initiating the development of European standards that consider and promote sustainable construction. The tasks and objectives of sustainable construction have been collated in an action plan (Action Plan for Sustainable Construction). Legislation, public procurement, standardisation, product labelling and certification are cited as political instruments for implementing the tasks at the building level.

The standards, legislation, guidelines and regulations discussed below are crucial for the building and construction industry and serve as rules or guidance for sustainable construction.

Sustainable Steel Buildings: A Practical Guide for Structures and Envelopes, First Edition.
Edited by Bernhard Hauke, Markus Kuhnhenne, Mark Lawson and Milan Veljkovic.

2.1 NORMATIVE BACKGROUND

In the construction sector there are standards that deal with the subject of sustainability, and various standards relating to sustainable construction have been published or are in preparation. Standards are a very important basis for construction and a short introduction to the standardisation activities will be given here. Reference will be made to important standards for sustainable construction and the relationships between them.

It is necessary to differentiate between the international, European and national committees for standardisation: the International Organisation for Standardization (ISO) is the global association of standards organisations. The European committee for standardisation (Comité Européen de Normalisation, CEN) prepares European standards (EN). The national institutions – for example, the German standards institute (Deutsche Institut für Normung, DIN) – represent individual country standardisation activities within CEN and ISO. CEN was formed in order to harmonise European standardisation activities, CEN standards therefore have to be adopted by all 30 countries represented in CEN. The standards are then introduced as 'DIN EN' standards in Germany or accordingly in any other European country. ISO standards can also be adopted as national standards directly or indirectly. They are then labelled as, for example, DIN EN ISO or DIN ISO in Germany. The preparation of a standard is a lengthy process of the strict procedures applicable and the challenge to find a consensus among various stakeholders. Thus despite the subject having been dealt with intensively by ISO and CEN, only a few standards for assessing the sustainability of buildings have actually been adopted so far. The planned future normative background for sustainable construction is shown in Table 2.1.

All the above-listed standards aim to define common principles for sustainability assessment of construction by defining suitable indicators and the methodology to calculate them. Developments are ongoing to have civil engineering works included in this framework (CEN/TC 350 WG6). The standards were initially focused on ecological performance. By including social and economic performance, the framework has been completed into a full sustainability assessment.

2.2 COMMENTS ON EN 15804 AND EN 15978

2.2.1 Modular life-cycle stages

EN 15804 [1] and EN 15978 [2] have been available since 2012, and in 2013 EN 15804+A1 was published, including an amendment. These specify the core rules for preparing environmental product declarations (EPDs) for construction products (EN 15804) and for the calculation of the environmental performance of buildings (EN 15978).

Both standards are closely interrelated insofar as they are based on a uniform modular presentation of the life cycle (Table 2.2). This means product environmental data from EPDs prepared in accordance with EN 15804 can be used

Table 2.1 Current and future normative background for sustainable construction.

<table>
<tr><td colspan="4">Sustainability in building construction
ISO 15392: Sustainability: Definition, goals and general principles for sustainability in building construction
ISO 12720: Guidelines for the application of the general principles on sustainability
ISO 21932: Terminology (in revision – adaption to EN 15804)</td></tr>
<tr><td colspan="3">Assessment of a building's sustainability
ISO 21929-1: Indicators: Definition, aspects to be considered when defining sustainability indicators, core set of indicators for assessing the sustainability performance of buildings
EN 15643-1: Goals of the assessment process, definition of a building life cycle, definition of the system boundaries of a building; ISO also describes criteria to be assessed
EN 15978: Instructions for general assessment and instructions for the assessment of the environmental performance, including substantial indicators</td><td rowspan="2">Sustainability of engineering structures
ISO 21929-2: Likely to be similar to 21929-1 but with special indicators for engineering structures</td></tr>
<tr><td>Environmental performance
EN 15978: See above
EN 15643-2: General principles and requirements for the assessment of environmental performance, including essential indicators
Environmental product declarations
ISO 14025: (not only for building products) Type III environmental declarations; integration of life-cycle assessment, general requirements, e.g. for the verification process
ISO DIS 21930:2016: (similar to 14025, but specifically for building products) purpose of declarations, information they should contain on building products, description of the building life-cycle phases, which documents must be drawn up during the preparation of environmental declarations of construction products and which information should be included
DIN CEN/TR 15941: Use of generic data (source of data and securing data quality)
EN 15804: Content and general validity of EPDs for construction products, e.g. determination of life-cycle phases and system boundaries for products, validity, ownership
EN 15942: communication format (rules for the structure and presentation)
Life-cycle assessment
ISO 14040: Description of the principles and framework for life-cycle assessment including basic definitions – background information for users of environmental product declarations
ISO 14044: Guidelines for compiling a life-cycle assessment including considerations and information from ISO 14040 – directed at life-cycle assessors</td><td>Social performance
EN 15643-3: Focus on service life, includes indicators for the assessment of social performance
EN 16309: Process for the assessment of social performance</td><td>Economic performance
EN 15643-4: Only describes the indicators 'intrinsic value' and 'life-cycle costs', so far; should later also provide details for the assessment process
EN 16627: Process for the assessment of economic performance</td></tr>
</table>

Table 2.2 Life-cycle stages of buildings and construction products for sustainable construction in accordance with EN 15804 and EN 15978.

Building Assessment Information				
Building Life-Cycle Information				
Product Stage	**Construction Process Stage**	**Use Stage**	**End-of-Life Stage** (Building)	**Benefits and Loads** Beyond Building Boundaries
A1: Raw Material Supply A2: Transport A3: Manufacturing	A4: Transport A5: Construction–Installation Process	B1: Use B2: Maintenance B3: Repair B4: Replacement B5: Refurbishment B6: Operational Energy Use B7: Operational Water Use	C1: Deconstruction C2: Transport C3: Waste Processing For Reuse, Recovery and Recycling C4: Disposal	D: Reuse–Recovery–Recycling Potential

directly to assess a building in accordance with EN 15978. The life cycle of a building thus starts with the extraction of the raw materials and the manufacture of the construction products, then covers the construction stage and finishes with demolition and waste management.

EN 15804 6.2.1: 'Only the declaration of the product stage modules, A1-A3, is required for compliance with this standard.' There is often no data available for the modules for the construction, use and disposal stages. At the end of a building's life, recyclable construction products, in this case steel products, are returned to the production loop. In order to completely describe the environmental impacts of a building, a fifth module should therefore be considered that describes, among other things, the credits and impacts resulting from recycling and reuse. This so-called Module D (Table 2.2) describes the recycling potential discussed in Chapter 3.11. The terms 'recycling' and 'reuse' are often used synonymously with steel and other materials. However, Module D considers all environmental benefits beyond the building system, such as exported renewable energy from integrated PV panels or potential of energy recovery, for example, from timber products.

2.2.2 Comparability of EPDs for construction products

EN 15804 5.3: 'In principle the comparison of products on the basis of their EPD is defined by the contribution they make to the environmental performance of the building. Consequently comparison of the environmental performance of construction products using the EPD information shall be based on the product's use in and its impacts on the building, and shall consider the complete life cycle (all information modules).' Comparisons between construction products – with or without an EPD – must always be conducted within the context of their application in the building. EPDs are therefore not suitable for comparing construction products outside the context of the building.

EPDs provide base data for assessing environmental performance. To assess the sustainability of buildings, it is also necessary to take social and economic performance into consideration. Suitable benchmarks are necessary to be able to actually assess the environmental performance of construction products within the context of the building – whereas EN 15804 or EN 15978 standards do not.

EN 15804 5.3: 'Comparisons are possible at the sub-building level, e.g. for assembled systems, components, products for one or more life cycle stages.' The complete building remains the basis for comparison for the description and assessment, because the same statutory and functional requirements of the principal have to be fulfilled. All assumptions and limitations - e.g. uncertainty of the data used - of such comparisons have to be presented transparently below the building level.

Comparisons between construction products or building materials based only on a declared unit of kg, m^3, etc. where the reference to function is missing are not acceptable.

2.2.3 Functional equivalent

EN 15978 7.2: 'Comparisons between the results of assessments of buildings or assembled systems (part of works) – at the design stage or whenever the results are used – shall be made only on the basis of their functional equivalency....If the assessment results based on different functional equivalents are used for comparisons, then the basis for comparison shall be made clear.' For comparisons at the level of building components, it is often not possible to avoid such comparisons, particularly at the development stage of buildings. Therefore comparisons have to be transparent and demonstrate functional equivalence in an appropriate manner. For example, for a single-storey building, when columns of different designs (articulated and fixed column base) are compared, the size of the foundation must be included in the comparison. The same applies to the comparison of beams of different materials and even span: the whole construction system has to be taken into consideration taking account of the resulting number and size of the columns and foundations.

2.2.4 Scenarios at product or building level

EN 15804 6.3.8: 'Scenarios shall support the calculation of information modules covering processes that deal with any one or all of the life cycle stages of the construction product except for the required modules A1 to A3....A scenario shall be realistic and representative of one of the most probable alternatives. (If there are, e.g. three different applications, the most representative one, or all three scenarios shall be declared.) Scenarios shall not include processes or procedures that are not in current use or which have not been demonstrated to be practical.

EN 15978 8.1: 'Scenarios for the building module D include information on reuse, recycling and energy recovery.'

The recycling system for steel scrap is an economic process that has existed for years. For structural steel (sections and plates), the current average recycling rate is close to 90%. There is a functioning market for steel scrap that is valuable economically. The reuse of steel construction products is possible in principle and is also practised to some extent. The rate of reuse is estimated as more than 10%, and there is still development potential. The greatest challenge is quality control and certification of the reused materials.

2.2.5 Reuse and recycling in module D

EN 15804 6.4.3.3: 'Module D recognises the "design for reuse, recycling and recovery" concept for buildings by indicating the potential benefits of avoided future use of primary materials and fuels while taking into account the loads associated with the recycling and recovery processes beyond the system boundary.' Design for recycling is also discussed in Chapter 4.3 in connection with recycling potential. Module D includes the benefit of avoiding the production of virgin steel. It also takes into account all impacts resulting from transport and subsequent processing of steel scrap after it was acquired by the steelworks. Collection, storage or sorting before the scrap is sold is considered the responsibility of the manufacturer producing the scrap, which therefore applies the polluter pays principle.

Only the net output flows are taken into consideration, that means, the input of scrap is deducted from the output flow that substitutes virgin steel. In the case more scrap is used in steel production than collected at end of life, a load is declared in Module D. The losses have to be replenished with virgin material in order to keep the mass balance in the product system.

In the case of downcycling when the output flow does not achieve the functional equivalence of the substitution process, a correction factor is used. Steel, and especially structural steel, can be recycled many times, with the quality of the steel remaining essentially unchanged. A correction factor is not necessary in this case.

EN 15804 6.4.3.3: 'Where a secondary material or fuel crosses the system boundary e.g. at the end-of-waste state and if it substitutes another material or fuel in the following product system, the potential benefits or avoided loads can be calculated based on a specified scenario which is consistent with any other scenario for waste processing and is based on current average technology or practice.'

EN 15978 7.4.6: 'Where a material flow exits the system boundary and has an economic value or has reached the end-of-waste stage and substitutes another product, then the impacts may be calculated and shall be based on: average existing technology, current practice, net impacts.'

This means that when calculating the benefits of avoided virgin production it is not possible to calculate, for example, the energy use from an outdated inefficient production process. The applied scenario must be made transparent in the report.

2.2.6 Aggregation of the information modules

EN 15804 7.5: 'The indicators declared in the individual information modules of a product life cycle A1 to A5, B1 to B7, C1 to C4 and module D as described in Figure 1 *shall not be added up in any combination of the individual information modules into a total or sub-total of the life cycle stages A, B, C or D. As an exception information modules A1, A2, and A3 may be aggregated.'* The primary concern here is transparency of the origin of the data and the corresponding transfer from the product level (EN 15804) to the building level (EN 15978). It is thus no longer permissible to add together modules A1–A3 and D, which used to be common for steel and other metals. However, by declaring Module D separately, the considerable positive contribution of steel recycling also becomes clear. Other building material sectors, which were initially rather critical of declaring Module D, have recognised this in the meantime as well.

EN 15978 12.6: 'Results shall be presented separately for all the building life cycle stages and for module D.' EN 15978 offers a method for calculating the environment-related performance of buildings. The principle of transparency also applies at the building level so that the results of the calculation have to be given separately for all modules. EN 15978 does not provide any benchmarks or method of evaluation for sustainable buildings. European building certification schemes, like BREEAM (United Kingdom), BNB and DGNB (Germany) and HQE (France) as well as the US scheme LEED do provide evaluation methods. For evaluation, the aggregation of the modules is common practice (e.g. DGNB), but the constitution of the modules must remain transparent. While the European schemes actually integrate the LCA results into their building assessment, LEED only grants credits for the fact of having made the calculations.

2.3 LEGAL FRAMEWORK

2.3.1 EU waste framework directive and waste management acts in European countries: product responsibility

The new EU Waste Framework Directive of 19 November 2008 formulates the long-term goal of having a circular economy without any waste and with high grade recycling und reuse. The first priority here are the products themselves, as these are far more important than recycling and recovery processes that already exist or are under development. Producers have to accept responsibility for their products, which also applies to construction products and processes.

Article 8 extended producer responsibility – *'1...any natural or legal person who professionally develops, manufactures, processes, treats, sells or imports products (producer of the product) has extended producer responsibility. Such measures may include an acceptance of returned products and of the waste that remains after those products have been used, as well as the subsequent management of the waste and financial responsibility for such activities. These measures may include the obligation to provide publicly available information as to the extent to which the product is re-usable and recyclable.'*

For example, in Germany, the new Waste Management Act (KrWG), whose legal obligations have been directly effective since 1 June 2012, implements the EU Waste Framework Directive in German law. Here the provisions are even more precise: §23 product responsibility, (1) Anyone who develops, manufactures, processes, treats or sells products shall have responsibility for fulfilling the goals of a resource-efficient life-cycle management of his products. Products should be designed to generate less waste during their manufacture and use and to ensure that the recovery or disposal of waste arising after their use causes less impact on the environment.

In particular, product responsibility includes

1 development, production and marketing of products that are suitable for multiple use, are technically durable and, after use, are suitable for proper, safe and high grade recovery and environmentally compatible disposal;
2 giving priority to use of recoverable waste or secondary raw materials in the manufacture of products;
3 labelling hazardous waste in order to ensure that waste remaining after use is recovered or disposed of in an environmentally compatible manner;
4 making reference via labelling of the products to the possible methods or obligations to return, reuse or recover and to deposit schemes;
5 accepting returned products and the waste that remains after the products have been used as well as their subsequent environmentally compatible recovery or disposal.

Additionally, a waste hierarchy is defined in Article 4 of the EU Waste Framework Directive and §6 of KrWG. It prioritises the measures for waste prevention and management as follows:

1 prevention;
2 preparing for reuse;
3 recycling;
4 other recovery, especially energy recovery and backfilling;
5 disposal.

These different waste routes are defined as follows:

- Reuse of products or components that have already been used in the same or a comparable function; KrWG §3(21): 'For the purpose of this act, reuse is any operation by which products or components that are not waste are used again for the same purpose for which they were conceived.'
- Recycling of secondary raw materials in the same material loop and enabling processing into products with similar or higher better quality (so-called recycling and upcycling); For example: standard steel product into advanced high strength steel product. KrWG §3(25): 'For the purpose of this act, recycling means any recovery operation by which waste materials are reprocessed into products, materials or substances whether for the original or other purposes; it includes the reprocessing of organic material but does not include energy recovery and the reprocessing into materials that are to be used as fuels or for backfilling operations.'

- Recovery of the material content of waste producing secondary (raw) materials for other functions. The original material quality is not reached – so-called down-cycling or 'open-loop'-recycling. (e.g; concrete crushed into road stone); KrWG §3(23): 'For the purpose of this act, recovery means any operation in the plant or in the wider economy the principal result of which is waste becoming – sometimes with further processing a secondary material which serves a useful purpose by replacing other materials which would otherwise have been produced to fulfil a particular function.'
- Recovery of the energy content of waste. Waste incineration with energy recovery is normally regarded as a disposal operation; KrWG §3(26): 'For the purpose of this act, disposal means any operation which is not recovery even when the operation has the reclamation of substances or energy as a secondary consequence.'

The goal is the prevention of waste by means of efficient product manufacturing and planning that is suitable for recycling and long lasting [KrWG §3(20)].

In addition, it is also of relevance to the construction and property industry that a recycling and reuse rate of 70% by 2020 is specified in the European Waste Framework Directive as well as in the German Waste Management Act. Unfortunately this target is not as progressive as it seems, because it includes also backfilling (downcycling) as part to reach the 70% target. To really achieve this objective and at the same time comply with the definitions of the waste hierarchy, building materials will have to be used efficiently and already selected today in such a way that they can be reused or recycled in as high end a manner as possible at the end of a building's life.

For steel construction products, published EPDs allow the recycling and reuse rates shown in Table 2.3 to be determined. It can be clearly seen that steel

Table 2.3 Average end-of-life scenarios for steel products.

	Average End-of-Life Scenario		
Product EPD	Recycling (per cent)	Reuse (per cent)	Collection Loss/ Disposal (per cent)
Bauforumstahl: Structural steel: open-rolled sections and heavy plate	88	11	1
Hot-dip galvanised structural steel: open-rolled sections and heavy plate [5]			
Akkon Steel Structure Systems Co.: light gauge steel profiles	70	0	30
IFBS: profiled steel sheeting for roofs, walls and ceiling constructions	90	0	10
Tata Steel: Colorcoat assessed cladding systems (trapezoidal profiled sheet for roofs and walls)	79	15	6
ThyssenKrupp Steel Europe AG: PLADUR, sheet, strip and single-skin construction products	90	0	10

and other metallic construction products vastly exceed the target of the Waste Management Act, which refers to both recycling and reuse. According to the waste hierarchy, the established system of recycling and reuse (of materials that are also valuable even as scrap) is ranked significantly higher than material recovery.

The collection rate of non-ferrous matals e.g. aluminium in the building sector can reach collection rates up to 98%.

2.3.2 EU construction products regulation

Council Directive 89/106/EEC relating to construction products came into force in Europe in 1988. The aim was to harmonise the different national technical regulations for construction products and to implement the free movement of goods within the EU.

The earlier directive has now been replaced by the new European Construction Products Regulation (EU 305/2011), which has been in force since July 2013, and construction products must now be in accordance with the new regulation. This applies above all to a declaration of performance and CE marking. A construction product may be covered by a harmonised standard or a European Technical Assessment that has been issued, but it must be accompanied by a declaration of performance in relation to the essential characteristics of the construction product in accordance with the applicable harmonised technical specifications. Thus the manufacturer assumes responsibility for the important characteristics of a construction product, which are specified in Annex ZA of the respective harmonised standard (EN 1090-1). Basic requirements for construction works that affect the important aspects of sustainability in many ways are given in Annex 1 of the Construction Products Regulation:

Basic requirement 1: Mechanical resistance and stability
Basic requirement 2: Safety in case of fire
Basic requirement 3: Hygiene, health and the environment
Basic requirement 4: Safety and accessibility in use
Basic requirement 5: Protection against noise
Basic requirement 6: Energy economy and heat retention
Basic requirement 7: Sustainable use of natural resources

Basic requirement 7 is new in the EU Construction Products Regulation: 'The construction works must be designed, built and demolished in such a way that the use of natural resources is sustainable and in particular ensure the reuse or recyclability of the construction works, their construction materials and parts after demolition.' Basically, steel structures and steel components always meet this requirement. Using long-span construction with open column-free areas and relatively small cross sections, steel structures are extremely economical, resource efficient and flexible. Steel buildings are capable of being adapted economically to another use, and so natural resources are used sustainably. By separating the supporting structure, facade and interior walls, the individual elements can be dismantled easily and replaced. The bolted joints commonly used in steel construction facilitate dismantling and provide an opportunity to reuse the components. Among the commonly used building materials, steel has an unsurpassed rate of recycling and reuse. EPDs prepared in accordance with ISO 14025 and EN 15804 provide the environmental impacts and energy consumptions for the specified building materials. Besides other product information, the impact on the environment during manufacturing and end of life including future potentials is presented (see Chapter 3.11). This makes it possible to use the EPDs to show that basic requirement 7 is fulfilled.

With respect to basic requirement 3, the EPD provides the possibility of transparency and thus showing compliance with respect to emissions to indoor air, soil and water during the use stage according to harmonised test provisions.

At present the standard committee CEN TC 350 is discussing if indicators for human and eco-toxicity, impacts from particles, radioactive radiation and impacts from land use should be added to the basket of indicators already given in EN 15978 and EN 15804. A technical report is expected soon that will provide the basis for a decision on such indicators. However, standardisation activity to adapt DIN EN 1090 to the Construction Products Regulation and thus to incorporate basic requirement 7 in Annex ZA is not expected to be completed before 2017.

2.3.3 EU building directive and energy saving ordinance

In May 2010, the European Parliament passed the new Energy Performance of Buildings Directive (EPBD). The implementation of the directive in Germany has to be adopted by amending the Energy Saving Ordinance (EnEV) and the Energy Conservation Act (EnEG). The revision of the EPBD makes an important contribution to achieving the EU's targets for energy consumption and climate protection. The overall target of the directives and ordinances is to achieve an almost climate-neutral stock of buildings by 2050. The new EnEV came into force in 2014. The changes regarding the requirements are listed below.

1. Greater access to energy performance data: The energy performance value of a property have to be published in any advertisements for commercial selling or renting. The energy performance certificate that hitherto had to be made available on request when re-letting property have to be handed over when finalising a sales contract or rental agreement. The energy performance certificate also have to include measures for comprehensive refurbishment as well as concrete proposals for 'smaller' modernisation tips in future. The obligation to prominently display energy performance certificates in public buildings is to be extended from buildings with a total useful area of over 1000 m^2 to those totalling 500 m^2 and in the long term even to those totalling 250 m^2.
2. Extension of possible controls: All member states are obliged to implement an independent control system for energy performance certificates and to randomly check the certificates awarded – certificates must then be presented to the responsible authorities upon request.

European Council and Council of the European Union

Location:	Brussels, Belgium
Architects and engineers:	Philippe Samyn and Partners architects & engineers, Lead and Design partner (with Studio Valle Progettazioni architects and Buro Happold engineers)
Building description:	Extension of the Residence Palace from 1927. On the north-east side two new facades transform its former 'L' shape into a cube. This outer area is converted into a glass atrium as protection from the urban dust. It covers the principal entrance as well as a new lantern-shaped volume incorporating the conference rooms.
Steel details:	Highly sophisticated steel structure for the inner and outer structural system.
Sustainability:	The council wishes this building to be from all points of view an example as far as sustainable development is concerned. This wish is displayed in many aspects of the architectural and technical design. As an example, an umbrella of photovoltaic panels for the electricity production covers both the new and the historical parts.

Figure 1 European Council and Council of the European Union. © Quentin Olbrechts.

Figure 2 Growing steel structure of the inner lantern-shaped volume. © Thierry Henrard.

Figure 3 The new double facade, made of a harmonised patchwork of reused old oak windows, provides the necessary acoustic barrier from the traffic noise and offers a first thermal insulation for the inner space. © Thierry Henrard.

Example provided by Philippe Samyn and Partners.

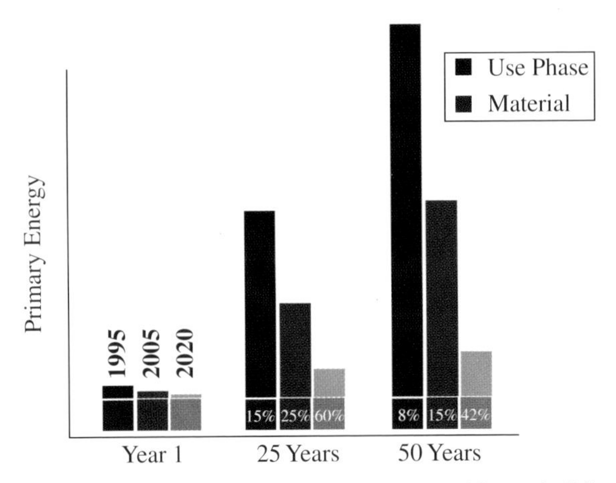

Figure 2.1 Effect of increasing energy efficiency of current and future buildings on the relationship of grey energy to the total energy used. Qualitative approach © bauforumstahl.

3 Increasing the energy efficiency of buildings and simplifying procedures for certain buildings: Only zero-energy new buildings (for heated or cooled buildings) will be permitted from 2021 onward – public buildings will already have to fulfil these standards from 2019 (exceptions are possible). The requirements for the energy efficiency of new buildings have been increased by 25% (primary energy requirement). Extensive refurbishment of building stock are also covered by the new regulations.

2.3.4 Focus increasingly on construction products

The so-called 'grey energy' contained in construction products, which is the energy needed to manufacture the construction products used, is important in achieving sustainable construction. However, it will become ever more important as the energy efficiency of buildings increases, whereas the operation of a building currently accounts for some 80% of a building's energy requirement today. The relative amount will decrease in future as a result of more thermal insulation measures and increased efficiency of household equipment, lighting, and so forth. The fraction of grey energy in the total energy needed during the life cycle of a building is expected to increase (Figure 2.1). Data on assessing the energy required to produce building materials is published, for example, in EPDs.

2.3.5 EU industrial emissions directive

The new European Industrial Emissions Directive (IED) came into force on 6 January 2011; it brings together seven directives including the previous directive on integrated pollution prevention and control from 1996. The IED affects numerous plants throughout Europe, including those for the production and processing of iron and non-ferrous metals, such as steelmaking. The best available techniques (BATs) should be used in industrial plants. The EU Commission will successively publish relevant descriptions in the form of BAT reference documents. These will contain, for example, binding requirements on emission values, which will have to be adhered to after a transition period of four years from the date of publication. The iron and steel industry is one of the first two branches of industry for which the BAT conclusions were published on 8 March 2012. Among others, the following measures for the production of iron and steel were proposed and detailed:

- A comprehensive environment management system should be set up and used in the plants.
- Among other things, the energy consumption should be reduced by using waste heat and process gases.
- The optimisation of material management, in other words in-house material flows, should facilitate reuse and recycling.
- Water resources should be conserved by introducing water recirculation systems and using rainwater.
- Flue gases must be filtered to prevent dust emissions.

European steelmakers have been practising most of these measures for many years. For example, in addition to being environmentally relevant, the use of waste heat, the efficient use of energy and water and optimal material management are factors that contribute to economic success.

Data on environmental impacts during the production of structural steel, on the complete recycling of production waste and on the future recycling potentials of structural steels are given in the EPD for structural steels (BFS-20130094-IBG1).

REFERENCES

[1] EN 15804. (2012) *Sustainability of Construction Works – Environmental Product Declarations – Core Rules for the Product Category of Construction Products.* Brussels: European Committee for Standardization.

[2] EN 15978. (2011) *Sustainability of Construction Works – Assessment of Environmental Performance of Buildings – Calculation Method.* Brussels: European Committee for Standardization.

[3] FIEC. (2014) *Annual Report 2014.* Brussels: European Construction Industry Federation.

[4] ISO 15392. (2008) *Sustainability in building construction – General principles.* International Organization for Standardization.

[5] ISO 12720. (2014) *Sustainability in buildings and civil engineering works - Guidelines on the application of the general principles in ISO 15392.* International Organization for Standardization.

[6] ISO 21932. (2013) *Sustainability in buildings and civil engineering works - A review of terminology.* International Organization for Standardization.

[7] ISO 21929-1. (2011) *Sustainability in building construction - Sustainability indicators - Part 1: Framework for the development of indicators and a core set of indicators for buildings.* International Organization for Standardization.

[8] EN 15643-1. (2010) *Sustainability of construction works - Sustainability assessment of buildings - Part 1: General framework.* Brussels: European Committee for Standardization.

[9] EN 15942. (2012) *Sustainability of construction works - Environmental product declarations - Communication format business-to-business.* Brussels: European Committee for Standardization.

[10] ISO 14025. (2006) *Environmental labels and declarations - Type III environmental declarations - Principles and procedures.* International Organization for Standardization.

[11] ISO/DIS 21930. (2016) *Sustainability in buildings and civil engineering works - Core rules for environmental declaration of construction products and services used in any type of construction works.* International Organization for Standardization.

[12] ISO 14040. (2006) *Environmental management - Life cycle assessment - Principles and framework.* International Organization for Standardization.

[13] ISO 14044. (2006) *Environmental management - Life cycle assessment - Requirements and guidelines.* International Organization for Standardization.

[14] EN 15643-3. (2012) *Sustainability of construction works - Assessment of buildings - Part 3: Framework for the assessment of social performance.* Brussels: European Committee for Standardization.

[15] EN 16309. (2014) *Sustainability of construction works - Assessment of social performance of buildings - Calculation methodology.* Brussels: European Committee for Standardization.

[16] EN 15643-4 (2012) *Sustainability of construction works - Assessment of buildings - Part 4: Framework for the assessment of economic performance.* Brussels: European Committee for Standardization.

[17] EN 16627 (2015) *Sustainability of construction works - Assessment of economic performance of buildings - Calculation methods.* Brussels: European Committee for Standardization.

[18] ISO/TS 21929-2. (2015) *Sustainability in building construction – Sustainability indicators - Part 2: Framework for the development of indicators for civil engineering works.* International Organization for Standardization.

[19] Directive 2008/98/EC of the European Parliament and of the Council on waste and repealing certain directives. *Official Journal of the European Union.*

[20] Waste Management Act (2012) *Gesetz zur Förderung der Kreislaufwirtschaft und Sicherung der umweltverträglichen Bewirtschaftung von Abfällen (Kreislaufwirtschaftsgesetz - KrWG).* Berlin.

[21] REGULATION (EU) No 305/2011 of the European Parliament and of the Council laying down harmonised conditions for the marketing of construction products and repealing Council Directive 89/106/EEC. *Official Journal of the European Union.*

Chapter 3
Basic principles of sustainability assessment

3.1 THE LIFE-CYCLE CONCEPT

Raban Siebers and Diana Fischer

3.1.1 What is the meaning of the life-cycle concept?

In the past, buildings were mainly designed to a construction cost budget. Because of growing environmental awareness, more emphasis was then put on high levels of thermal insulation of building envelopes based on the Energy Performance of Buildings Directive (EPBD). As a result, the energy use over the life cycle of a building was reduced. Nowadays, buildings are required to achieve a high economic, environmental and social performance in order to be classified as sustainable buildings.

Planning and constructing sustainable buildings requires a holistic assessment. It is important that the assessment focuses on the life cycle of a building. The production of greater thicknesses of thermal insulation may require more material, which means that a higher amount of embodied energy is needed for the building. Over the building's life, however, the higher level of thermal insulation results in considerable energy savings. These savings may outweigh the higher energy requirement for the production after a relatively short time. Thus, it is important to consider the investments required to achieve operational savings during occupancy.

3.1.2 Life-cycle phases of a building

The call for a 'life-cycle assessment' brings up the question of how to define a building life cycle. Setting the system boundaries is of crucial importance for all aspects of sustainability assessment. The basic approach is called 'from cradle to grave'.

Sustainable Steel Buildings: A Practical Guide for Structures and Envelopes, First Edition.
Edited by Bernhard Hauke, Markus Kuhnhenne, Mark Lawson and Milan Veljkovic.

Table 3.1 Life-cycle stages of buildings and construction products for sustainable constructions in accordance with EN 15804 and EN 15978.

Building Assessment Information				
Building Life-Cycle Information				
Product Stage	**Construction Process Stage**	**Use Stage**	**End-of-Life Stage** (Building)	**Benefits and Loads** Beyond Building Boundaries
A1: Raw Material Supply A2: Transport A3: Manufacturing	A4: Transport A5: Construction–Installation Process	B1: Use B2: Maintenance B3: Repair B4: Replacement B5: Refurbishment B6: Operational Energy Use B7: Operational Water Use	C1: Deconstruction C2: Transport C3: Waste Processing for Reuse, Recovery and Recycling C4: Disposal	D: Reuse–Recovery–Recycling Potential

The term 'cradle' denotes the acquisition of raw materials needed for the construction products. The 'grave' is reached when (following dismantling) the building materials are either being disposed of (landfill/energy recovery) or are being reused or recycled. In the latter case, the term 'from cradle to cradle' is used. This reflects that certain building materials and products have no end of life in the normal sense but are available for future construction works, because of their ability to be reused or recycled.

The building life cycle consists of the following phases as shown in Table 3.1.

3.1.2.1 *Design stage*

The construction project starts with the design and planning process. Even though this process does not directly form part of the building life cycle, it is nevertheless important, because it determines the basic aspects that will influence the sustainability performance of the building. The building owner should take advantage of this potential and take account of all the building's relevant requirements during the planning stage. Here design for deconstruction, recycling and reuse plays an important role (see Chapter 4.3.3).

3.1.2.2 *Product stage*

Resource extraction, transport of the resources, the manufacturing process and completion of the finished products at the factory 'gate' all form part of the production stage of the construction materials. Thus, the production stage comprises all processes from 'cradle-to-gate'. For structural steel in the form of sections and heavy plates, this stage includes the mining of iron ore and steel production in the blast furnace or production in the electric arc furnace from scrap. For each production route, the further processing in rolling mills is included.

3.1.2.3 *Construction process stage*

The construction process stage consists of all processes that are necessary for the completion of a building that is ready for occupancy. The construction stage in general starts with the transportation of the building materials from the manufacturer's factory to the construction site and ends with the completion of the construction work on site. An important characteristic for construction in structural steel is the prefabrication of components in workshops prior to the actual on-site assembly.

3.1.2.4 *Use stage*

The time in which a building is used is called the 'operational phase' or 'use stage'. From a temporal, environmental and social point of view, the use stage accounts for most of a building's life cycle. From a financial standpoint, it is the period that enables amortisation of investment costs. As requirements for buildings can vary over time, it is not possible to predict the duration of an entire life cycle. When it comes to office and administration buildings, an average service life of 50 years is often used in their evaluations. The DGNB certification system (Deutsche Gesellschaft für Nachhaltiges Bauen) and the German assessment system for sustainable federal buildings BNB (Bewertungssystem Nachhaltiges Bauen) use this period of time. LEED (Leadership in Energy & Environmental Design) from the U.S. Green Building Council and the British BREEAM (Building Research Establishment Environmental Assessment Methodology) assume 60 years for consideration (see Chapter 5).

3.1.2.5 *Conversion/refurbishment*

Refurbishment of existing buildings is of major importance and will continue in the future. The aim of sustainable design is to ensure that present-day buildings will still be usable in 50 years. Depending on the evolution of lifestyles, buildings are likely to be converted several times. The cost for adaption and extension can be reduced considerably by taking measures during the design and construction phase to facilitate the future work, for example, by design to higher loads or use of wider spans and choosing a structural floor system that accommodates a number of mechanical service distribution schemes based on different occupancies.

Refurbishment is an important aspect of the conservation of natural resources, because buildings can be put to new use. Most or all of the original structure of a building can be preserved. The environmental and financial cost-saving opportunities are significant: the refurbishment of an existing building can be up to 80% less cost-intensive than demolition and construction of a new building [1]. Thus, the building stock has a huge potential from an environmental and economic point of view. Furthermore, existing buildings have a social purpose: they create local identity and are witnesses to the past and to the culture. Therefore, the preservation and expedient utilisation of existing buildings meet the criteria for sustainable development.

The optimisation of the energy use of existing buildings is an important area. While improvements have been achieved in the field of energy needs for new buildings, large parts of the building stock still have to be modernised. Today, old buildings account for about 90% of CO_2 emissions from private households [2]. The existing building stock is a priority for action in order to achieve long-term climate protection goals.

3.1.2.6 *Deconstruction and end-of-life stage*

Addressing the deconstruction of a building that has not even been constructed seems not to be a pressing requirement given the timescales involved. During the building design phase, few planners and building owners will give thought to the possibility that 'their' building will have to be demolished or, better, deconstructed. Nevertheless, this phase has to be taken into account when adopting a holistic approach. At its end of life, a building can be disassembled into its component elements or constituent material. The remaining materials are classified as follows:

1 structural elements and building materials that can be reused, recycled or used otherwise (e.g. for energy recovery);
2 structural elements and building materials that have to be disposed of. This includes destructive incineration in which energy is not recovered and landfilling of waste materials.

Taking into consideration the 'cradle' at the beginning of the production phase, materials of the first category have a life cycle 'from cradle to cradle'.

Building materials of the second category have reached their grave at the building's end of life. This corresponds to a life cycle 'from cradle to grave'.

3.2 LIFE-CYCLE PLANNING

3.2.1 Building Information Modeling in steel construction

Billie Kaufman, Lorenz Erfurth, Michael Huhn and Ronald Kocker

Increasing pressure to manage time and resources more efficiently has moved the subject of Building Information Modeling (BIM) to the centre stage. The planning method with the goal to efficiently manage the entire life cycle of buildings – from design to execution and finally operation and maintenance – is increasingly a focus for construction companies, architects, engineers and other planners. At the same time, government and institutions are increasingly encouraging the 'digitalisation of construction' and are trying to lay the foundation for a widespread use of BIM methods and technologies. The central goal is to make construction projects more economical, resource efficient and sustainable through use of integrated planning systems and information exchange.

3.2.1.1 *Better planning with BIM*

BIM refers to a method that aims to model, optimize and communicate the structure of a building's components and systems with all its relevant information throughout its entire life cycle. This involves the 3D, highly detailed visualization of the structure as will be built. With BIM, buildings are virtually created long before their structures actually exist. This enables planners and construction companies to determine early on if the project can be realistically and efficiently executed in terms of time, materials and costs, thus reducing waste and making projects more viable and sustainable.

At the core of successful BIM lies an extensive database that contains all relevant information on the building, including detailed information on all building components. However, not all building models are true BIM models. Models, which only include visual 3D data but whose 'objects' do not carry any information on the attributes of components, do not satisfy the goals and possibilities of BIM. These models do not contain the relevant information to efficiently support modeling, fabrication and procurement of the project. In addition, models should be parametric, with building components correlating with each other in terms of measurements and connections and also enabling designers to add or eliminate elements without difficulty. Finally, the BIM model can be enriched through valuable project management information, such as data regarding logistics, time or cost scheduling.

The advantage of compiling comprehensive data for a construction project over its entire life cycle is clear. When using traditional planning methods, information has to be accumulated again and again in every new project phase, and valuable data is often lost from one phase to the next. With BIM, the amount of data on a project increasingly builds up, enabling project parties to make well informed decisions at key times. The path for sustainability is set with the first designs, as better planning can significantly decrease the waste of materials and other resources in later project phases and also includes the efficient operation and maintenance of the building in the future. With a BIM model, all parties involved are constantly up to date regarding the status of the project. The improvement of communication and collaboration is therefore often cited as one of the most important advantages of BIM for construction projects (see Figure 3.1).

3.2.1.2 *Integrated planning with BIM*

The integrated planning process begins with the architectural design. The different project parties work on their specific designs and components of the building. The architect designs the structure in an architectural-led BIM solution, in which

Figure 3.1 BIM is the process that aims to model, optimise and communicate the structure of a building throughout its entire life cycle, in order to make construction projects more efficient and sustainable. © Trimble.

the 3D planning of walls, ceilings, columns, doors or windows is created. The structural engineer then concentrates on the design of the structure with all its relevant information. Windows, doors or lightweight partition walls are not taken into consideration. The structural model is then enhanced through static loads, load combinations and so forth in a structural analysis solution. The structural designer then checks the cross sections and the model is optimized according to the building requirements, before it is again passed on to the architect to confirm the final acceptance of the suggested changes and to further detail the model. Simultaneously, the MEP (mechanical, electrical and plumbing) contractor designs the building's HVAC (heating, ventilation and air conditioning) system according to the architectural design. Possible change requests, such as adding additional wall breakthroughs, are communicated and coordinated at this stage.

The integrated planning process is never linear. A large number of different participants are in constant dialogue with each other, in order to harmonise their designs and expectations for the project. The cycle of design, coordination and optimisation is therefore continuously repeated. This additional effort in early project phases pays off in terms of the construction phase: the digital building model is up to date at all times, thereby enabling project parties to make well-informed decisions, based on a solid data foundation. Costly mistakes, such as clashes between models of different trades, are identified before they reach the construction site. The actual material requirements can be accurately calculated based on the 3D model, which reduces waste. In addition, design alternatives can be easily compared and possible changes are integrated efficiently.

Apart from the coordination of following trades, the 3D model provides valuable information for fabrication, logistics and installation. Modern BIM software solutions allow the direct integration of production machinery and CNC (computer numerically controlled) machines. In addition, drawings and transportation lists can be automatically generated. With the help of specific scheduling tools or through the integration of project management solutions, different project phases can be simulated and controlled. The model is enhanced through additional information on time, resources and costs. Information on the current status of the project can be continuously delivered from the construction site and incorporated into the BIM model, in order to visualise the progress of the project and to make additional calculations and predictions in the project management solution (see Figures 3.2–3.5).

3.2.1.3 *Interoperability and open BIM*

The description of the integrated planning process shows that, contrary to widespread beliefs, BIM is not characterized by the creation of a single, huge building information model that all project parties share and simultaneously alter. Rather, the different disciplines work on their specific models, such as architectural model, structural steel model or MEP model. BIM forms a bridge between all parties involved, enabling them to continuously coordinate their designs.

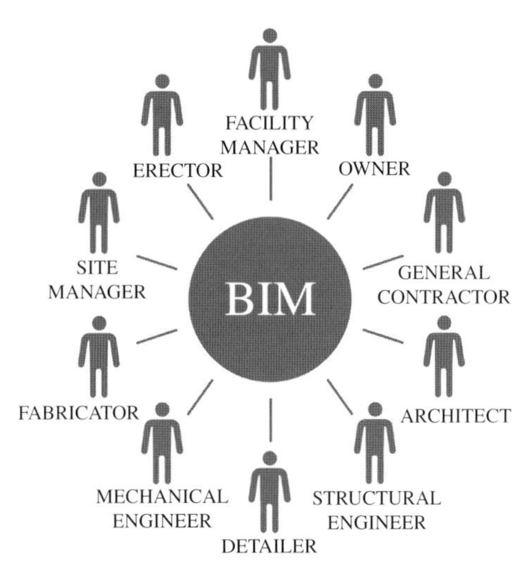

Figure 3.2 Through integrated planning, BIM brings all project parties closer together. © Trimble.

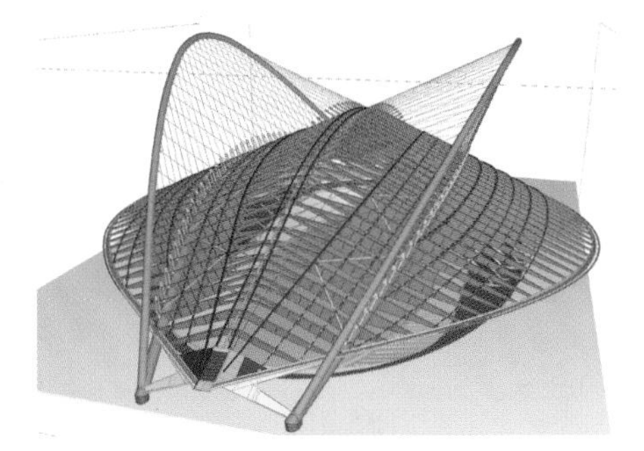

Figure 3.3 BIM model of the Velodrome of the 2004 Olympics with a total of 26,000 elements and 5740 single-part, assembly and general arrangement drawings. © Trimble.

Figure 3.4 With BIM, buildings can be planned down to the smallest level of detail. Due to parametric components, changes can be incorporated quickly and efficiently. © Trimble.

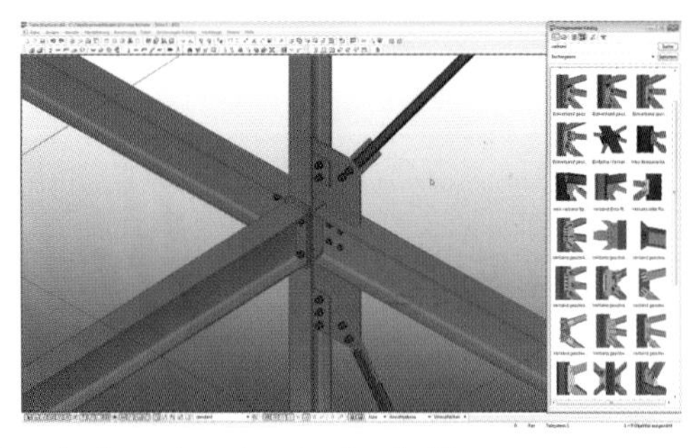

Figure 3.5 Detailing of a braced steel node. © Trimble.

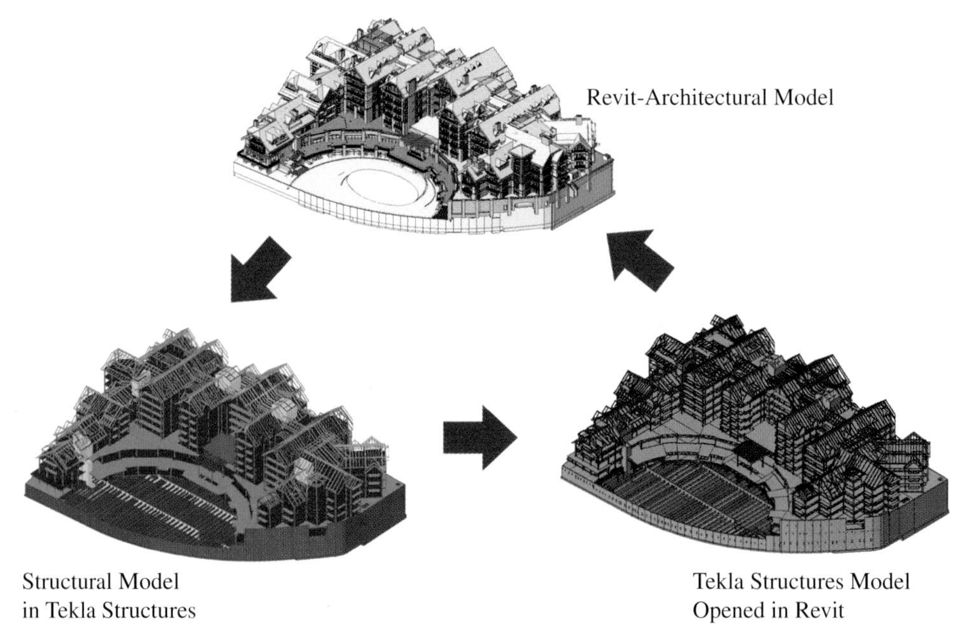

Figure 3.6 Urban renewal project Solaris in Vail, Colorado. With open BIM, all project parties could work with the tools best for their individual needs. Revit architectural model and structural model in Tekla Structures and Revit pictured. © Trimble.

A fundamental requirement for these processes is a high degree of interoperability between different software solutions, enabling project partners from structural engineer to MEP contractor to work with the tools and solutions that are best suited for their individual needs. The key to this is open BIM. The concept of open, model-based data exchange is based on the data format IFC (Industry Foundation Classes), which all software solutions involved need to be able to read and write. The respective designs of the other parties are imported as a reference model and can thereby be used as a point of orientation for their own model. The coordination between all planning processes within a central BIM software solution is possible as well (see Figures 3.6 and 3.7).

Figure 3.7 Clash checking in Tekla BIMsight. Possible errors are identified before they reach the construction site, ensuring timely and sustainable project execution. © Trimble.

3.2.1.4 *BIM in practice: Vienna railway station*

BIM offers substantial possibilities to optimise construction projects and increase their overall sustainability. Many successful projects today can illustrate the added value of BIM. A good example is the recently completed roof of the new Vienna railway station. A total of 145,000 passengers and 1,000 trains are expected to use the open, modern building daily. The Unger Steel Group was responsible for planning, fabricating and assembling the partially transparent, architecturally stunning roof of the new railway station. Fourteen diamond-shaped trusses, each 76 m long, form a unique structure, 15 m above platform level, with no element being exactly like any other and none of the components arranged horizontally.

From the beginning of the project, the steel contractor consequently relied on BIM to optimize design and construction. The engineers imported the architects' model as 3D DWG files to the BIM software solution in order to design and detail the diamond trusses. The company created constructable steel structures, modeled and detailed the main structure and substructures like cable channels, and extracted data and drawings for production and assembly, also simulating welding sequences of the complex components. Up to 10 designers worked on the model in multiuser mode. For coordination between project parties, the team utilized the free BIM collaboration software Tekla BIMsight, enabling all project parties to continuously access the building information and efficiently work together, including the teams from fabrication and on site. Apart from relevant data for design and fabrication, the BIM model also included information on the exact positioning of the prefabricated steel elements on site. The data was transferred to

Figure 3.8 Integrated planning played a central role in the construction of the new roof of the Vienna railway station with its diamond-shaped trusses (simulation). © ÖBB/Stadt Wien.

Figure 3.9 The complex steel structure of the roof of the Vienna railway station was planned in the BIM software solution. © Trimble.

a Trimble total station, and the team utilised the Total Station to mark the building axes and precisely position and assemble the elements.

The consistent use of BIM and the exemplary information managed during the construction of the Vienna railway station majorly contributed to the timely, cost-efficient and sustainable execution of the project (Figures 3.8–3.10).

Figure 3.10 With the help of the Trimble total station and the data from the BIM model, the roof elements were precisely positioned. © Trimble.

3.2.2 Integrated and life-cycle-oriented planning

Lamia Messari-Becker and Ingo Schrader

Depending on the climate, the form and function of a building, urban integration and the use of specific materials always have to be considered. As it is the case for area or energy consumption, it is also necessary to optimise and reduce the amount of materials used. Minimising material usage is limited on the one hand by safety aspects and standardisation related to the primary structure (load-carrying capacity, fire protection). On the other it has to satisfy requirements relating to user comfort and health (e.g. thermal insulation, moisture proofing and sound absorption, acoustic and lighting comfort) as well as energy efficiency. The latter requirements are defined mostly by the efficiency of the building shell or facade as secondary structure. If environmental aspects (ecology, waste prevention, recycling) are added to these considerations, the process of material selection, use of resources and optimisation becomes iterative. When life-cycle-based methods of planning and evaluation are used, aspects such as dismantlability, modularity, flexibility and low maintenance are now key factors in determining the long-term economic stability of an object. There are several interactions, so this process can only provide useful solutions (also in the long term) if the parties work together in an integrated manner at all stages of the project.

Sequential Planning	Integral Planning	Life-Cycle Engineering
• Planning and Operation Separated • **No** Interaction Between Planners (Architect, Engineers) and the Operator/User	• Planning and Operation Linked • Interaction Between Planners and Transfer **to** the Operator/User	• Interaction Between Planners (Architect, Engineers) and Transfer **to and from the** Operator/User • Decisions Depending on Life-Cycle Costing (LCC) and Life-Cycle Assessment (LCA)

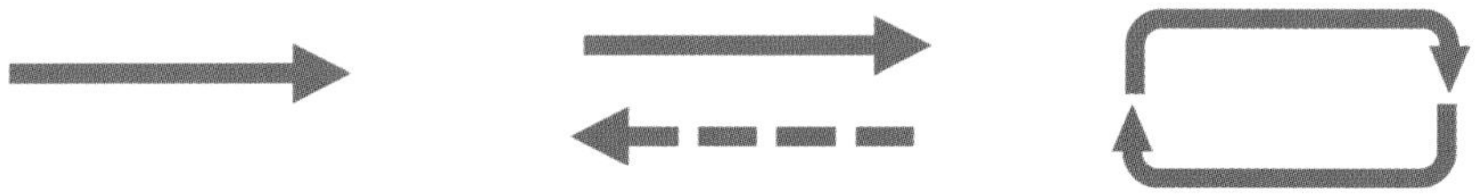

Figure 3.11 From sequential planning to life-cycle engineering [3].

3.2.2.1 *A change in planning culture – integrated planning processes today*

Twenty years ago, the construction and operation of a building did not constitute a planning entity. The planning phases were a series of steps one after another. There was barely any interaction between the planners and the user or operator of the building. For the most part, no consideration was given to the integration of the user's interests, for example, concerning operational costs. As part of integrated planning, manufacturing and operating aspects became an entity later. Cooperation took place both within the planning team and between the planning teams and the future user or operator. The parties became ever more conscious of the operation/use of a building being a decisive phase. In the meantime, the subsequent user and/or operator has become a partner of the planning team. Decisions that have far-reaching consequences during the life cycle of the object, for example, energy supply, convertibility of use, flexibility of the ground plan or implementation of energy and sustainability standards, can now be reached based on feedback between the operator or user and the planners. The implementation of several sustainability labels and political guidelines for sustainable building are further advancing resource and life-cycle-oriented planning (see Figure 3.11).

3.2.2.2 *The life cycle of a building*

The life cycle of a building includes all phases that a building experiences during its lifetime (product manufacturing and construction of the building, use, demolition and dismantling, recycling or landfilling) (see Section 3.1). Both environmental impacts and costs occur during the life cycle. The environmental impacts arise in the form of gases that are damaging to the environment as a result of the extraction of the raw materials, manufacturing, replacement (during refurbishment or conversion) and demolition as well as the conditioning of buildings (heating, cooling, ventilation) during operation. In an analogous manner, after the construction of a building additional costs also arise for the operation, use, conversion, refurbishment and demolition. Keeping these (subsequent) environmental impacts and costs as low as possible or optimising them ensures the environmental efficiency of the object and its value retention and is the task of life-cycle-oriented planning.

3.2.2.3 *Methods used in life-cycle-based planning*

A *life-cycle assessment* (LCA) analyses the whole life cycle, for example, of a building, the relation to ecological impacts and the resultant environmental impacts (pollution). A life-cycle assessment can address various environmental categories depending on the need and relevance. Life-cycle assessment is now an important constituent of proof of performance, such as in the case of integrated sustainability assessments or certifications (see Chapter 5).

For *life-cycle cost analysis*, the calculation of the life-cycle costs is based on the (dynamic) net present value method. The present value of all payments that are made during the whole life cycle are included. Present values are the amounts that would have to be invested today to pay for all costs incurred during a certain time period and at a specific rate of interest. In classical project planning, only the production costs were taken into account and continually determined with ever-greater precision to keep within the budget. The property sector has been familiar with comprehensive life-cycle cost analyses for longer, albeit in a modified form. They are used as a basis for making decisions or for optimisation. The accounting periods for a life-cycle analysis can vary markedly. When a building is being assessed more for real estate funds, particular attention might be given to the maintenance phase. Life-cycle costs for buildings are continuing to make their way into planning. They are becoming increasingly important for the public sector, municipalities and towns, including with regard to the maintenance costs of the infrastructure. Certification schemes like DGNB, BNB, LEED and BREEAM (see Chapter 5) define some parameters for their specific purposes, but there is no general standard. A harmonization is desirable, especially with respect to the accounting periods and maintenance costs [4].

3.2.2.4 *Resource efficiency via integral planning: Oval roof of north entrance to the Frankfurt trade fair grounds*

Since August 2013 an oval roof has covered the north entrance gate and the security post building at the Frankfurt trade fair grounds. Architect Ingo Schrader created a striking architectural ensemble. The aesthetics of the construction rely essentially on the sophisticated design and the clarity and simplicity of the details. The design of the roof, which is located on an existing road bridge, is composed of a girder grid made of irregularly arranged flat steel girders. The girders vary in height (up to 600 mm high) depending on the distribution of the forces. This design was developed in close cooperation with the structural engineers of Bollinger + Grohmann. The algorithms defining the optimized shape and layout of the grid are similar to the principles found in natural growing processes as in trees, leaves etc. The steel columns have a triangular cross section (see Figure 3.12).

The project was awarded a special prize by the BMUB (Federal Ministry for the Environment, Nature Conservation, Building and Nuclear Safety) in 2014 for sustainable steel architecture. The high level of resource efficiency of the construction, due in part to optimisation and prefabrication, was particularly commended as a reason for the special award. Selected sustainability aspects of this prize-winning construction project are presented below.

Figure 3.12 Oval roof of the north entrance to the Frankfurt trade fair grounds, Architecture: Ingo Schrader Architekt BDA, Structural Design, Physics and Sustainability: Bollinger + Grohmann. Photo: © Christian Richters.

The construction is in steel, used for its material and resource efficiency. Use of steel creates a delicate appearance that meets the demanding architectural requirements and allows the structure to have a special lightness with a free projection of up to 10 m. The individual supports are made of 20 or 40 mm thick plates in S355 grade and thus from a starting material that only has to be cut to size and joined together. Simplicity is pursued as the guiding principle for details and connections. The use of complex profiles and connections was abandoned in favour of a uniform and simple method of construction.

The roof covering consists of plywood panels, which are coated with polyurethane on the upper face. A low solvent paint was used for the components. If the whole life cycle is taken into consideration, the recyclability of steel complements the material efficiency in producing the construction. Returning the steel to the material loop minimises the total grey energy and the CO_2 emissions that occur during the whole life cycle. The susceptibility to corrosion is reduced significantly by the small surface area of the flat steel supports and the fact that they are exclusively vertical, which in turn contributes to low maintenance costs.

For integrated technology, lighting, heating cables and cameras are integrated in channels milled into the plates. There is a downpipe in the area of the security post building to drain the water from the flat roof. The housing of the pipe also contains the electrical connection between roof and guardhouse. The installations for domestic engineering are thus naturally part of the shape (see Figure 3.14).

For prefabrication and construction, as a result of the parametric design the use of materials is minimised, which enhances resource efficiency. The roof was

Figure 3.13 The roof was prefabricated in the workshop, then divided into seven transportable segments and assembled on the ground immediately next to the final location. © Ingo Schrader Architekt BDA.

prefabricated in the workshop, then divided into seven transportable segments and assembled on the ground immediately next to the final location. The division into transportable elements reduced the number of welds that had to be carried out on site to a minimum. The almost complete prefabrication on the ground meant that the closure of the bridges and the time that the entrance to the trade fair grounds had to be closed was kept short (Figure 3.13).

This reduced the transport cost, the construction time and emissions of dust and noise at the construction site. These are important aspects of environmental protection in urban construction while operations of the facility continue. The optimised construction process also lead to improved occupational safety.

3.2.2.5 *Guardhouses at Frankfurt trade fair grounds*

By using a family of different security post buildings at the Frankfurt trade fair grounds, the differing space requirements and sight lines were taken into account by means of different sizes depending on the location and function. The buildings are relatively small and have a similar appearance, which ensures they are readily recognisable at the busy fair grounds. The variation of the types and the quasi-uniform appearance were made possible and supported not least via use of the architecturally driven steel solution.

Regardless of the different size of the buildings, the steel construction, materials and main details are very similar. This approach not only facilitated the construction process but also emphasizes the formal relationship between the different guardhouses (Figure 3.15). The steel structure also serves as the support

Figure 3.14 Oval roof of the north entrance of the Frankfurt trade fair grounds, Architecture: Ingo Schrader Architekt BDA, Structural Design, Physics and Sustainability: Bollinger + Grohmann. © Ingo Schrader Architekt BDA.

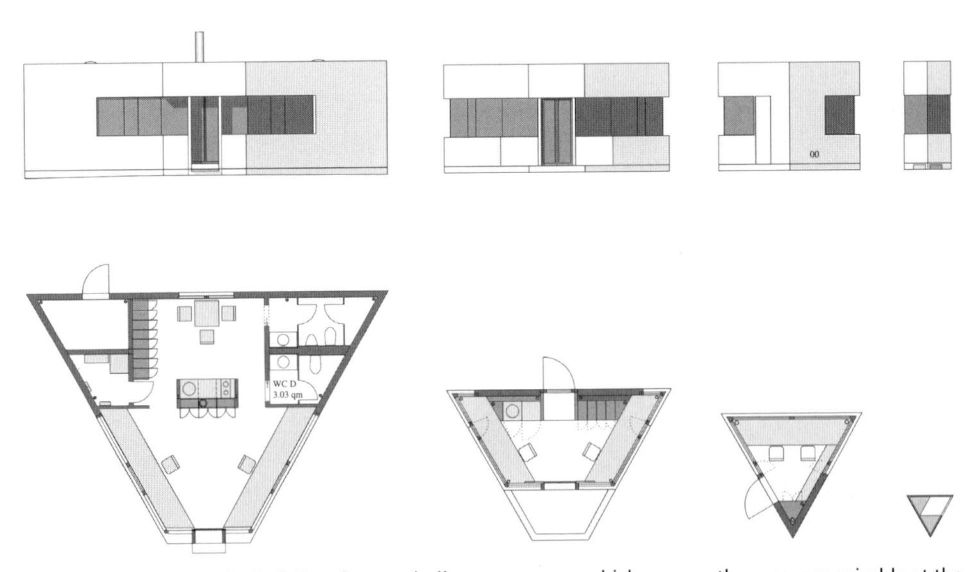

Figure 3.15 The security buildings have a similar appearance, which ensures they are recognisable at the busy fair grounds. Regardless of the different size of the buildings, the steel construction, materials and main details are very similar. © Ingo Schrader Architekt BDA.

for the secondary design level and ensures proper thermal insulation, sound insulation and fire protection, and thus satisfies the comfort, health and safety of the users. For the floor plan, the triangular shape provides an optimized overview for the guards and a recognizable contrast to the mainly rectangular large volumes of exhibtion halls, containers or trucks on the fairground.

This type of construction ensured the efficient use of space for the specific functions of the security post buildings. The similarity of these buildings was further enhanced by the arrangement and dimensional coordination of the elevations; the height of the parapet, ribbon window and fascia is identical for all buildings. The facades were clad using red-lacquered aluminium sheet.

3.2.2.6 Integrated design

It is important to involve all specialist disciplines at an early stage when planning a building or structure that conserves resources. Regional, climatic, cultural and statutory framework conditions play an important role when choosing construction methods and materials that are suitable from an ecological, economical and sociocultural point of view. For planning, the first tasks are to define space requirement, size and shape of the building. It is only during the planning process and after several interactions, ideally using estimated life-cycle-oriented methods of planning, that a single suitable solution becomes clear that is resource efficient, flexible in use and economically viable.

However, there is already potential for optimisation during the production phase. Optimised integrated planning of the supporting structure allows the materials to be used in a stress-oriented (load-bearing-oriented) and thus efficient manner (bionics principle: highly stressed beams are used efficiently while low stressed ones are removed).

Simple details and connections allow further material savings (and a harmonious and sophisticated appearance even in an aesthetic sense). If the integrated technique is successful, it contributes to the ease of maintenance. Furthermore, steel allows a high degree of prefabrication off site depending on the shape and function. Thus, the material loss due to the production process is minimised. In addition, construction times are shortened, which means lower noise and disturbance on site. Life-cycle-oriented methods of evaluation, for example, LCA or LCC, are becoming increasingly important but require simplification and harmonization. The standardization of characteristic and reference values contributes to comparability and to transparency with respect to a scientific and practice-oriented discourse.

3.3 LIFE-CYCLE ASSESSMENT AND FUNCTIONAL UNIT

Raban Siebers

Life-cycle assessment (LCA) is a method for determining and assessing ecologically relevant processes. Originally developed primarily for assessing products, it is also used today to assess processes, services, behavioural patterns and complete buildings.

The principles and rules governing the implementation of LCAs are defined internationally in the ISO 14040 and ISO 14044 standards. According to these standards, an LCA involves four elements:

- definition of the goal and scope of the assessment;
- life-cycle inventory;
- impact assessment;
- interpretation.

In defining the goal and scope of the assessment, it is first necessary to define what the LCA is to be used for. This affects all subsequent decisions and is thus an important step. The use and functions of the object and the basic life cycle are then defined, starting with the extraction of the raw materials and finishing with the end-of-life respective recycling, reuse or disposal. In addition, interactions are taken into consideration, assumptions and limitations are specified, and the preliminary limits of the assessment, the so-called 'cut-off' criteria, are defined. Another important point is the definition of the 'declared unit'. This is taken to mean the product-specific size on which environmental impacts will subsequently be based (e.g. a square metre of gross floor space).

In the life-cycle inventory, quantitative statements are made regarding the product life cycle that has previously been documented. For this, the consumption of resources (inputs) is compared with the use (functional unit) or the correlated emissions (outputs). The life-cycle inventory is in itself a purely descriptive model without any assessment whatsoever.

The impact assessment subsequently divides the results of the life-cycle inventory into different impact categories that comply with scientifically based qualitative criteria and show, for example, the relevance of different emissions to the greenhouse effect or the formation of the hole in the ozone layer. The result of the impact assessment is a number (typically 5–10) of quantitative environmental impacts caused by a product. In addition to renewable and nonrenewable primary energy consumption, the environmental impacts detailed in the following sections are usually considered.

In the subsequent evaluation, important parameters (e.g. individual life-cycle stages or impact categories) are identified and analyses of consistency, completeness and sensitivity are then carried out in order to obtain a result. Based on this result, conclusions are drawn and recommendations developed, and a report is compiled.

An LCA for construction also requires a holistic approach. Here it is important that the analysis considers the whole life cycle of a complete building or at least a functional unit. This means, for example, using a comparable supporting structure including foundations, building shell or other components that have to be looked at independently. When comparing different types of construction, materials or methods of construction, it is necessary to ensure that the options being compared fulfil the same technical purpose or that the system boundary has been chosen in such a way that possible impacts from any differences can be documented. The quality of a building can only be determined once savings and improvements during use have also been taken into consideration, in addition to the effort required for the production and disposal or reuse.

3.3.1 Environmental impact categories

3.3.1.1 Global warming potential

The accumulation of greenhouse gases in the atmosphere leads to air layers near ground level becoming warmer (the greenhouse effect). The global warming potential (GWP) of a substance is always compared to the GWP of carbon dioxide (CO_2), which means that emissions that will potentially cause global warming are expressed in carbon dioxide equivalents (CO_2eq). As the greenhouse gases in the atmosphere have differing half-lives, the GWP value has to be based on a specific time period. A time period of 100 years is used to characterise the contributions to GWP. Impact factors are also used to describe the extent to which different substances contribute to global warming potential. Over a period of 100 years, methane (CH_4) has an impact factor that is 25 times greater than that for the same mass of CO_2.

3.3.1.2 Ozone depletion potential

Ozone is only present in the atmosphere in a small concentration, but it is of major importance for life on earth. It is capable of absorbing short-wave UV radiation and emitting this again independent of direction with a longer wavelength. The ozone layer shields the earth against a large part of the UV-A and UV-B radiation from the sun, prevents excessive warming of the earth's surface and protects flora and fauna. The accumulation of harmful halogenated hydrocarbons in the atmosphere contributes to damage to the ozone layer. The consequences include the development of tumours in humans and animals as well as disturbance of the process of photosynthesis. The ozone depletion potential (ODP) is expressed in kg R11eq; the ODP values refer to the reference substance chlorofluorocarbon CFC-11, which is also known as R11. All substances with a value less than one are less ozone depleting than CFC-11 and values above one are more strongly ozone depleting.

Photochemical ozone creation potential

The photochemical ozone creation potential (POCP) describes the harmful trace gases expressed as an equivalent based on mass. In combination with UV radiation, these trace gases, such as nitrogen oxides and hydrocarbons, contribute to the formation of ozone at ground level. This contamination of the air layers at ground level due to a high ozone concentration is also referred to as summer smog. Summer smog attacks the respiratory organs and causes damage to flora and fauna. The concentration of ozone at ground level is determined regularly at air measuring stations and recorded in ozone air-pollution maps.

Acidification potential

The acidification potential (AP) indicates the effect of acidifying emissions; it is expressed in sulphur dioxide (SO_2) equivalents. Airborne contaminants such as sulphur and nitrogen compounds react with water in the air to form sulphuric or nitric acids; these then fall to earth as so-called acid rain and enter the soil and surface water. Acid rain causes damage to living creatures and buildings. For example, nutrients in acidified soil are quickly digested chemically and are thus leached out more rapidly. Likewise, toxic substances can form in the soil and attack root systems and damage the water balance of plants. In total, the many individual impacts of acidification have two serious consequences: damage to forests and fish mortality. Also acid precipitation attacks buildings, and in particular it attacks the sandstone of historical buildings.

Eutrophication potential

'Eutrophication' (overfertilisation) is the term used to describe the process that transforms water or soil from being nutrient poor (oligotrophic) to nutrient rich (eutrophic). It is caused by the ingress of nutrients, especially phosphorus and nitrogen compounds. These can enter the environment during the manufacture of construction products and by leaching due to emissions caused by combustion. If the concentration of nutrients present in the water rises, there is also an increase in the growth of algae. This can lead, inter alia, to fish mortality.

3.4 LIFE-CYCLE COSTING

Gerry O'Sullivan, Christian Stoy and Christopher Hagmann

The main factors that lead to a final building form are

the client's perception of what is needed;
the architect's interpretation of those needs in relation to the building's functions;
the planner's concept to reach the set goals.

These participants are influenced by

regulations and standardization;
the local circumstances and environment.

Added to these is the experience of the available construction market and the benefit that the design solution can give to property values. As-built costs are clearly one of these factors. Timing in terms of completion of the project is another factor, as the building may be a key part of the client's business priorities. In this case, the choice of material may be an important factor that allows the client to achieve that particular business opportunity. Therefore, any additional capital costs arising is balanced by the benefits achieved when meeting the target. Clients are often prepared to pay a premium to achieve early completion.

When a building is constructed purely for rent or sale on the property market, then timing is critical in order both to reduce the cost of borrowings earlier and to recoup the investment. The concept of valuing the benefits accruing from the project is consistent with these aims.

Time is also a factor for the contractor in reducing overhead and management costs on a project. In construction the focus is generally on the initial expenditure, known as the 'capital costs', without due reference to the whole-life costs of the building. With the rise of procurement arrangements such as the public private partnership (PPP) and the private finance initiative (PFI) and their variants in both public and private sectors, the life-cycle costs of buildings has come more into focus. Furthermore, in real estate discerning property investors seek to invest in or rent buildings that can demonstrate lower operational and maintenance costs and display a pleasing aesthetic image.

In the late 1990s, a survey was carried out into the long-term costs of owning and using office buildings [5]. This survey, based on a London city office, established a ratio of the costs over a period of 20 years, which is frequently quoted in many articles on life-cycle costing and facilities management.

The ratio C 1 : F 5 : S 200 compared:

C = Capital costs (excluding land purchase);
F = Facilities management or operational costs;
S = Costs of providing final services by occupying and using the facility (staffing).

The results of the survey have been seriously challenged by a more recent analysis, where the ratio was reduced to C 1 : F 3 : S 5. The operating costs over the life of a building for the operator or owner is certainly significant, and it is strongly influenced by the choices made at the design and construction stages.

An equivalent exercise for a typical office building in the United States demonstrates an even lower ratio for operational costs, nearer 1:1. It further demonstrates the distribution of capital and operational costs over the elemental functions of the building, in which the superstructure and external elements represent 29% of construction costs but only about 20% of the operational costs.

Higher specification and quality products can minimise maintenance costs and reduce operating costs throughout the building's life, particularly in relation to energy consumption. The added value of the 'better building' can be demonstrated by lower employment costs or higher gross output (lower recruitment/retention staff costs, reduced absence or higher productivity) [5].

It is generally accepted that some form of life cycle costing methodology has to form the basis for evaluating the sustainability of building projects. Various methodologies for measuring sustainable indicators such as BREEAM (UK), LEED (United States), CASBEE (Japan), DGNB (Germany) and HQE (France) have been developed, and many others are under development, even though the different valuation systems are not always easily compared. These labels are gaining importance for the marketing and commercialisation of buildings, office space and houses. In addition, the reputation that accompanies certification plays an increasingly important role in this context. Sustainability has become a pressing issue for all governments. Construction activities, including the manufacture and transport of construction products, have a major contribution to make in reducing a nation's carbon footprint. Life-cycle analyses and economy as major integral parts of sustainability are therefore very important for assessing this topic.

The European Commission has identified the use of life-cycle costing tools and criteria in all key phases of the construction process as one of the main ways of improving the competitiveness of the construction sector, and it was a core recommendation of the Sustainable Construction Working Group [6].

By taking into account not only the initial capital costs but also all subsequent operational costs, clients are able to undertake proper assessment of alternative ways of achieving their requirements whilst integrating environmental considerations. Life-cycle costing (LCC) has been defined as 'an economic evaluation method that takes account of all relevant costs over the defined time horizon (period of study), including adjusting for the time value of money'. This equally applies to whole-life costing (WLC), in which, according to the cost breakdown structure in ISO 15686-5 Building and constructed assets - Service-life planning - Part 5: Life-cycle costing the costs of a building asset can be subdivided as follows:

LCC:

construction (capital costs): construction cost including design fees, site costs, statutory charges, taxes associated with foregoing costs, finance charges and development grants;

maintenance: planned maintenance, replacement and emergency repairs so that the building meets the required levels of quality and functionality;

operation: cleaning, energy consumption (heating, cooling, electricity, water and drainage) waste management, property management (administration, insurance, etc.) and occupancy costs (security, information and communication technology, laundry, cleaning, car parking, etc.);

end of life (disposal, demolition, refurbishment, etc.) to meet change of use requirements or upgrade to new demands of the client/operator.

and

WLC:

operational costs (leases, rents, taxes, etc.);

income (rent and service charge payments, etc.);

externalities (costs associated with the asset but not included in foregoing).

The environmental costs could also be a consideration under LCC or WLC and are dependent on the national statutory and legal requirements, including planning, and the environmental policies of the client.

Underlying all LCC are environmental and social costs including those of sustainable development. An optimal sustainable development is one that balances the total economic costs, social change and the environmental consequences, but current design approaches do not address this balance. The position on carbon costing remains confused, ranging from renewable obligation certificates between electricity generating companies to international green certificate trading. Accounting for embodied energy in construction products is an ongoing process, but it is dependent on the local energy sources used in production. Several databases have been developed, mainly on a national basis (see later in this chapter), and processes like the ECO-EPD, an environmental product declaration with European consistency, have been introduced to assess this aspect (see Section 3.8).

Statements that sustainable design will add 10%–15% to the total WLC are oversimplifications, and it is advised that only verifiable facts from reliable sources should be included in the calculations. However, the challenge is to balance the initial construction expenditure for more sustainable buildings with the benefits that accrue from the investment in the WLC cost of the building. Sustainable design is likely to increase construction costs slightly but will be more economical in the long run in terms of either WLCs or LCCs.

3.4.1 Life-cycle costing – cost application including cost planning

Not all these costs will be relevant to all LCC calculations. Some will have negligible value whilst others will not affect the design decisions. The aspects in a construction programme and the requirements and priorities of the investor, the main purposes for undertaking a life cycle study, are

- to predict a cash flow over a fixed period of time for budget, cost planning, cost or audit purposes;
- to compare cost assessments of design options (value engineering) or to appraise tenders.

In all cases, the earlier the exercise is carried out, preferably at an early planning or design stage, the greater the benefits. In the context of a life-cycle study, there are five important criteria to define:

- lifespan or service life;
- discount rate and future inflation value;
- value of future costs and incomes (nominal costs);
- current costs (real costs);
- residual and terminal values.

Real costs are those when all capital expenditure is complete and revenue expenditure commences. Nominal costs are the future costs discounted to bring them to their value at the base date. The important issue is that they should be established as accurately as possible and preferably in the early planning and design stage. Without early accurate cost advice arising from effective cost planning, there is no raw data to complete even a basic cost study exercise. Reasonably accurate estimates of the anticipated construction costs must be available to validate any LCC. The period of appraisal depends on the purpose of the exercise.

An example of a simple exercise would be a project where a short-term target is involved, such as achieving an opportunity cost. It may involve a comparison of alternative construction systems or materials to see which can best achieve the target completion date. If there are maintenance and operational cost differences, they need to be included in the calculations, as the opportunity gain must also be compared with the long-term costs to the client. Considering structural frame options into these exercises can be fruitful. Where the size and scope of a project requires meeting a retail tenant's market deadline, then the best choice will normally be the option that involves as much off-site prefabrication and reduced on-site installation as possible.

This was the case in respect of Europe's largest shopping development in Dundrum, Dublin. With PPP or PFI procurement methods, the client will need a longer review period to cover not only the time over which the building or service is procured and constructed. For a typical school programme of, say, six schools, this could be between 9 and 18 months for a construction period of 18 months. The overall economic life is often as high as 60 years for a school, even though the franchise period may be much shorter, say, 25 years. The service life of the various building features is typically

substructure/superstructure (say 50–90 years);
components (30–35 years);
services (10–20 years, depending on the system);
finishes (5–20 years, depending on the finish);
external ground works and the expected service life of the components.

The following periods are defined in the construction sector [7]:

economic life: the period of occupation, which is considered to satisfy a required functional objective;
functional life: period until the building achieves its function for the purpose for which it was built;
legal life: period until it no longer satisfies legal or statutory requirements;
physical life: time when deterioration may lead to loss of safety of parts of the building;
social life: time reached when the building fails to meet its social functions;
technological life: time when it no longer suits the client requirements in terms of future technological developments.

The actual periods of the evaluation are often less than these limits, and for many commercial developments, the payback/break-even period for the investment will dictate the period of evaluation. Two methods of discounting can be used, which are annual equivalent and net present value. The latter is preferred for construction assets.

3.4.2 Net present value method

The net present value (NPV) method compares the value of money now with the value of revenue or expenditure in the future, taking account of inflation and a discount rate. This can be expressed as the formula below:

$$NPV = \sum_{t=0}^{T} C/(1+r)^{t}$$

t = period of the project over which the costs/income are assessed
C = annual sum to be discounted
r = discount rate

The discount rate takes account of the prevailing inflation rate. The discount rate set by the UK government for public sector projects is 3.5% and is suitable for cost study periods of up to 30 years [8]. In Germany for this section 5.5% and 50 years is set in DGNB and BNB [9]. For considerations in the private sector, options should be tested for the sensitivity of the result to varying discount rates.

3.4.3 Life-cycle cost analysis

Many factors contribute to the global competitiveness of steel solutions. The price of steel itself is only one factor, and in respect to the structural frame, the prime aspect to be considered is usually the potential lifespan of the building.

3.4.3.1 Steel structures

Steel framed structures for buildings are normally considered as an economic option when the designers wish to create long spans for large open spaces, for example, in industrial buildings, large retail areas or open plan offices. Usually the external wall system is nonstructural, and this can give the freedom to include large expanses of glazing or uninterrupted space. With a greater number of storeys, framed structures become the only viable option and transfer the weight of floors and internal walls and live loads through the framed structure to the foundations. For instance, the slim steel columns of high rise buildings do not take up much space and increase usable space. Steel also provides possibilities to speed up the construction stage, if this is a factor in costs and in achieving a client's goals for the project.

Even with non industrial buildings of one to four storeys, where the common view is that traditional loadbearing walls are more economic than a framed structure, the latter justifies any additional capital costs by giving greater flexibility over the lifespan of the building and ultimately reducing the maintenance and operational costs. For example, loadbearing internal walls lack flexibility in use in buildings such as offices, schools and universities, which can be expected to undergo many future changes to their internal layout. A framed structure can provide greater flexibility in relation to room layouts by allowing future removal of partition walls. The LCC analysis here shows that the additional initial capital expenditure buys the client the option of future functional flexibility. These options have to be considered particularly with public buildings, which are expected to have lifespans of 50 years or more but should allow for future changes in needs.

The design phase of hospitals, for instance, often has many changes between the feasibility stage and final construction. Throughout the life of the building, medical science and procedures advance, which lead to new space, equipment and loading requirements. A hospital designed using a framed structure would provide open spaces to accommodate operating theatres with special clean room partitioning. Efficient connecting 'clean' and 'dirty' corridors and allowing large areas required to service the theatres from above is possible. Steel framed structures are seen as providing the greatest flexibility by allowing internal changes of layout with minimal future costs. Many common problems with the current building stock, especially in the public sector, stem from its inflexibility, which results in an inability to meet functional or environmental changes without expensive refurbishment.

Given the considerable initial investment that is involved in land purchase and building construction, the frame option should be considered early as part of the LCC exercise. Where the steel structure option is considered, it will usually be evaluated against the reinforced concrete (cast in situ or precast) and, less frequently, timber. Commonly the choice is based on initial capital cost rather than LCC, and the steel solution has to be considered in relation to the fire safety and weathering options, together with the appropriate aesthetic requirements. If a comparative LCC analysis is required, the lifespan of the additional treatments must be considered together with the residual values of the steel frame compared with the alternatives. Generally, the disposal and demolition costs of a steel structure are considered as nil because it is completely recyclable, whereas concrete is only partially recyclable and will have a disposal costs. It must also be considered that steel structures are often refurbished for a new use, such as the conversion of offices into apartments or hotels. However, if future options for alternative uses are to be considered as part of the client's needs, then the comparative flexibilities of the frame materials should also be considered, and a steel frame may prove more adaptable if the design anticipates this need for flexibility. The comparison of cost alternatives can be extended to cover the full range of the different structural systems steel has to offer.

Steel in S235 grade was once the standard grade used in building construction. It has been largely replaced by the higher strength S355 steel, which is used for rolled wide-flanged sections. This higher strength steel can reduce the overall weight of the structure by 10%–15%, giving material and on-site welding cost savings and improved weldability. For high rise multistorey structural frames where weight reduction is of paramount importance, high strength steels with low alloy content complying with S460 grade have been developed that can reduce the construction weight further (see Chapter 4.5). For a moderate increase in cost alternatives to the standard I- and H-sections can be found in the form of frames using hollow sections of various shapes that can offer alternative solutions. The LCC evaluation must relate to the client's initial brief, which must include any added aesthetic value required to reflect its corporate image or that is essential to sell the client's business. Some initially considered more costly solutions may give greater value to the client.

Steel structures are no longer confined to heavy frames. Light steel frames of galvanized and cold-rolled profiles manufactured and designed for residential buildings and hotels are competitive against both timber framing and concrete construction. Steel structures that are inherently lightweight reduce loads on foundations and are beneficial when adding additional storeys to existing structures.

Figure 3.16 Kraanspoor project in Amsterdam harbour. Three storeys of lightweight steel structure on top of an existing concrete frame. © Initiative and design: Trude Hooykaas, architect: OTH architects, photographer: Rob Hoekstra.

A similar option was taken on the Kraanspoor project in Amsterdam harbour (Figure 3.16). A three-storey office building had to be constructed on top of an existing concrete frame structure, on which two rail-mounted harbour cranes used to run. Due to the limited loadbearing capacity of the existing structure, the third floor was possible only by using a steel structure and a specially designed floor system. The 30% increase in rentable floor space provided by the additional floor has also made the project more profitable for the investor.

Where soil conditions are poor, lightweight steel structures may prove to be the only option to reduce the loads on foundations. Floors were traditionally formed either with cast in situ concrete or precast concrete. Developments in steel technologies include decking systems that replace temporary formwork and lead to a thinner concrete floor. Furthermore, the use of a steel structure can in itself result in slimmer floors with a consequential reduction in facade heights and reduced cladding area. The introduction of asymmetric beams in new slim floor concepts has provided cost-effective flat slab solutions for steel construction.

These integrated floor systems, often called slim floor beams, simplify service integration and avoid the problem of having an additional service zone below the beam level that otherwise would increase floor to floor height (see Chapter 4.14). Another solution with minimal floor depth for multistorey framed buildings is to use cellular beams with large web openings. The openings accommodate the service zone within the beam space, thus reducing the floor height and, likewise, the facade area and column heights (see Chapter 4.12) (Figure 3.17). These long

Figure 3.17 Loading of a cellular beam – here a roof girder in industrial building. Cellular beams are also solutions with minimal floor depth for multistorey framed buildings. © ArcelorMittal.

span options can offer suitable exposed soffits as well as integrated services and are inherently fast to install because of the fewer structural members.

In industrial buildings, steel floor plates offer cost-effective solutions with slip-resistant durable surfaces for walkways, stairs and platforms. Returning to the basic structural steel frame option, attention should be paid to the following points in order to optimise the LCC of a steel structure [11]:

- select available steel sections with the lowest cost;
- select steel sections with the lightest weight;
- select the minimum number of different types of sections;
- select sections with the minimum total exposed perimeter to reduce fire protection costs

The sustainable characteristics of structural steel that offer economic benefits are

- 100% recyclable – good end-of-life value;
- uses minimum volume of material;
- clean, dust-free construction process;
- minimal site wastage;
- off-site fabrication in a controlled environment;
- adaptable and flexible to suit changing lifetime requirements.

Steel buildings of modular form are commonly manufactured for such purposes as medical units to house operating theatres, clean rooms etc. All of these off-site prefabricated options are dependent on the ongoing demand to sustain the costs of industrial-type manufacturing facilities that achieve 'economy of scale' by a certain level of standardisation.

3.4.3.2 *Cladding/roofs*

Steel cladding was once just the preferred material for agricultural buildings, industrial warehouses and stores. However, with the introduction of insulated sandwich panels with a high quality external finish, the possibilities of using it for external walls and roof cladding has increased, both as a material that can compete with the other traditional solutions and as an aesthetic choice when seeking a more modern, technological look for buildings. Table 3.2 compares the life-cycle costs of two possible cladding systems.

Looking at the thermal performance of existing buildings and their envelopes and at rising energy costs in recent years, thermal retrofitting by means of over-cladding or replacing existing cladding might also be reasonable from an economic point of view, taking into account energy losses, which can be calculated for various options. Numerous options for protective coatings with various lifespans are available from the major steel manufacturers. There are also combinations incorporating other energy efficiency options, such as photovoltaic (PV) cells or using stainless steel and weathering steel for the cladding. Recent LCC models

Table 3.2 Life-cycle cost analysis example: Comparing two alternative cladding systems.

Options	Steel Cladding A	Steel Cladding B
Capital costs	€40,000	€55,000
Recoat frequency	10 years	20 years
Cost of recoating	€10,000	€20,000
Lifespan of building	40 years	40 years
Discount rate	3%	3%
Service life	20 years	40 years

Assumed annual maintenance and cleaning costs and residual value the same

Life-cycle costing	Steel Cladding A	Steel Cladding B
Capital costs	€40,000	€55,000
Recoat A yr. 10 €10,000 × 0.744 PV	€7,440	-
Replacement A yr. 20 €40,000 × 0.554 PV	€22,150	-
Recoat B yr. 20 €20,000 × 0.554 PV	-	€11,080
Recoat A yr. 30 €10,000 × 0.412 PV	€4,120	-
Totals	€73,710	€66,080

Conclusion: Cladding B is 12% less expensive over the lifespan though the capital outlay was 38% more expensive.

published in the UK [12] compared steel cladding products with common alternative materials. Even though it was a general exercise rather than relating specifically to a particular building, it gave an indication of how cladding products compare. The aim was to produce a working LCC model that would demonstrate the WLC difference between various options. A number of assumptions were included as a basis for these life-cycle models:

the capital costs of the cladding system include its substructure – the supporting steel rails or purlins;
routine maintenance – includes annual inspection and repairs;
exceptional maintenance costs include recoating the cladding after 20 years;
the end-of-life value of a steel-based system is estimated zero as the scrap value will equal the demolition costs;
demolitionand disposal costs in masonry (one of the alternative materials);
other maintenance costs;
every five years, roof weather sealing and repainting at industry-standard Ievels;
the period to repaint decision (PRD) frequency is based on data from the manufacturer;
the costs exclude energy consumption during the building's lifetime.

As an example, options for a roof of 1,000 m² in area with an assessment period of 30 years are considered. The steel metal profile roof with a PRD of 30 years was compared with a concrete tile roof and a single-ply pitch polymer system on galvanised decking (flat roof). The LCC for the polymer system was higher due to the need to replace it every 20 years, even though it had the lowest capital costs. The concrete tile proved to have a continuous LCC mainly because of a long PRD. However, factors such as aesthetics and speed of construction, the effect of self weight loads and degree of adaptability were not considered. In addition, the cost model demonstrated that the coated steel sheeting compared favourably against the other materials in the lifespan of 30 years (Figure 3.18).

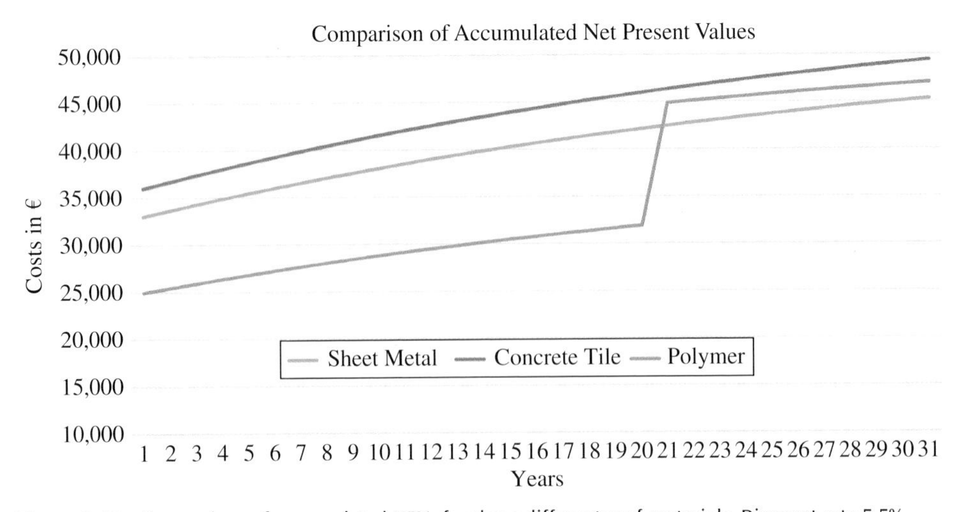

Figure 3.18 Comparison of accumulated NPVs for three different roof materials. Discount rate 5.5%.

The question of costs always has to be considered in the context of the 'value proposition' for the customer, not just the contractor who buys products or solutions but the final user or future owner of the building.

3.5 ENERGY EFFICIENCY

Raban Siebers and Markus Kuhnhenne

Minimising energy consumption and limiting harmful environmental impacts are central aspects of a sustainability assessment. Energy saving has been the subject of legislation and standards for decades on national levels. Later the European directive 2002/91/EC on energy performance of buildings and the succeeding directive 2010/31/EU and 2012/27/EU on energy efficiency provide guidelines that have to be realized in national regulations. Increasing energy efficiency is one of the key features of sustainable construction. On the one hand, the energy requirement itself is assessed, and on the other, the environmental impacts and costs associated with the energy requirement are determined.

When auditing energy flows, for example, in Germany in accordance with the former Heat Insulation Ordinance (WärmeschutzV 1995), only the heating requirement was recorded; the effort needed to produce and distribute the heat and losses due to nonideal control were not recorded. In addition, the effect of thermal bridges on the heat losses due to transmission were not taken into consideration.

These aspects were subsequently taken into consideration in the calculation method presented in energy saving ordinances such as the German EnEV 2002, which allowed the primary energy requirement for heating (including hot water) to be determined. With the introduction of the revised EnEV 2007, which was an update to meet the regulations of European directive 2002/91/EC, additional energy required for lighting and cooling was included. However, in principle the levels of performance were no more stringent than in EnEV 2002. The significant changes in EnEV 2007 were that the use of the building within prescribed types of use was taken into consideration, and this affected in particular lighting, cooling and the necessary air changes. A target set for the permitted energy requirement was no longer required, but a 'reference building' was used to determine the target.

In the next step of energy saving ordinance EnEV 2007/2009, it became necessary to calculate the annual primary energy requirement for all nonresidential buildings when at least one of the conditioning systems of heating, cooling, ventilation, humidification, lighting and domestic hot water systems is present. In addition to the requirements with respect to primary energy and limiting the heat transfer by thermal transmission, generally applicable requirements are also required for

- air-tightness, minimum air-change rate;
- minimum thermal protection and control of losses at thermal bridges;
- equipment for heating, cooling and ventilation and air conditioning;
- examination of alternative systems of energy supply;
- use of energy performance certificates (as required by EU legislation).

When EnEV 2014 came into force the requirement on the overall energy performance becomes stricter of about 25% compared to the level of EnEV 2009.

The energy requirement is affected by various building-related and use-related factors such as thermal insulation standard, type of ventilation, losses during heat generation, lighting concept and cooling system. EnEV 2014 takes into consideration all parameters that affect the energy requirement of a building during the operation phase, and this is apparent from the complexity of the method of calculation.

With its demands on new buildings, EnEV 2014 also limited the heat transfer by transmission via the building envelope. In areas where thermal bridges have an influence on the overall performance, proof of minimum thermal protection has to be provided, and the 'additional' heat transfer has to be taken into consideration in the calculation of the primary energy requirement.

The demands placed on the air-tightness of the building shell are aimed at avoiding unnecessary heat transfer and issues such as condensation. The airtight layer should prevent warm moisture-laden air flowing through components, as leaks in the airtight layer can lead to condensation damage in the building fabric.

3.6 ENVIRONMENTAL PRODUCT DECLARATIONS

Raban Siebers and Diana Fischer

Many certification schemes have been developed in recent years that enable the ecological or even the sustainable quality of a building to be determined. Examples are the DGNB label developed by the German Sustainable Building Council or the LEED scheme from the United States. Certification schemes place different demands on the environmentally relevant performance of buildings. In order to undertake a comparable assessment of the quality of a building, it is also necessary to take the construction products used into consideration in the certification. Due to the large number of production processes involved for each construction product (such as the mining of raw materials, transport to the plant and to the construction site, processing, etc.), it is not possible for the users of products to collect and process the complete product data for a building. The manufacturers of the products have to undertake this task because they have the specialist knowledge as well as the insight into the required company data on the source of the raw materials and the production and manufacturing processes or the energy consumption.

Different types of environmental declarations circulate on the market. Some of them evaluate specific products and are used especially for marketing purposes in business-to-consumer communication. The type discussed here, environmental product declarations (EPDs) for construction products, are so-called Type III environmental declarations. The purpose is business-to-business communication without evaluation or comparison on the product level (Table 3.3).

When sustainability assessments of buildings first began in Germany, LCAs were prepared for many products based on the average data available, which was

Table 3.3 Different environmental declaration types (EPDs are type III declarations).

	Type I Certified labels	Type II Self-declarations	Type III EPDs
Auditing by independent external auditors	Yes	Possible	Yes
Parameters	Several external set criteria – evaluation	By the company itself defined environmental standards	Environmental and technical data along the product life cycle – no evaluation
Example	EU Ecolabel	CO_2 Footprint	EPD 'Structural Steel'

published in the ÖKOBAUDAT database of the Federal Ministry of Building and Urban Development (see Section 3.10). In recent years, this database has been expanded and updated and was being adapted to meet the requirements of EN 15804. The development of the basic database was an important step towards considering construction from an ecological point of view. However, when using the data it should be noted that ÖKOBAUDAT only includes the results of the LCA but no other product information over and above that. The data it contains are average data for Germany so it also takes data into account on a pro-rata basis from producers outside Europe that also supply their construction products to Germany. This means manufacturers from Germany and Europe with above-average eco-friendly production are at a disadvantage.

For these reasons, each individual or an association of various manufacturers can prepare an EPD. Manufacturers are using EPDs to show that they are particularly eco-friendly or more resource efficient compared with manufacturers of similar products. In addition, EPDs can include further eco-relevant information about the product, for example, on recyclability or even technical information such as service intervals and energy consumption during use. Planners and owners of a building can consult this information. As EPDs are producer related, they may only be applied if the product actually used is also from the producer listed in the declaration. If no EPD is available, average data from the ÖKOBAUDAT database has to be used.

Several producers of similar products can also prepare a joint EPD. The advantage of such declarations is that the customer does not have to choose a specific producer at the outset but can use better than the average data for calculation purposes. When it comes to finally ordering the materials, the customer can take advantage of competition between the EPD participants without having to change the LCA previously prepared.

To prepare an EPD, it is first necessary to collect the information needed for the audit, such as details on the quantity and type of raw materials required. A manufacturer-specific LCA is then prepared using this data and the results are published in an EPD. These declarations also contain additional product information, such as on properties associated with the building physics performance. To ensure the quality of EPDs, certification authorities in Europe appoint authorised experts to check the correctness of the basic principles and the data in an EPD independently. This ensures adherence to the rules on eco-auditing with regard to the content of an EPD given in ISO 14025 and EN 15804.

An EPD involves carrying out a life-cycle inventory (LCI) and a life-cycle impact assessment (LCIA) and presentation of other indicators, for example, type and quantity of waste produced. The LCI contains data on resource consumption, for example, energy, water, renewable resources and emissions into the air, water and soil. The impact assessment is based on the results of the LCI and specifies concrete environmental impacts. These include the greenhouse effect (CO_2 emissions), damage to the stratospheric ozone layer, acidification of water and soil, eutrophication (overfertilisation), formation of photochemical oxidants (smog) and exhaustion of fossil energy resources and mineral resources (see Section 3.3.1).

3.6.1 Institute Construction and Environment (IBU) – Program Operator for EPDs in Germany

Burkhart Lehmann

Since 1982, the Institut Bauen und Umwelt (IBU, Institute Construction and Environment), which was formerly Arbeitsgemeinschaft Umweltverträgliches Bauprodukt (AUB, a working group for eco-friendly construction products), has been aiming to achieve higher levels of sustainability in the building and construction industry. Since 2004 it has acted as the programme operator of an intersectoral declaration scheme in Germany for EPDs, based on ISO 14025. Thus IBU is described here as a good example for a national EPD program operator.

EPDs are based on independently verified data from LCAs and have become established across Europe, particularly in the building and construction sector. The reason is that different demands are placed on construction products depending on the type and use of a building, as well as its location. As semifinished products, the technical capability of construction products usually only manifests itself when they are used in components and in construction. This means the environmental impacts of a construction product can only be properly evaluated once the actual installation and the building context are known.

This viewpoint has become established across Europe, not least because of the standardisation activities of the European Committee for Standardization on the sustainability of construction works (CEN/TC/350) initiated in 2005. The main outcome of these activities is that the assessment of the ecological quality of a building should be based on an LCA with the environmental declarations of the construction products providing the base data.

As part of CEN's standardisation work, rules for LCA calculations in environmental production declarations over and above those specified by ISO have been specified. The European standard EN 15804 introduced in April 2012

Figure 3.19 Logo for products with an EPD from IBU.

provides the core product category rules (core PCRs) for environmental declarations for construction products and all kinds of construction services. Uniform indicators and methods of calculation form the basis for the EPDs, which are valid throughout Europe.

The IBU was the first EPD programme in Europe to successfully implement the new guidelines given in the European standards for the preparation of EPDs. All 1300 EPDs that have been verified and published by IBU are following EN 15804 since it became valid in 2012.

The institute has an independent expert advisory board (SVR) that oversees its technical activities. This body of independent third parties includes experts from science and standardisation as well as representatives of the building and environmental authorities, testing facilities and nature conservation organisations. The SVR ensures that the product-specific requirements (PCR instructions) for certain groups of construction products prepared by product group panels conform to the normative requirements of ISO and CEN standardisation, and it is responsible for the quality assurance of the verification process in the IBU EPD programme. Further information is available at www.ibu-epd.com (Figure 3.19).

3.6.2 The ECO Platform

Christian Donath

The ECO Platform, based in Brussels, was founded in 2013 to support the development of a common core EPD for construction products in Europe. The founding members included programme operators from Germany, Sweden, Norway, Great Britain, France, Spain, Poland, Portugal, Slovenia and the Netherlands. Construction Products Europe (CPE, formerly CEPMC) is also represented.

EPDs serve as important sources of information for the consumption and environmental impacts of construction products. EPDs are needed for life-cycle-based assessments of sustainability on the building level. Only by balancing initial 'efforts' (costs, emissions, consumptions, effects) with later 'impacts on the building-level during the full life-cycle', can an assessment of sustainability be reasonable and accurate.

Up to now, the absence of a common harmonized solution for product-specific life-cycle data in a common format required enormous effort for the industry and hindered a comprehensive way of communication to the market. The fact that 'national' EPDs for products from construction product manufacturers were not recognised in other markets also created a problem. As a consequence, producers had to provide several EPDs with similar content for different markets.

The ECO Platform is an umbrella organization for the different European EPD programme operators and aims for mutual recognition of EPDs throughout Europe by finding a common way of interpreting the relevant standards. This 'European core EPD' is strictly based on the European standard EN 15804 (*Sustainability of Construction Works – Environmental Product Declarations – Core Rules for the Product Category of Construction Products*). In addition to the programme operators, various European trade associations for construction products, green building councils or operators of green building rating schemes as well as providers of LCAs are actively involved in the ECO Platform.

The ECO Platform is working on the harmonization in three working groups that deal with technical aspects, quality management and certification as well as communication. The working group for communication is particularly important for contact and coordination with the European Commission, CEN and the product technical committees, in which the product-specific scenarios covering the life cycle of the product will be defined (PCR). The various activities are closely linked with respect to organisation and personnel to ensure a harmonized approach. All ECO Platform members can participate in the working groups. The involvement of the trade associations for construction products ensures that product-specific knowledge and the requirements of the industry are taken into consideration. Green building councils and the rating schemes are increasingly relying on information from EPDs in their ratings for buildings and products and are interested in the wider use of a European construction product EPD of uniform quality that provides a well-founded basis for building assessments.

The harmonization process is ongoing. However, as an important step toward the final goal the first ECO EPDs based on common verification guidelines were introduced to the market in 2014. By this step, all EPDs prepared by one of the member EPD programme operators in accordance with the ECO verification guidelines are recognised by the other ECO members, meaning that an EPD that is prepared and verified by one of the member programmes (e.g. IBU in Germany) is also recognised as being prepared and verified in accordance with the jointly defined ECO Platform standard by all other member programmes.

The full comparability of EPD, however, is not given yet. To achieve the goal of a fully comparable core EPD, several issues, often out of the influence of the EPD programme operators (such as background databases or national regulations; see Section 3.9), have also to be aligned. The ECO Platform acts to bring all relevant parties together. By making the differences transparent and providing solutions to overcome the gaps, the ECO Platform contributes to a common harmonzed approach for EPDs in Europe (Figure 3.20).

Figure 3.20 Logo of the Eco Platform.

3.7 BACKGROUND DATABASES

Raban Siebers

When preparing LCAs, for example, for publication in environmental product declarations, experts resort to background databases. Software facilitates access to these databases, and background data can be inserted into a model on aspects such as production processes and compositions. This means an LCA for a product can be prepared by modelling a manufacturing process with raw materials and consumables, energy flows, waste and transport. The well-established programs include the GaBi software from thinkstep and the SimaPro software. They are capable of accessing many different background databases and are not only suitable for preparing LCAs but enable the user to simulate different scenarios and thus strive to achieve optimisation. This approach is called 'design for the environment' or 'ecodesign'. It allows various ways of processing raw materials or different production processes to be compared and optimised.

In the background databases, numerous datasets on raw materials can be found starting with products, production and disposal processes as well as energy production and conversion. They complement and complete the environmental data collected by the company itself. The datasets are designed in such a way that measurable parameters such as power consumption, material masses and distances can be used to determine the environmental impacts requested by the LCA, such as global warming potential, acidification potential, ozone depletion potential and renewable and nonrenewable primary energy (Section 3.3.1). For example, the power consumption for a specific production chain can be determined. This information is then combined with the dataset for the respective power mix from a background database, and the environmental impacts of the power consumption for the production chain being considered are then calculated.

The ecoinvent database and GaBi database from thinkstep are often used as background databases for LCA of construction products. There are other specialist databases as well for foodstuffs, agricultural products, chemicals and transport as well as country-specific databases that are based on common background databases.

The ecoinvent database was originally established in Switzerland and is currently the leading data system worldwide with over 4500 users in over 40 countries. Ecoinvent is often used for internal purposes because it is currently mainly based on data from the literature that has not been verified externally. With its large volume of data for almost all fields it also satisfies the prerequisites for academic research. Its use is somewhat more complicated when preparing an EPD as a Type III declaration that complies with ISO 14025 and EN 15804. This has to be developed with the involvement of an independent third party and checked independently (see Section 3.6), which means verification and external examination of the background data used has to be carried out.

The GaBi database is increasingly being used in this field. A four-step review process is required to access certified datasets, which have mainly been collected from and by industry itself. The subsequent verification of an EPD is considerably easier because it uses datasets that have already been checked. The quality of the data from the GaBi database is held in high regard and offers greater specialisation than ecoinvent. The developers and distributors of this database also offer to prepare datasets on request. This allows datasets that are currently not available to be acquired additionally, or available datasets can be adapted to the specific requirements of the project.

3.8 EUROPEAN OPEN LCA DATA NETWORK

Tanja Brockmann

Sustainability considerations in the construction sector are nowadays established in many European member states, and many of them are using certification schemes to evaluate sustainability at building level. Within these, the LCA is a central instrument for the evaluation of the ecological quality of the building, for example, global warming, ozone depletion, photochemical ozone creation, acidification and eutrophication potential (see Section 3.3).

In Germany, as part of the BNB [9], the Federal Ministry for the Environment, Nature Conservation, Building and Nuclear Safety (BMUB) provides important infrastructure for choosing suitable building products. There is a complete 'tool chain' offered for use. It starts with the basic material data from EPDs, which are imported in the online database ÖKOBAUDAT [10], exported from there to the calculation tool 'eLCA', and used for the final evaluation of a sustainable building resulting in a bronze, silver or gold certificate (Figure 3.21).

3.8.1 ÖKOBAUDAT

The German online database ÖKOBAUDAT has been provided within BNB by BMUB since 2009, and it is available for anyone interested in an ecological evaluation of building with a consistent database. Its main users are planners and architects, who analyse the environmental indicators of the products, services and processes as integrated in the database.

The webpage (www.oekobaudat.de) offers user-friendly direct access to the database. Basis information, as well as interesting links, is given. Different versions

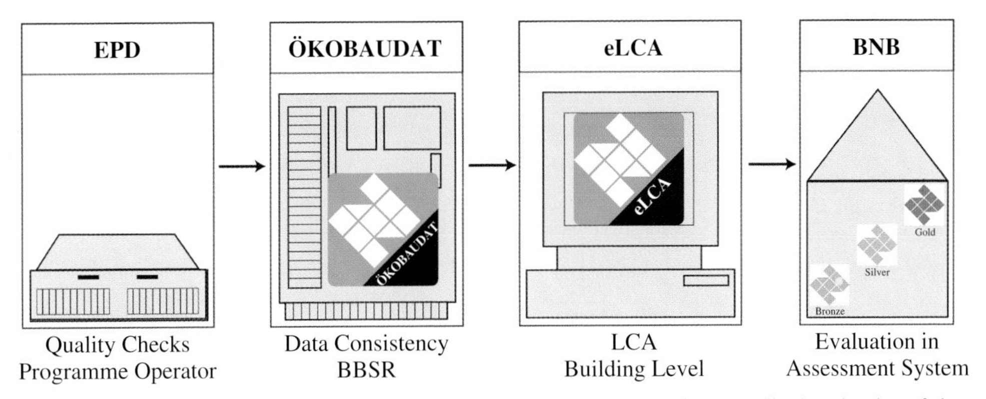

Figure 3.21 LCA within the German BNB – from product and basis data via LCA to final evaluation of the sustainability of the building.

of the ÖKOBAUDAT are archived. This can be relevant for ongoing projects as well as for reproducing former LCA results. Search and filter functionalities allow finding relevant datasheets for chosen materials or products directly in the online database. A comprehensive English version of ÖKOBAUDAT has been provided since June 2015.

ÖKOBAUDAT contains generic basic datasets that provide suitable averages of the environmental indicators for the building materials, as well as product-specific datasets that are determined within EPD. Using the generic datasets allows sustainability studies of buildings already in early planning stages when architects or planners do not yet work with product-specific but with generic building product information. In a later stage, the generic data in the model is then substituted with specific data representing actual construction products.

Since 2013, ÖKOBAUDAT comprehensively meets the demands of European Standard EN 15804 (see Chapter 2.2). It was one of the key developments to offer a data format that meets EN 15804 but also follows an extended International Reference Life Cycle Data System (ILCD) format. As an internet-aware data format, it has been designed to explicitly allow publishing and linking of data as resources via the internet (Figure 3.22). The advantage of this new approach is that existing software tools with built-in support for the ILCD format can be easily enabled to support the new EPD dataset as well, with only minor changes to their internal information structures. ÖKOBAUDAT is running on an open-source programme software platform (soda4LCA), which allows the development of further modules that use or may add new features to the procedures.

In Germany, the IBU as an important EPD program operator has equipped its own database application with facilities to directly import its data online into ÖKOBAUDAT (see Section 3.7). Not all institutions that may offer suitable building materials-related data will be able to generate data with a tool of their own. Thus, a further project has been set up to modify the widely adopted open-source LCA modelling tool 'openLCA' accordingly to allow creating suitable EPD data that subsequently can be imported into ÖKOBAUDAT, even online directly from openLCA. This indirect data transfer is currently carried out by German EPD

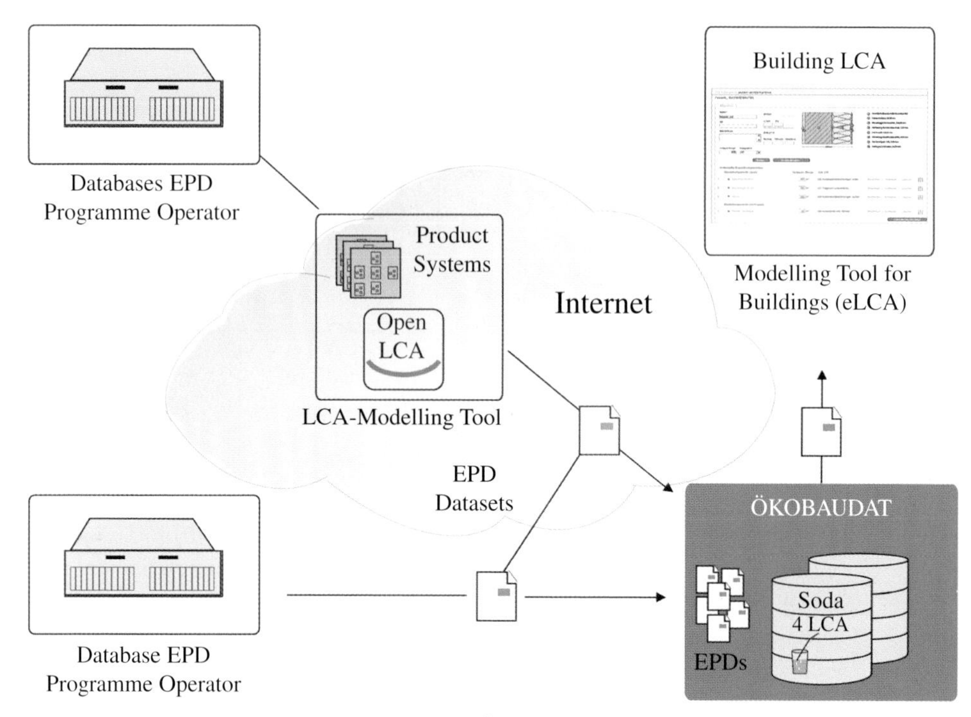

Figure 3.22 LCA data import into and export from ÖKOBAUDAT.

program operator ift Rosenheim, and the Austrian EPD program operator Bau-EPD, as well as by the German Thünen Institute for its LCA data, that is, averaged LCA data derived from industrial background data regarding wooden materials. All this newly imported data has been published with the release of ÖKOBAUDAT in summer 2015.

All data transfer procedures are administrated by BMUB/BBSR (Federal Institute for Research on Building, Urban Affairs and Spatial Development). Thus, an approval of the data transfer into ÖKOBAUDAT is required. BMUB/BBSR always checks the adherence to the 'requirements for the acceptance of LCA data in ÖKOBAUDAT' [10], for example, the compatibility with EN 15804, EPD program rules, verification of data by an independent third party, period of validity of data, data format and some additional requirements for ÖKOBAUDAT. All data are imported to an 'inbox', where they are quality checked for plausibility and completeness before being released in ÖKOBAUDAT.

3.8.2 eLCA, an LCA tool for buildings

ÖKOBAUDAT delivers LCA data of building products that are factored into the LCA at building level. Consequently, a further BMUB project was initiated to develop a BNB-compliant LCA tool for buildings. eLCA is an open-source online tool, available since January 2015 (www.bauteileditor.de). It is user friendly and

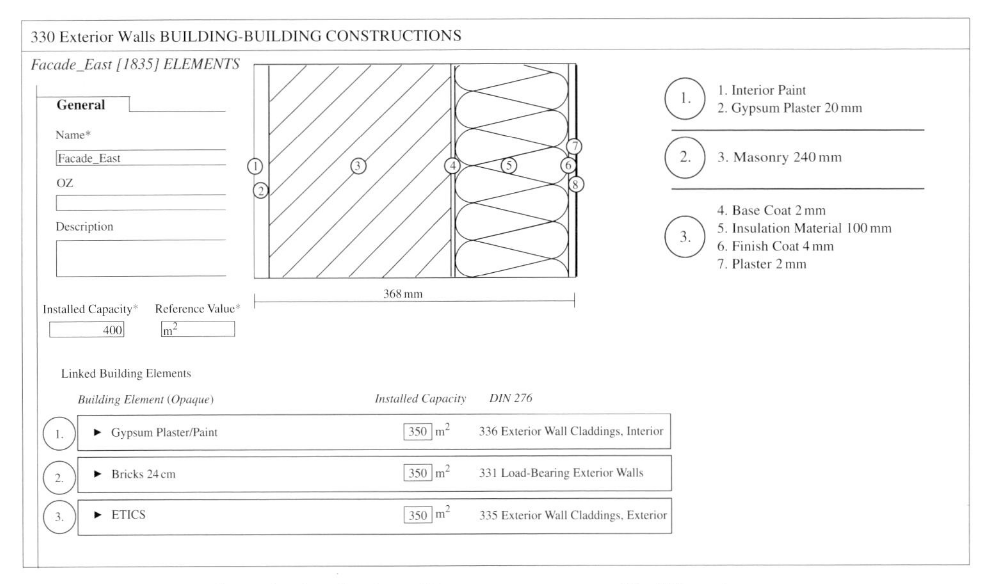

Figure 3.23 eLCA – dynamic visualisation of input parameters and building elements.

allows for a consistent and comparable LCA at building level. Motivation for the development of eLCA was to gain independence from commercial tools and to adapt an LCA tool to the needs of BNB. Furthermore, the tool can be used for the derivation of benchmarks, and it allows adapting the tool to any required changes with a high level of independence.

Within eLCA the building structure is created by the components and construction elements with the associated materials. The underlying data of materials is given in the ÖKOBAUDAT – in eLCA the contribution of all materials to the environment has to be calculated in the amount of the materials as used in the real construction, according to the German standard DIN 276 (costs in construction sector). A specific feature is that the creation of elements is associated with dynamic graphs that show the thickness of different material layers – this helps to prove the created building components and elements (Figure 3.23).

The product and construction stage, use stage and end-of-life stage are considered within eLCA. The evaluation of the project's LCA can be presented as a total score result, which is relevant for the evaluation of the addressed sustainability criteria within BNB, but also separated into construction elements (cost groups according to the German standard DIN 276) or relative to the life-cycle stages (Figure 3.24).

With eLCA the calculation of new construction and refurbishment is possible. It is used for evaluation of different sustainability criteria within BNB and may influence the final result of the building. As the choice of construction and products may have a significant influence on these ecological results within BNB, eLCA is used in the different planning stages. Today, eLCA is used by many engineering offices as well as by numerous universities. An English version of eLCA is planned.

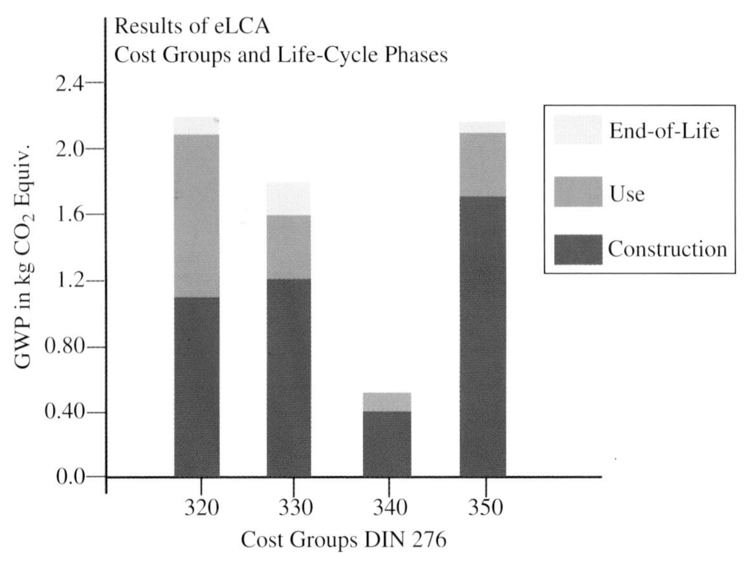

Figure 3.24 eLCA – graphic results of GWP in cost groups according to German standard DIN 276 (costs in the construction sector) and life-cycle stages – 320 foundation, 330 exterior walls, 340 internal walls, 350 ceilings.

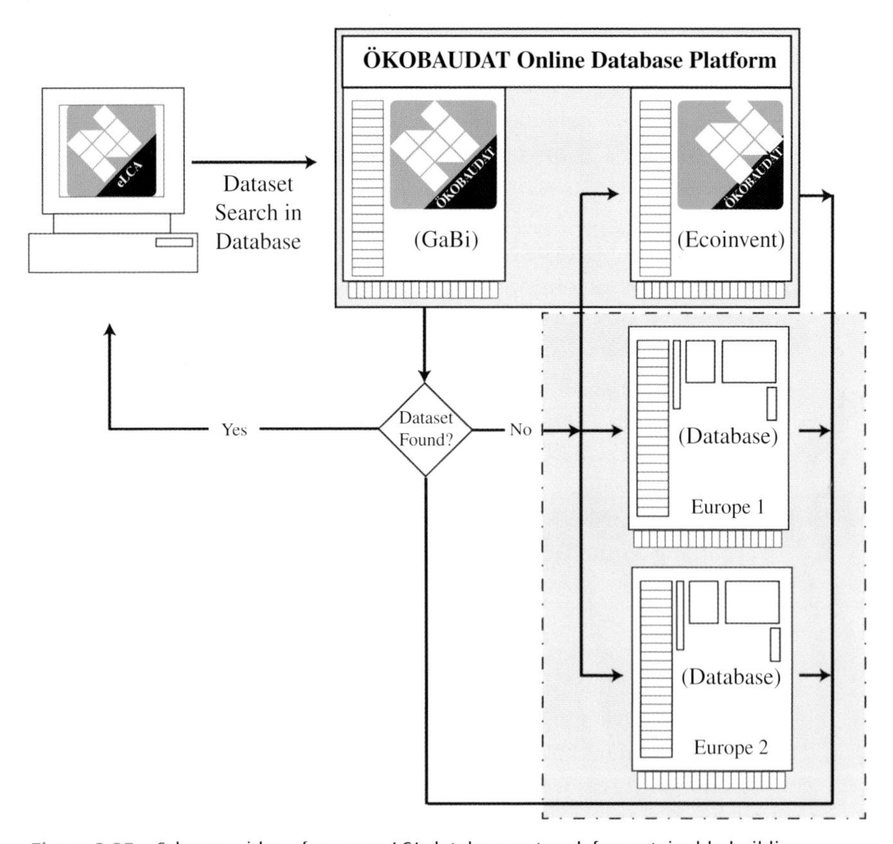

Figure 3.25 Scheme – idea of an open LCA database network for sustainable building.

3.8.3 LCA – a European approach

The possibility to directly import data into an online database (this could be a European joint database, or it could also be the ÖKOBAUDAT), or linking databases with a given harmonized data format that follows the generally accepted European standards, is a great chance for the idea of a consistent and harmonized way of using material and product-relevant LCA data or EPD data, respectively, for LCA at the building level within Europe. The idea behind the described developments is the vision of an open-data network. The main concept is an open-data network with the idea that there exist independent national databases that are linked to each other and due to a common standard allow open search and use of data (Figure 3.25). Each state will have the opportunity to set up its own national rules for the use of data in subsequently used LCA tools or the like.

Currently, there are the following international cooperations with ÖKOBAUDAT: transfer of data into ÖKOBAUDAT (Austria), external use of ÖKOBAUDAT (Denmark), and the planned linking of the Spanish database 'opendap' with ÖKOBAUDAT (Figure 3.26). These examples are a good starting point for further developments, and more European countries are already interested

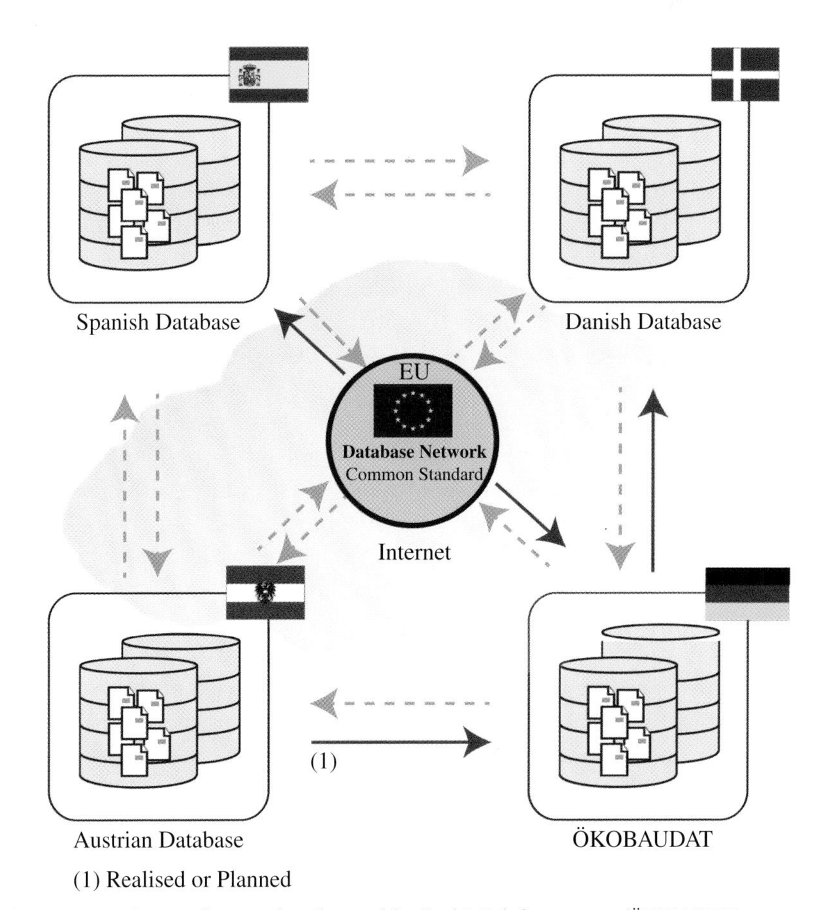

Figure 3.26 Current (international) use of (online) LCA infrastructure/ÖKOBAUDAT.

in using the developed data format and finding a way of linking databases. Still, there are important aspects that need to be discussed and analysed in further projects. For example, harmonization of background databases, harmonization or mapping of different product category structures, development of a common minimum standard of required information, determination of levels of data quality, administration and help desk.

The idea of an open-data network would support European ideas. In order to bring forward this idea, in general, political programmes and support are very helpful, also for the definition of responsibilities. It is very helpful to run one central national database for EPD data rather than having different databases by several programme operators. Experience in Germany has shown that the centrally bundled provision of data and tools by the government is a highly supportive instrument for a wide application of LCA at building level and the realisation of sustainable buildings.

3.9 ENVIRONMENTAL DATA FOR STEEL CONSTRUCTION PRODUCTS

3.9.1 The recycling potential concept

Bernhard Hauke, Johannes Kreißig and Markus Kuhnhenne

A large proportion of the structural steel used in Europe, profiles, merchant bars and heavy plates, is made from secondary raw material scrap steel. This means that when considered in isolation, making steel from scrap in an electric arc furnace (EAF) saves primary energy because it dispenses with ore preparation and the extraction of pig iron from the ore. However, blast furnace (BF) steel also makes an important contribution to the industrial cycle. Every tonne of primary steel (which is what BF steel is referred to because of its production from iron ore – a primary raw material) creates de facto a construction material that can be recycled an infinite number of times. Steel construction leads to long service life and has experienced high growth rates in Europe in recent decades. This means large quantities of steel are currently used in products and there is insufficient scrap available via the scrap market to satisfy current demand. Primary steelmaking is therefore necessary in order to ensure there is a fresh supply of new products. Regardless of whether it is obtained by the primary or secondary route, it is important that the valuable used steel (scrap) is collected after use and returned for recycling. This 'cradle-to-cradle' concept reduces the use of primary raw materials and improves the LCA of structural steel. Recycling must therefore also be taken into consideration when preparing LCA for construction products made of steel. There are basically two approaches: recycled content and recycling potential.

With the recycled content approach, only the production process 'cradle to gate' in question is considered. What is important for the evaluation is the proportion of material (e.g. scrap) made available without environmental burden. It aims to encourage in particular the use of secondary material; in contrast, no consideration is given to collection rates or recyclable construction. However, end-of-life steel, in other words scrap, is a material that is already being actively traded as a commodity. The recycled content approach encourages uneconomical material flows for scrap but neglects the far more important aspect of conservation of materials by achieving the highest possible collection rates.

In contrast, the recycling potential approach takes into consideration a material's complete industrial cycle. By considering the current collection rate (reuse rate + recycling rate), it takes into account the fact that secondary production reduces the need for the corresponding amount of primary production and the associated impacts. This is also called the 'end-of-life' approach to recycling. What is important is the net amount of material conserved, which reduces the overall environmental impact. It also encourages design for recycling.

If a production mix of 40% BF steel and 60% EAF steel is assumed and around 25% steel scrap is used in the BF route, this will result in 110 kg of reusable products and an additional 180 kg of scrap (Figure 3.27). These products and the additional

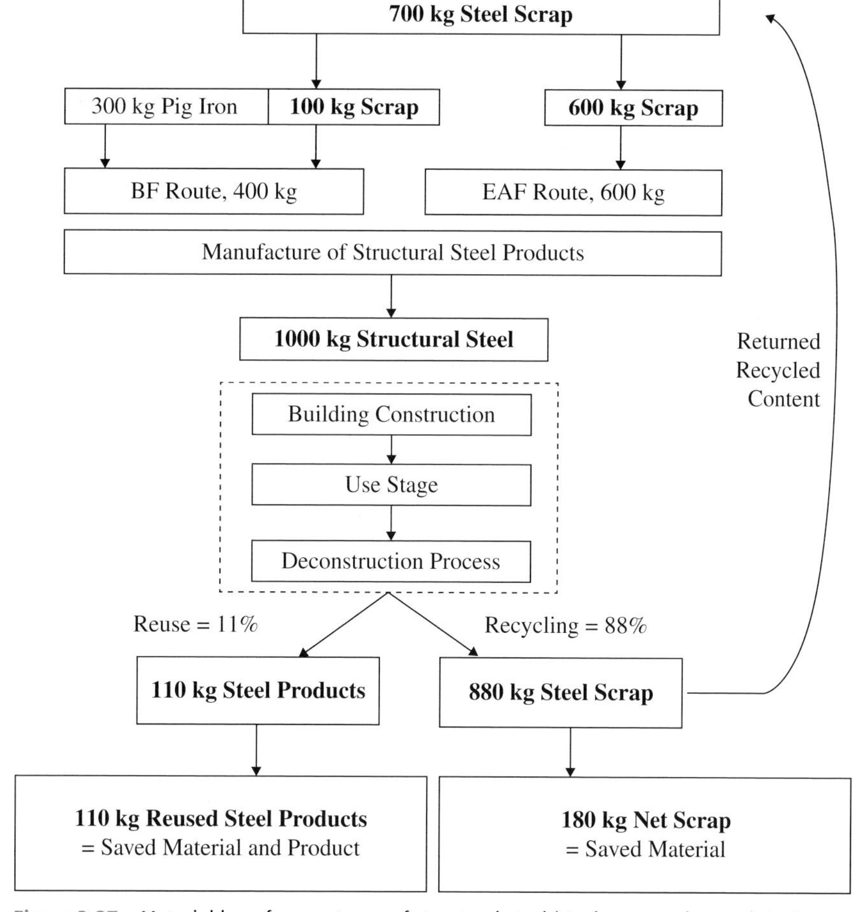

Figure 3.27 Material loop for one tonne of structural steel (steel scrap and reused steel products replace production from iron ore). © bauforumstahl.

scrap minimise the demand for primary raw materials and thus conserve natural resources. This data corresponds approximately to the EPD for structural steels from bauforumstahl members (Section 3.11.2) [13].

As mentioned above, if a material can be recycled there is a reduction in the consumption of raw materials and energy and in CO_2 emissions. The scrap needed for production (e.g. 700 kg steel scrap per tonne) has to be deducted from the 88% of the steel scrap that is recycled. The remaining scrap (180 kg steel scrap per tonne) and the steel products that can be reused directly (110 kg) avoid the need for production from primary raw materials. This is referred to as the 'recycling potential'. In this example, the recycled content, based on a simple calculation (only postconsumer scrap), would be 700 kg.

Even though evaluating the amount of recycled material in the construction materials used appears to be easier, it does not achieve the desired objectives of resource efficiency and avoidance of waste. The collection rate after deconstruction of a building and the quality of the construction materials, for example, whether they are reused or subjected to additional recycling operations, are not taken into consideration. This is the advantage of the recycling potential approach. Current average market data are used in the calculation, as required by EN 15804. This means collection and recycling rates are based on currently available facts and technologies. Differing service lives, which in the case of steel construction products can be very long, play a role in this approach. Changes in the average collection rates, the market shares of the BF or EAF routes or possibly even improvements in recycling technology can lead to an adjustment of the respective current recycling potential.

The basic idea of the recycling potential approach is to assign environmental impacts to each material cycle as a net balance in a 'cradle-to-cradle' accounting framework. For steel production, the first production step needed is the BF route with iron ore as the most important primary raw material, in which the energy requirement and emissions are relatively high. Steel made from iron ore is often used for consumer products such as cars or washing machine, and also for construction products, for example, heavy plates for bridges. Assuming that the collection rate is 100% when a washing machine is no longer used, or a bridge is dismantled, the resulting valuable steel scrap can then be recycled into new construction products without any loss in quality. Only the difference in effort and emissions between, for example, a construction product (here a heavy plate) and the secondary material (the scrap from the first life cycle of the steel) has to be taken into account. If all of the scrap is then collected and starts a second life cycle by being remelted in an EAF, the energy requirement and the emissions are considerably less than in the BF process. This is possible because the secondary material does not have any environmental burden from its primary production. The newly manufactured construction product from the same original material, this time possibly a rolled beam, can now be used in a building. Here, too, it is assumed that demolition results in a collection rate of 100%. This means that even after the second life cycle the steel scrap is completely available. Only the effort required to produce the new construction product (rolled beam) from the steel scrap has to be taken into account in the second life cycle.

If, however, a collection rate of zero has to be assumed – in other words the steel is lost completely – the full effort necessary to make steel from iron ore via the BF route would have to be taken into account for each new (construction) steel product. This would be independent of whether the scrap that was lost had been manufactured via the BF or the EAF route.

The recycled content approach only takes into consideration the fraction of secondary material used in the production process – the simple use of a material or a construction product is regarded as an environmental impact. In contrast, the recycling potential approach takes more precisely into account the environmental impact of any material that is lost and any production steps that have to be repeated (via the collection rate or the manufacture of the product using the EAF route, respectively). From the point of view of a methodology that considers the complete life cycle, the recycling potential approach is more suitable for steel construction products.

With a simplified LCA calculation according to the above-described recycling potential approach and the recycled content approach, the benefits and shortcomings are illustrated. For this purpose only the total required energy is regarded, but the results may be transferred to other efforts or emissions. The following highly simplified assumptions were made: the required energy for the BF process purely from iron ore (100% pig iron) is assumed with 26 MJ/kg structural steel product. Here the manufacture of pig iron as a material extracted from ore is assumed with 14 MJ/kg and the subsequent steelmaking and rolling process resulting in a product (e.g. a heavy plate) is assumed with 12 MJ/kg. For the EAF process based on the existing material scrap with a material value of 14 MJ/kg as explained above, the required energy to manufacture a product (e.g. a rolled beam) is assumed as 12 MJ/kg.

Figure 3.28 shows that for the BF process both approaches result in the same required energy if a collection rate zero is assumed. The input of cooling scrap at the BF steelmaking process, which is typically between 20% and 30% and may be considered by reducing the material effort accordingly, is neglected here for simplicity. Figure 3.29 shows that for the EAF process both approaches result in the same required energy if a recycling rate of 100% is assumed. In both basic cases the recycling potential is zero because the considered effort of the regarded life cycle is lost at the end of life (e.g. of the building). For the BF process the product and the material values are lost (collection rate zero) and for the EAF process the product value is lost (recycling rate 100% but no reuse).

If now for the BF process the recycling rate is assumed 100% as shown in Figure 3.30, the recycled content approach does not reflect this whilst the recycling potential approach results in a reduced energy requirement because only the product effort is considered and the remaining material effort is not. And if for the EAF process instead of recycling now 100% reuse is assumed as shown in Figure 3.31, the recycled content approach leads again to unchanged results whilst the recycling potential approach results in zero energy required because material

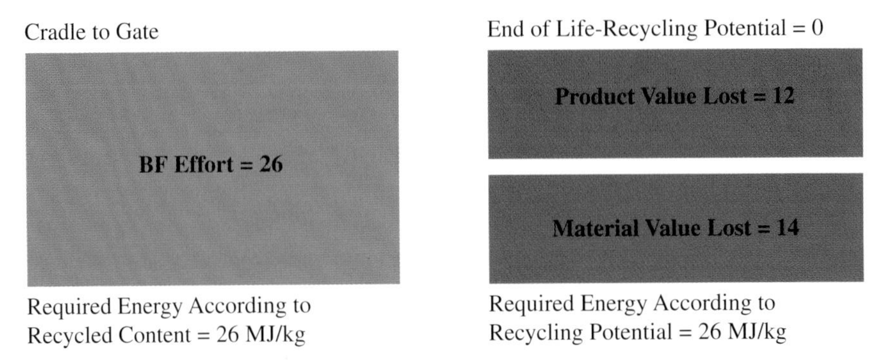

Figure 3.28 BF, no cooling scrap, 0% collection. © bauforumstahl.

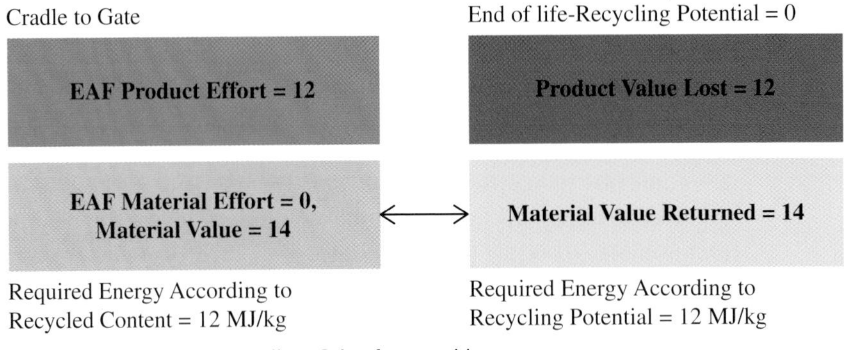

Figure 3.29 EAF, 100% recycling. © bauforumstahl.

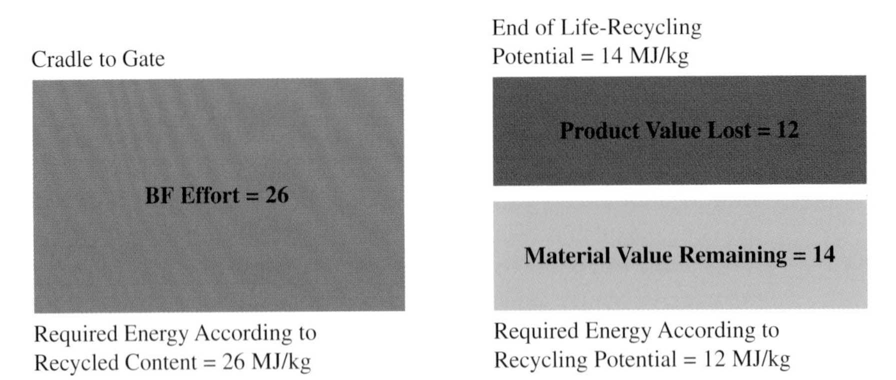

Figure 3.30 BF, no cooling scrap, 100% recycling. © bauforumstahl.

and product effort are remaining. This is of course a principle lowerbound consideration disregarding any efforts of deconstruction, material testing and so forth for the reuse of, for example, rolled beams.

Figures 3.30 and 3.31 show the same scenario of 100% recycling for the EAF and BF routes, respectively. Comparing the required energy illustrates that the

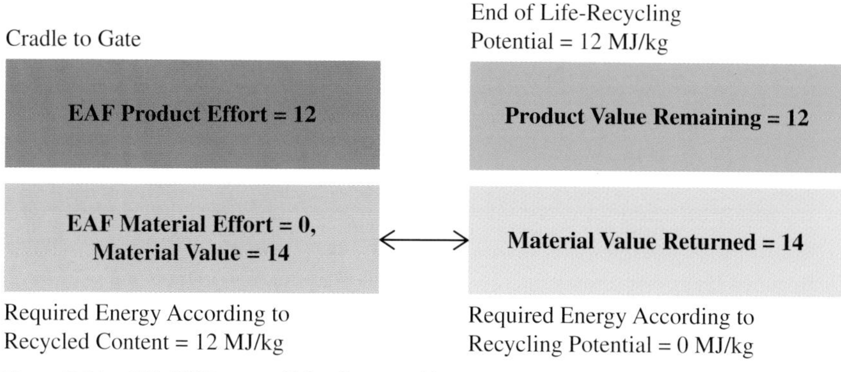

Figure 3.31 EAF, 100% reuse. © bauforumstahl.

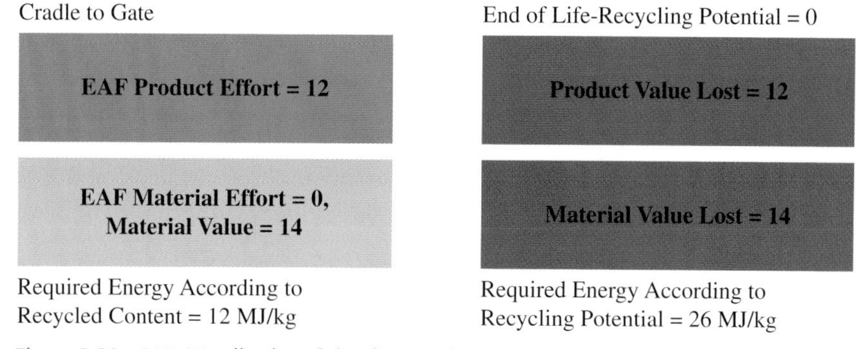

Figure 3.32 EAF, 0% collection. © bauforumstahl.

recycled content approach favours the EAF route with its 100% scrap content whilst the recycling potential approach honours the preservation of the material value (of scrap) equally for the EAF and BF routes.

So far, it seems that the recycling potential approach always results in favourable results for steel compared to the recycled content. However, let us now assume for the EAF process a zero collection rate. As illustrated in Figure 3.32 the recycled content approach still results in low energy demand despite the fact that not only the lost product value but also the lost material value must be replaced at the end of life with a fresh material supply from a BF process with iron ore. The recycling potential approach does reflect this complete loss for the EAF process with the same high energy demand as for the BF process (Figure 3.28).

Figure 3.33 shows then the typical market situation for structural steel according to Figure 3.27 with a mix of BF (cooling scrap considered) and EAF routes and a mix of recycling and reuse. Thanks to a collection rate of 100% the calculated energy demand according to the recycling potential is lower than for the recycled content approach. For reinforcing bars, which are exclusively produced with the EAF route, the results would be less favourable assuming only a typical collection rate of 70% [14].

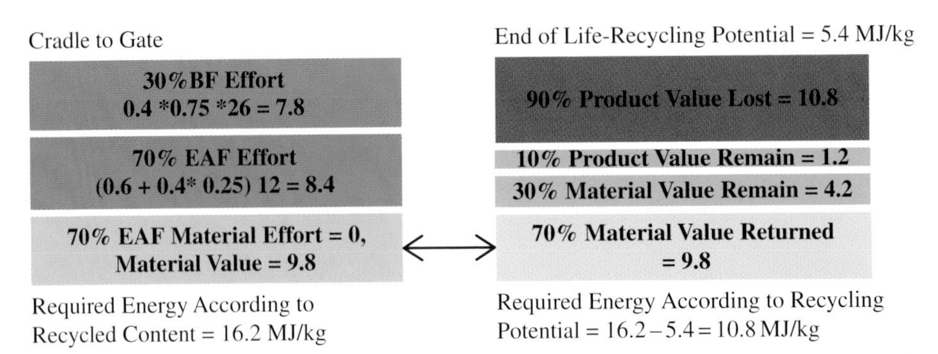

Figure 3.33 60% EAF and 40% BF with 25% cooling scrap, 90% recycling, 10% reuse. © bauforumstahl.

3.9.2 EPD for structural steel

Under the management of bauforumstahl, the steel profile manufacturers ArcelorMittal, Tata Steel, Peiner Träger GmbH and Stahlwerk Thüringen GmbH and the heavy plate manufacturers Dillinger Hütte and GTS Industries, Tata Steel and Ilsenburger Grobblech joined forces to disclose their environmental data in an EPD that complies with EN 15804. IBU (Section 3.7) commissioned independent specialists to verify the EPD Structural Steel: Sections and Plates (EPD-BFS-20130094-IBG1) [13]. In addition to the results of LCAs for manufacturing and recycling potential, it also contains product definitions and data associated with the building physics, data on base materials, source of materials, descriptions of manufacturing the product, information on product processing and standardisation, and data on the conditions of use, exceptional impacts and the postutilisation phase. It covers steel structural sections, steel bars and heavy plate in the steel grades S235 to S960. For structural steel, there are two different production routes: in the BF process, steel is essentially produced from iron ore, coking coal, coal and scrap (up to about 35%) and subsequently hot rolled to form the required shape. In the EAF, steel scrap is melted down and also rolled to new steel products. Both processes are used at the participating steelworks so that a mix of the BF and EAF routes is shown in the EPD (see Section 3.11.1).

An LCA was carried out and presented in accordance with ISO 14040 and the requirements of EN 15804 and thus conforms to EN 15978, the standard for evaluating the sustainability of construction. The IBU guidelines for Type III declarations and the PCRs for structural steels constitute the formal framework. The environmental data for the structural steels covered by the EPD (Table 3.4) is more than 37% less than the data from ÖKOBAUDAT and even significantly less for individual environmental parameters (Figure 3.34). This reflects the fact that the participating European manufacturers of structural steel profiles have continuously invested in modern, eco-friendly steel production and to high social standards.

As a basis, specific data for the products investigated were obtained from all the plants together with data from the GaBi 6 database (see Section 3.9). The LCA covers the life-cycle stage's raw materials and energy supply and consumption

Table 3.4 Environmental impact and resources for 1 ton of structural steel.

Results of the LCA – environmental impact and resources: 1 ton of structural steel

Impact category	Unit	Production A1–A3	Recycling Potential D	Total A1–A3 & D
Global warming potential (GWP)	[kg CO_2–Äq.]	1735	-959	776
Total use of renewable primary energy resources (PERT)	[MJ]	840	92.4	932.4
Total use of nonrenewable primary energy resource (PENRT)	[MJ]	17,800	-7210	10,590
Ozone depletion potential (ODP)	[kg CFC11–Äq.]	1.39E-7	6.29E-9	1,45E-07
Acidification potential (AP)	[kg SO_2–Äq.]	3.52	–1.32	2.2
Eutrophication potential (EP)	[kg $(PO_4)^{3-}$– Äq.]	3.7E-1	–1.26E-1	2.44E-1
Photochemical ozone creation potential (POCP)	[kg Ethen Äq.]	6.98E-1	–4.14E-1	2.84E-1
Abiotic depletion potential for nonfossil resources (ADPE)	[kg Sb Äq.]	2.85E-4	–1.11E-4	1.74E-4
Abiotic depletion potential for fossil resources (ADPF)	[MJ]	17,000	–7450	9550

Source: EPD Structural Steel: Sections and Plates (EPD-BFS-20130094-IBG1) [13].

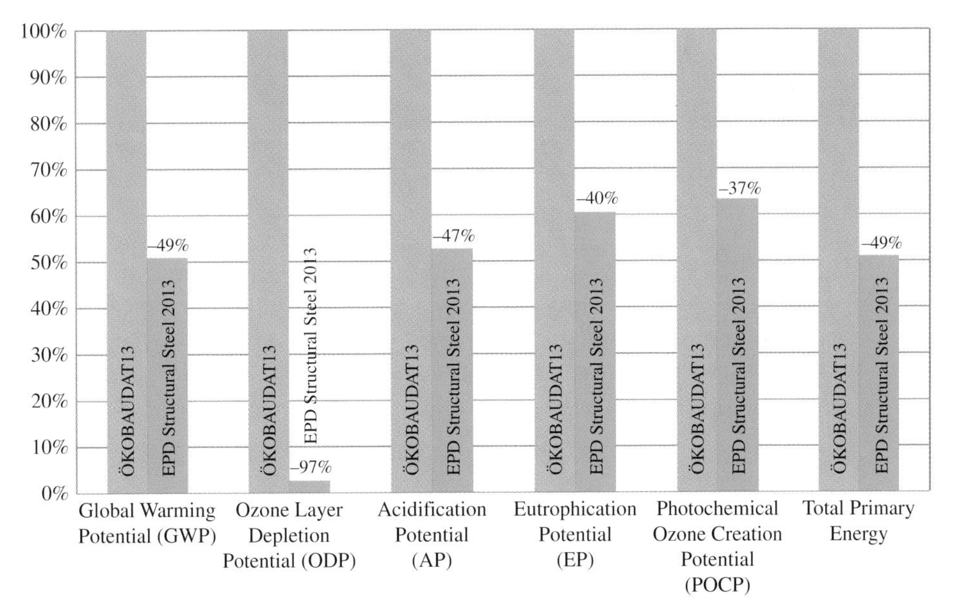

Figure 3.34 Comparison of ÖKOBAUDAT (dataset for nongalvanised structural steel) with the EPD for structural steel. The recycling potential is taken into consideration in all datasets. © bauforumstahl.

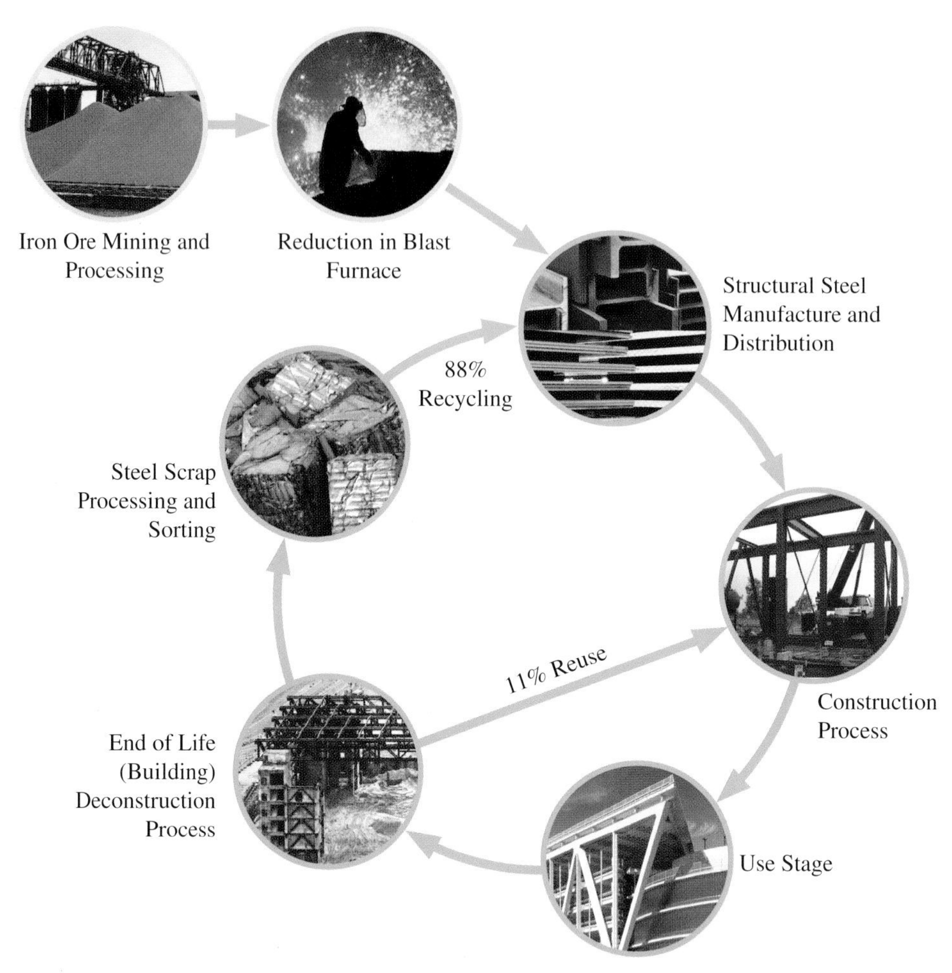

Figure 3.35 Life cycle of structural steel covered by the environmental product declaration for structural steel EPD-BFS-20130094. © bauforumstahl.

including the transport of the raw materials, the production of the structural steels and recycling at the end of the life cycle, including consideration of the recycling potential. With these quantitative statements, the declaration is targeting planners, architects, property developers, property companies, facility managers and companies involved in production and services along the value chain to the finished construction (Figure 3.35).

3.9.3 EPD for hot-dip galvanized structural steel

Raban Siebers

A further EPD was produced in cooperation with the German association for hot-dip galvanising (Industrieverband Feuerverzinken, IVF). The EPD Structural Steel: Open Rolled Profiles and Heavy Plates [13] described above served as

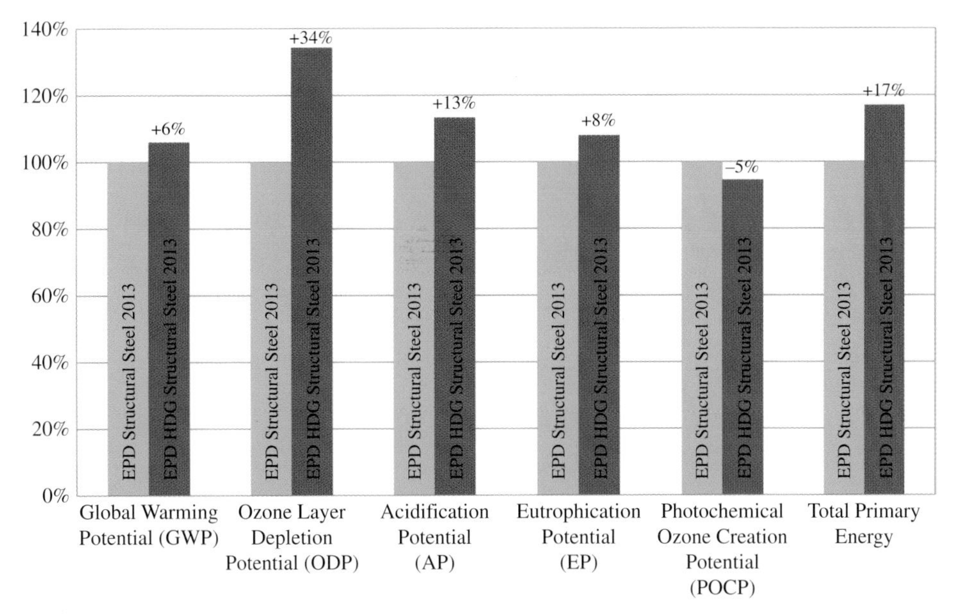

Figure 3.36 Comparison of EPDs for structural steel and hot-dip galvanised structural steel. The recycling potential is taken into consideration in all datasets. © bauforumstahl.

basis [15]. It applies to the members of IVF that are located only in Germany and bauforumstahl members. It provides objective data and facts on the effects of hot-dip galvanised structural steels on humans and the environment. The data of the EPD issued by IBU shows that hot-dip galvanised steel is an optimal material from a sustainability point of view. The environmental product declaration for hot-dip galvanised structural steels (EPD-BFS-20130173-IBG1) is a Type III declaration that complies with ISO 14025 and EN 15804, which was developed and checked with the involvement of an independent third party. It conforms to the international standards for eco-auditing.

The environmental data for hot-dip galvanised structural steel is at least 35% lower than even the figures given for non-galvanized structural steel by ÖKOBAUDAT. Compared to the EPD for nongalvanised structural steel the environmental impacts are slightly higher (Figure 3.36 and Table 3.5). For further information on environmental data for batch hot dip galvanized steel, on European level see 4.6.4.

3.9.4 EPDs for profiled sheets and sandwich panels

Kai Kahles

In 2012, members of the International Association for Lightweight Metal Construction (IFBS) worked together to develop industry-wide EPDs for trapezoidal sheets, liner trays, corrugated and standing steel seams, sandwich panels with polyurethane and mineral wool cores, and aluminium trapezoidal sheets,

Table 3.5 Environmental impact and resources for 1 ton of Hot-Dip Galvanized Structural Steel.

Results of the LCA – environmental impact and resources: 1 ton of structural steel

Impact category	Unit	Production A1–A3	Recycling potential D	Total A1–A3 & D
Global warming potential (GWP)	[kg CO_2–Äq.]	1847	−1019	828
Total use of renewable primary energy resources (PERT)	[MJ]	1384	18.7	1402.7
Total use of nonrenewable primary energy resource (PENRT)	[MJ]	20,030	−7939	12,091
Ozone depletion potential (ODP)	[kg CFC11–Äq.]	1.96E-7	−1.03E-9	1.94E-07
Acidification potential (AP)	[kg SO_2–Äq.]	4.01	−1.51	2.5
Eutrophication potential (EP)	[kg $(PO_4)^{3-}$–Äq.]	4.06E-1	−1.43E-1	2.63E-1
Photochemical ozone creation potential (POCP)	[kg Ethen Äq.]	7.12E-1	−4.43E-1	2.69E-1
Abiotic depletion potential for nonfossil resources (ADPE)	[kg Sb Äq.]	1.44E-1	−3.86E-2	1.05E-1
Abiotic depletion potential for fossil resources (ADPF)	[MJ]	18,710	−8099	10,611

Source: EPD Hot-Dip Galvanized Structural Steel: Sections and Plates (EPD-BFS-20130173-IBG1) [15].

corrugated and standing seamsystems. These industry-wide EPDs were first launched in 2013. The evaluation was performed by thinkstep and the EPDs are published by IBU.

The industry-wide EPDs enable companies to assess and calculate the environmental impact of a variety of lightweight metal construction products, covering profile types and profile weights. The environmental impact of each product is described in detail and recorded as a figure in the form of weight per square meter (kg/m^2). The EPDs also show the linear correlation between weight and environmental impact, enabling companies to perform reverse calculations if required.

The environmental impact of each product was assessed by a number of IFBS member companies, the names of which can be found in the EPDs themselves. The environmental impact analyses can also be applied to products from other manufacturers if no other reliable data is available. The EPDs also include an end-of-life analysis for recycling purposes, based on a worst-case scenario. However, it is possible to assess local recycling processes and strategies to obtain more accurate results.

3.9.4.1 *Profiled steel sheets*

The majority of the production stage A1–A3 relates to the production of raw materials in stage A1 with 97%–99% in all impact categories. The cold forming process for steel profiles in stage A3 contributes just 1%–3%. There is a linear correlation between the weight and environmental impact of the raw material.

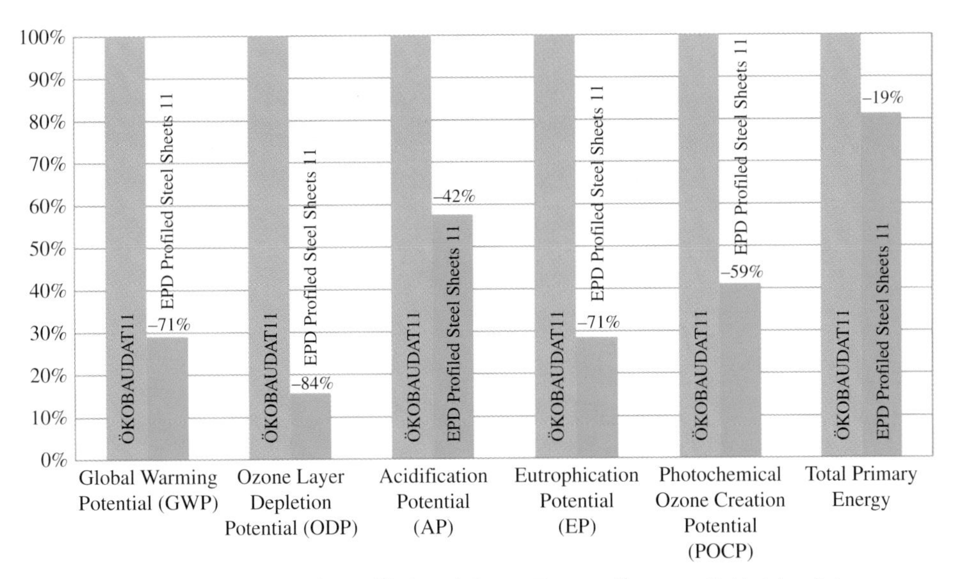

Figure 3.37 Comparison of EPDs for profiled steel sheets. The recycling potential is taken into consideration in all datasets. © IFBS.

The end-of-life credit for the steel scrap following the usage stage (with a collection rate of 90%) helps to reduce (negative value) the majority of the environmental impact category results [16].

In comparison to the ÖKOBAUDAT, the Association EPD data is better in all relevant categories (between −19% and −84%) (Figure 3.37). The comparison is based on 2011 data. Newer datasets may deliver different results, depending on new default energy and raw material datasets, like European electricity grid mix data or Worldsteel LCI.

3.9.4.2 *Profiled sheets made of aluminium*

The majority of the production stage A1–A3 relates to the production of raw materials in stage A1 with 97%–99% in all impact categories. Metal makes the greatest contribution at 98%. The cold forming process for aluminium profiles in stage A3 contributes just 1%–3%. As above, there is a linear correlation between the weight and environmental impact of the raw material. The end-of-life credit for the aluminium scrap following the usage stage (with a collection rate of 90%) helps to reduce (negative value) the majority of the environmental impact category results [17].

3.9.4.3 *Double skin steel-faced sandwich panels with a polyurethane core*

The environmental impact in the product stage is mainly determined by the raw material extraction and processing in stage A1. The steel sheets (68%) provide the main part of the primary energy, determined for a sandwich panel of 40 mm thickness. The value for ridged PU core amounts to 32%. For a sandwich panel of 160 mm thickness, the PU core (66%) provides the main part for primary energy.

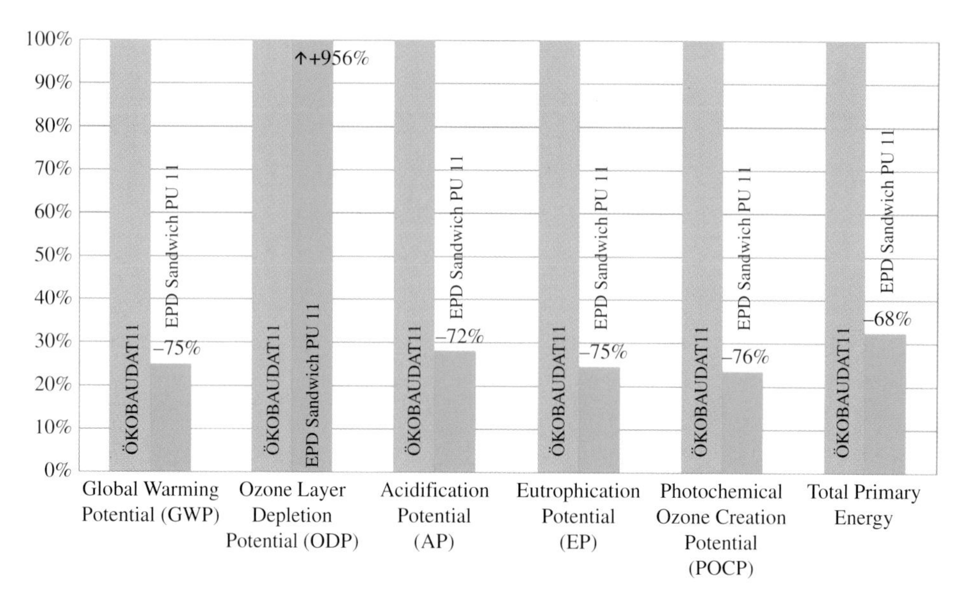

Figure 3.38 Comparison of EPDs for double skin steel-faced sandwich panels with a polyurethane core. © IFBS.

The steel sheets amount to 34%. The absolute results for the steel sheet remain comparable for all thicknesses in all declared products. The absolute results for the PU core clearly increase with thickness. A linear correlation between environmental impact and the amount of PU core material can be identified. The rating for the next product system includes the rating for steel recycling as well as substitution of primary fuels for electricity and steam generation from waste incineration plants for the PU core [18].

In comparison to the ÖKOBAUDAT, the Association EPD data is better in most of the relevant categories (between −68% and −76%) (Figure 3.38). The ODP seems to be tremendously poor, but the values are very low, between $1{\cdot}10^{-6}$ and $1{\cdot}10^{-7}$. Therefore, also little deviations result in high percentage values. It must also be considered that the ÖKOBAUDAT does not contain any end of life for the core material. The comparison is based on 2011 data. Newer datasets may deliver different results, depending on new default energy and raw material datasets, like European electricity grid mix data, Worldsteel LCI or ISOPA (European trade association for producers of diisocyanates and polyols) data.

3.9.4.4 *Double skin steel-faced sandwich panels with a mineral wool (MW) core*

The environmental impact for the product stage is mainly determined by the raw material extraction and processing in stage A1. The steel sheets (70%) provide the main part of primary energy for a sandwich panel of 40 mm thickness. The mineral wool amounts to 30%. For a sandwich panel of 160 mm thickness, the MW core (57%) provides the main part for primary energy. The steel sheets amount is 43%. The absolute results for the steel sheet remain comparable for all thicknesses in all declared products. The absolute results for the MW core increase with

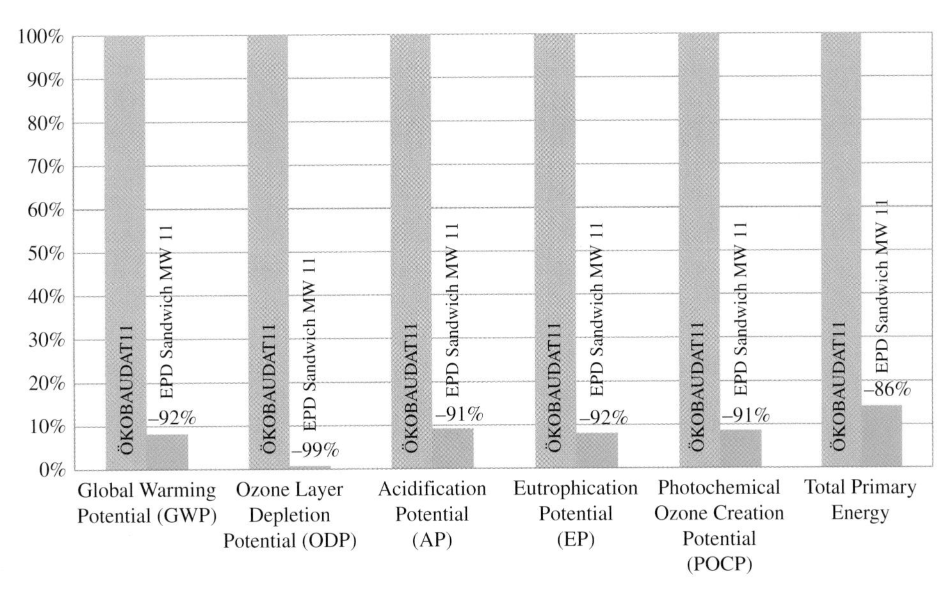

Figure 3.39 Comparison of EPDs for double skin steel-faced sandwich panels with a mineral wool core. © IFBS.

thickness. A linear correlation between environmental impact and the amount of MW core material can also be identified. The rating for the next product system includes the rating for steel recycling [19].

In comparison to the ÖKOBAUDAT, the Association EPD data is better in all relevant categories (between –86% and –99%) (Figure 3.39). The default data of ÖKOBAUDAT is based on global data. Therefore, it is clear that collected energy data during the production process results in better values in all categories. The main contributor here seemed to be better data for the core material than the default values. It must be considered that the ÖKOBAUDAT does not contain any end of life for the core material. The comparison is based on 2011 data. Newer datasets may deliver different results, depending on new default energy and raw material datasets, like European electricity grid mix data, Worldsteel LCI or global material data.

The EPDs for structural steel presented above and other EPDs for steel construction products, such as the EPDs prepared by IFBS for profiled sheets and sandwich panels, are freely available at Institut Bauen und Umwelt IBU (www.bau-umwelt.de).

3.10 KBOB-RECOMMENDATION – LCA DATABASE FROM SWITZERLAND

Rolf Frischknecht and Franziska Wyss

An increasingly significant part of the overall energy consumption during the life of buildings is caused by the production of construction materials and the building technology (heating, ventilation, electric and sanitary equipment). Primary energy consumption, greenhouse gas emissions and other relevant environmental impacts get more and more into the focus of building owners and architects. These environmental impacts from construction to usage and to the end of life of buildings can strongly be influenced with mindful planning by engineers, architects and construction planners.

Instruments based on life-cycle thinking and LCA data are crucial when aiming toward sustainability in the construction sector. In Switzerland the most commonly used data source is the KBOB-recommendation 2009/1:2014 'Life Cycle Assessment Data in the Construction Sector' [20], which contains environmental information about construction materials, components of building technology, energy supply, transport and waste management services. KBOB is the Swiss coordinator of federal, cantonal and communal building and real estate authorities.

3.10.1 KBOB-recommendation as a basis for planning tools

The KBOB-recommendation is a data pool of environmental information of various construction materials (steel sections and sheets, concrete, wood products, insulation materials and many more), building technology components (ventilation, heating, hot water supply and heat distribution, sanitary systems, electrical equipment, energy supply (e.g., light fuel oil, natural gas, wood, biogas, electricity (based on various sources) and district heat) and transport and waste management services. It is similar to the German ÖKOBAUDAT (Section 3.10). However, the KBOB-recommendation differs in a few key aspects: Firstly, full transparency is ensured by providing all necessary underlying data. Secondly, environmental impacts are quantified using a single score method wich allows for comprehensive and distinct results. Thirdly, the underlying methoological approch is identical for all product groups thus guaranteeing full consistency across all construction materials.

The KBOB-recommendation is not only used in planning tools but serves also for standards and codes. Figure 3.40 illustrates the usage of the KBOB-recommendation and shows where LCA data is applied in the Swiss regulatory construction system SIA (Swiss society of engineers and architects). For example, the KBOB-recommendation builds the basis for the technical bulletins about energy certificates of buildings, SIA 2031 [21], about 'grey energy' of buildings, SIA 2032 [22], about mobility induced by buildings, SIA 2039 [23], about the SIA energy efficiency path, SIA 2040 [24], the building certificate MINERGIE-ECO and more.

Transparency, independence and data consistency are among the KBOB-recommendation's key features and of great importance since the data are widely used. Data acquisition and modelling guidelines were established to ensure harmonized modelling and data quality [25], [26]. The KBOB-recommendation is updated and enlarged in regular intervals. The last update was conducted and published in July 2014 and is by and large based on the ecoinvent data v2.2+ [27].

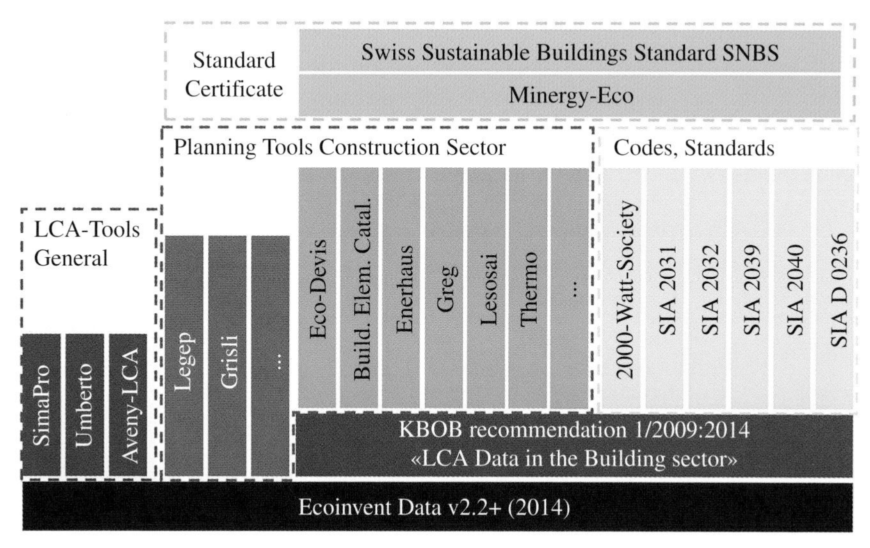

Figure 3.40 How LCA data is used in Swiss construction planning tools and codes, standards and certificates.

Within the last update, the LCI of hot-rolled steel sections was updated and newly modelled. It could be shown that the present environmental impacts of hot-rolled steel sections are actually substantially lower compared to the environmental impacts reported in the previous version.

As the KBOB-recommendation supports an efficient and consistent scheme of tools, standards and certificates, it is widely accepted and used in Switzerland by all parties involved in the construction process (building owners, construction managers, producers of building materials, public authorities, planners, architects).

3.10.2 Environmental impact assessment within the KBOB-recommendation

The KBOB-recommendation 2009/1:2014 serves three environmental indicators: environmental impacts according to the ecological scarcity method 2013 (environmental impact points [EIPs]), cumulative energy demand (CED in MJ oil-eq., [28]), and total, nonrenewable and greenhouse gas emissions 2013 (kg CO_2-eq., [29]). While CED and greenhouse gas emissions are rather well-known indicators, the environmental impacts indicator is described here.

The ecological scarcity method [30] makes it possible to assess the impact of pollutant emissions and resource extraction activities on the environment (impact assessment) as part of an LCA. The key metrics of this method are eco-factors, which measure the environmental impact of pollutant emissions or resource extraction activities in EIPs per unit of quantity. The ecological scarcity method evaluates the pollutant emissions and resource extraction activities on a distance to target principle. The calculation of the eco-factors is based on one hand on the actual emissions (actual flow) and on the other hand on the targets defined by Swiss environmental policy and legislation (critical flow). The ecological scarcity method allows for an environmental optimisation within the framework of a country's environmental goals.

The current version of eco-factors according to the ecological scarcity method covers 19 environmental topics:

water resources;
energy resources;
mineral resources;
land use;
global warming;
ozone layer depletion;
main air pollutants and particulate matter;
carcinogenic substances into air;
heavy metals into air;
water pollutants;
persistent organic pollutants into water;
heavy metals into water;
pesticides into soil;
heavy metals into soil;
radioactive substances into air;
radioactive substances into water;
noise;
nonradioactive waste to deposit;
radioactive waste to deposit.

The constantly changing emission levels and the new regulatory situation as well as new scientific findings were taken into account in the recently updated version. The databases for existing eco-factors were updated and new eco-factors for traffic noise, persistent organic pollutants and metallic and mineral primary resources were introduced.

3.10.3 Environmental impacts of hot-rolled steel products

In Switzerland about 1.3 million tonnes of steel scrap is produced yearly [31]. Two Swiss steelworks transform it into new steel products and supply the Swiss steel market together with the imported steel products.

In 2013–2014 the Swiss centre for steel construction (SZS) commissioned an LCA and environmental product declaration, EPD according to EN 15804 (see Chapter 2.2) of hot-rolled steel profiles (sections, bar steel, flat steel, wide flat bar steel > 600 mm) produced in Europe and consumed in Switzerland. These data was imported to the KBOB-recommendation 2009/1:2014.

3.10.3.1 *Hot-rolled steel profiles manufacturing*

The material and energy consumption and the emissions of five Swiss and European steelworks were analysed. Hence the main data originate from the steel producers and stems from the years 2008–2013. Background processes such as energy supply or transport services were modelled using ecoinvent data v2.2+ [27] (see also Section 3.9). Missing data was completed by expert estimation and literature values.

The life-cycle phases analysed cover the provision of raw materials and energy, the transport of the raw materials to the steelworks, the production of the hot-rolled steel profiles, their transport to the Swiss regional storage and the end-of-life phase. Transport to the construction site, construction itself and the use phase were excluded.

1.1 kg of scrap steel is needed to produce 1 kg of hot-rolled steel profiles. According to the information from the steelworks supplying the Swiss market, only secondary steel is used for the production of sections. At its end of life the steel is fully recycled. Eleven per cent can directly be reused after the end of life as steel profiles [32]. Eighty-eight per cent goes back to the steelworks and is recycled. Production and end of life have a loss of 1%.

3.10.3.2 *Environmental impacts of hot-rolled steel profiles*

The environmental impacts of hot-rolled steel profiles are quantified with the three environmental indicators described in Section 3.12.2. It becomes evident that the production phase causes the main impact for all environmental indicators. Because the hot-rolled steel profiles are recycled at the end of life, their environmental impacts at the end-of-life stage are negligible. Table 3.6 summarizes the results for the total environmental impacts, the CED (primary energy demand) and the greenhouse gas emissions of 1 kg of hot-rolled steel profiles.

The environmental impacts of the product stage are dominated by manufacturing, which has a share of 65%, followed by the supply of raw materials with a share of 23% (scrap and auxiliary materials). The manufacturing itself is dominated by the energy supply (mainly electricity) and the process specific emissions. Transport from the steelworks to the Swiss regional storage causes 10% of the overall impact, whereas

Table 3.6 Total environmental impacts, primary energy demand (total and nonrenewable) and greenhouse gas emissions of 1 kg of hot-rolled steel profiles supplied to the Swiss market.

	Total Environmental Impact	Primary Energy Total	Primary Energy Nonrenewable	Greenhouse Gas Emissions
Unit	EIP	MJ oil eq	MJ oil eq	kg CO_2 eq
Product stage	999	13.34	12.43	0.73
End-of-life stage	0.00	0.00	0.00	0.00
Total	999	13.34	12.43	0.73

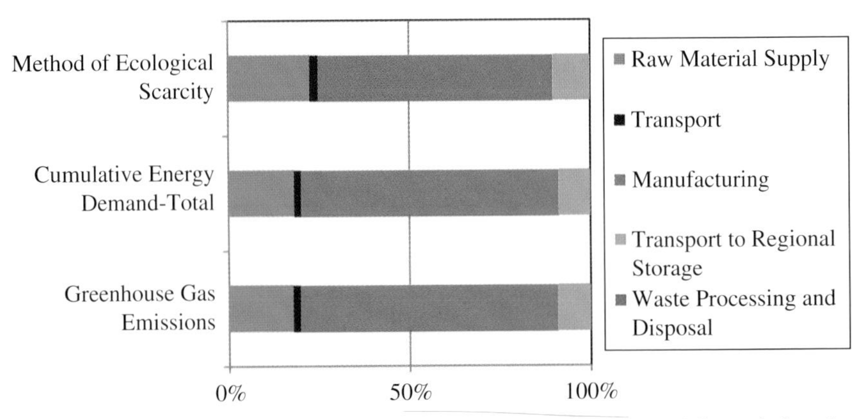

Figure 3.41 Shares of raw materials supply, manufacturing, transport and disposal of total environmental impacts according to the ecological scarcity method 2013, CED total and greenhouse gas emissions of 1 kg hot-rolled steel profiles.

transport of the raw materials to the steelworks contributes a share of 2%. Figure 3.41 shows the shares of the life-cycle stages of all indicators analysed.

Primary energy consumption is dominated by the consumption of electricity and natural gas in manufacturing. Manufacturing causes 71% of the primary energy consumption, while transport to the regional storage has a share of 9% and transport of raw materials to the steelworks 2% (total CED and nonrenewable CED). The supply of raw material causes 17%–18% of the primary energy demand. The greenhouse gas emissions add up to 0.73 kg CO_2 eq/kg and the shares of the life phases are equal to those of the primary energy demand.

The LCA of hot-rolled steel profiles shows that the main impact of the steel profiles is caused by the manufacturing, followed by the supply of raw materials. The electricity consumption and the origin of the raw materials dominate the assessment and have a large influence on the environmental performance. Auxiliary materials and water consumption within the production as well as supply transport have a minor influence on the overall environmental impacts. At the end-of-life stage almost no waste occurs, as 99% of the hot-rolled steel profiles are either recycled or reused.

3.10.4 Example using data from the KBOB-recommendation

Patric Fischli-Boson

It is possible to compare anything from individual structural elements to eco-balances of entire buildings. Using the example of a building column, the effects of different materials on its eco-balance can be determined under consideration of loads, used materials and buckling load. The following data from the KBOB-recommendation is used for this example (Table 3.7).

Table 3.7 Excerpt from the KBOB-recommendation (July 2014) [33]. This data forms the basis of the eco-balance calculations.

Eco Balance Data for Construction Materials						KBOB / eco-bau / IPB 2009/1:2014								
Construction Materials *[Bibliographie treeze, version 2.2+]*	**Weight per Volume**	**Reference**	**EIP'13** *UBP*			**Primary energy**						**Greenhouse Gas Emissions**		
						Total			**Non-renewable**					
			Total EIP	Production EIP	End-of-Life EIP	**Total** MJ Oil-eq	Production MJ Oil-eq	End-of-Life MJ Oil-eq	**Total** MJ Oil-eq	Production MJ Oil-eq	End-of-Life MJ Oil-eq	**Total** kg CO_2-eq	Production kg CO_2-eq	End-of-Life kg CO_2-eq
Concrete (no reinforcement)	**kg/m³**													
Unreinforced concrete CEM II/A (cement content 290 kg/m³)	2,420	kg	**95**	**68**	**27**	**0.781**	0.580	0.201	**0.723**	0.529	0.194	**0.097**	0.0867	0.0105
Unreinforced concrete CEM II/B (cement content 290 kg/m³)	2,420	kg	**91**	**64**	**27**	**0.739**	0.538	0.201	**0.683**	0.489	0.194	**0.093**	0.0826	0.0105
Bricks	**kg/m³**													
Clinker bricks	900	kg	**199**	174	25.4	**2.94**	2.75	0.188	**2.67**	2.48	0.182	**0.247**	0.238	0.00903
Sand-lime bricks	1,400	kg	**155**	130	24.7	**1.57**	1.39	0.179	**1.44**	1.27	0.173	**0.139**	0.130	0.00869
Cement bricks	1,700	kg	**132**	108	24.7	**1.01**	0.832	0.177	**0.924**	0.754	0.171	**0.130**	0.121	0.00885
Metallic construction materials	**kg/m³**													
Reinforcing steel	7,850	kg	**2,850**	**2,850**	**0**	**13.5**	13.5	0	**12.7**	12.7	0	**0.681**	0.681	0
Bright steel sheet	7,850	kg	**3,560**	**3,560**	**0**	**28.9**	28.9	0	**27.8**	27.8	0	**1.83**	1.83	0
Steel profile, blank	7,850	kg	**999**	**999**	**0**	**13.3**	13.3	0	**12.4**	12.4	0	**0.733**	0.733	0
Timber and timber materials	**kg/m³**													
Three-ply laminated (PVAc glue) solid timber panel	470	kg	**1,290**	1,210	79.6	**35.1**	34.9	0.194	**10.9**	10.70	0.191	**0.679**	0.569	0.111
Glue-laminated (UF glue) timber panel, for dry spaces	470	kg	**950**	863	86.8	**34.4**	34.2	0.212	**8.13**	7.92	0.208	**0.545**	0.424	0.121
Glue-laminated (MF glue) timber panel, for humid spaces	470	kg	**995**	908	86.8	**35.1**	34.9	0.212	**8.86**	8.65	0.208	**0.583**	0.463	0.121
Solid timber beech/oak, kiln dried, raw	700	kg	**512**	485	26.9	**22.9**	22.8	0.125	**2.11**	1.99	0.123	**0.119**	0.109	0.0101
Solid timber beech/oak, kiln dried, planed	700	kg	**611**	584	26.9	**24.1**	24.0	0.125	**2.90**	2.78	0.123	**0.162**	0.152	0.0101
Solid timber spruce/fir/larch, kiln dried, planed	450	kg	**496**	469	26.9	**25.0**	24.9	0.125	**3.62**	3.49	0.123	**0.138**	0.128	0.0101

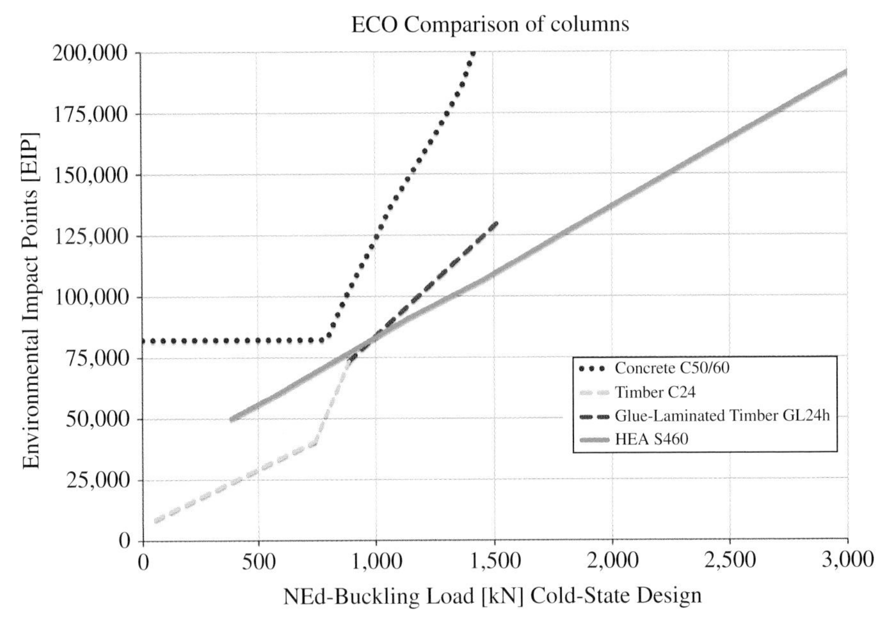

Figure 3.42 Comparison of eco-balances of building columns made of steel, concrete and timber. © SZS.

The results of the calculations can be seen in Figure 3.42: for small loads, a timber column (timber grade C24) is the best choice, followed by a steel column (steel grade S460) and finally by a concrete column (concrete grade C50/60, reinforcing steel grade B500B). The concrete column is the least favourable choice here, because it requires a minimum amount of reinforcement even for very small loads. The results look completely different if a glue-laminated timber column has to be used instead of a solid timber column due to higher design loads. In that case, the steel column achieves similar results and starts to outperform the glue-laminated timber column as the design loads increase (Figure 3.43). A further important advantage of using a steel column is its smaller cross-sectional area, a fact that was not considered here.

A comparison of bending elements shows a similar pattern. Again, the steel element achieves significantly better eco-balance values than a glued laminated timber beam of the same structural capacity. The ecological advantage of a steel-beam compared to other materials makes it ideal for industrial construction. By using steel as a building material for structures such as industrial halls, the fundamental idea of environmentalism is put into practice: to use materials in the most efficient and environmentally friendly way possible. To realise this idea, it is also possible to combine different materials, provided this makes sense from a structural and ecological point of view. An interesting and promising novel construction method that adheres to this fundamental idea of environmentalism consists of using steel for the main structural elements, while employing timber panels as wall and floor elements. This hybrid construction method combines the advantages of both materials.

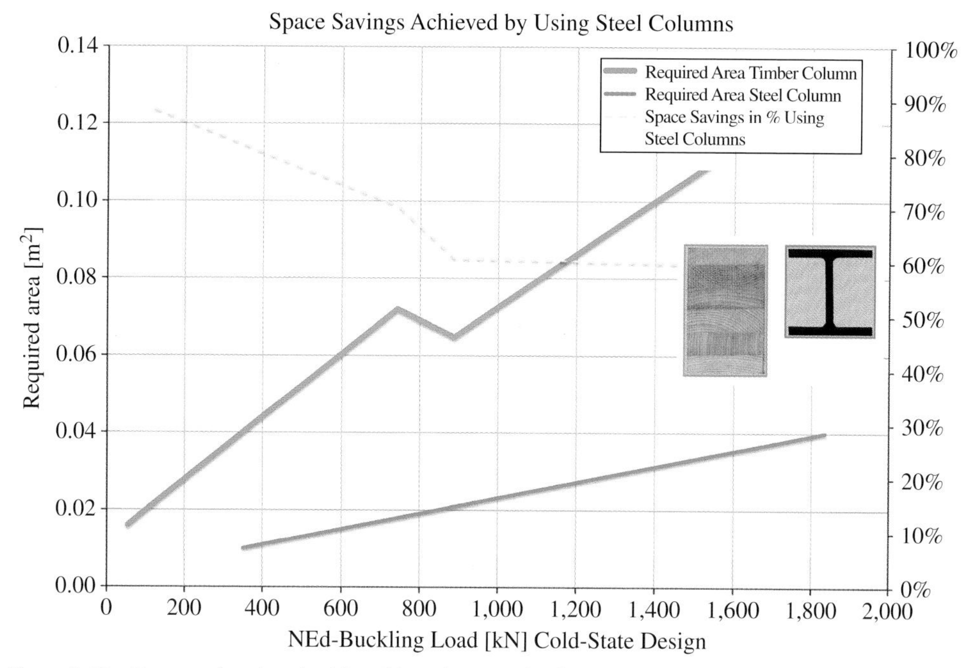

Figure 3.43 Space savings in m^2 achieved by using a steel column instead of a timber column. © SZS.

REFERENCES

[1] Schulze Darup, B. (2010) *Deutsche Bundesstiftung Umwelt: Energetische Gebäudesanierung mit dem Faktor 10*. Osnabrück: DBU.

[2] Messari-Becker, L. (2008) Nachhaltiges Sanieren? Zur ökologischen Effektivität und ökonomischen Effizienz energetischer Sanierungsmaßnahmen im Altbau am Beispiel von Dachsanierungen. Bauphysik 30(5).

[3] Mösle, P. (2010) Entwicklung einer Methode zur Internationalisierung eines ganzheitlichen Zertifizierungssystems zum nachhaltigen Bauen für Bürogebäude. Forschungsergebnisse aus der Bauphysik 3.

[4] Pelzeter, A. (2009) Lebenszykluskosten von Immobilien im Praxistest, Journal of Practical Business Research 9, doi:10.3206/0000000026.

[5] Graham, I. (2006) Re-examining the costs and value ratios of owning and occupying buildings. Building Research & Information 34(3), pp. 230–245.

[6] European Commission. (2004) *Final Report - Working Group Sustainable Construction Methods & Techniques January 2004*. Brussels.

[7] Royal Institute of Chartered Surveyors. (1986) *Quantity Surveying Division Report 1986*.

[8] HM Treasury. (2015) Green book guidance on public sector business cases; using the five case model. Available at https://www.gov.uk/government/uploads/system/uploads/attachment_data/file/469317/green_book_guidance_public_sector_business_cases_2015_update.pdf. Accessed 29 October 2015.

[9] Bewertungssystem Nachhaltiges Bauen (BNB). (2011) BMUB Version V 2011_1A1 BNB_BN 2.1.1 Büro- und Verwaltungsgebäude, Berlin.

[10] Bundesministeriums für Umwelt, Naturschutz, Bau und Reaktorsicherheit (BMUB). (2015) ÖKOBAUDAT 2015. Available at www.oekobaudat.de. Accessed 22 October 2015.

[11] Sarma, K. C., Adeli, H. (2002) Life-cycle cost optimization of steel structures. International Journal for Numerical Methods in Engineering 55(12), pp. 1451–1462.
[12] Oxford Brookes University Dept. of Architecture with the assistance of Sworn King & Partners. (2003) InsiteTM on Life Cycle Costs. London.
[13] EPD-BFS-20130094. (2013) *Environmental Product Declaration: Structural Steel: Open Rolled Profiles and Heavy Plates.* Berlin: Institute Construction and Environment (IBU).
[14] Steel Recycling Institute. (2013) Steel recycling rates 2013. Available at https://www.recycle-steel.org. Accessed 18 November 2015.
[15] EPD-BFS-20130173. (2013) *Environmental Product Declaration: Hot-Dip Galvanized Structural Steel: Open Rolled Profiles and Heavy Plates.* Berlin: Institute Construction and Environment (IBU).
[16] EPD-IFBS-2013211. (2013) *Environmental Product Declaration: Profiltafeln aus Stahl für Dach-, Wand- und Deckenkonstruktionen.* Berlin: Institute Construction and Environment (IBU).
[17] EPD-IFBS-2013111. (2013) *Environmental Product Declaration: Profiltafeln aus Aluminium für Dach-, Wand- und Deckenkonstruktionen.* Berlin: Institute Construction and Environment (IBU).
[18] EPD-IFBS-2013172. (2013) *Environmental Product Declaration: Double Skin Steel Faced Sandwich Panels with a Core Made of Polyurethane.* Berlin: Institute Construction and Environment (IBU).
[19] EPD-IFBS-2013171. (2013) *Environmental Product Declaration: Double Skin Steel Faced Sandwich Panels with a Core Made of Mineral Wool.* Berlin: Institute Construction and Environment (IBU).
[20] KBOB, eco-bau and IPB. (2014) KBOB-Empfehlung 2009/1:2014: Ökobilanzdaten im Baubereich, from April 2014. Koordinationskonferenz der Bau- und Liegenschaftsorgane der öffentlichen Bauherren. Bern: KBOB.
[21] Schweizerischer Ingenieur- und Architektenverein (SIA). (2009) Merkblatt 2031: Energieausweis für Gebäude gemäß SN EN 15217 und SN EN 15603. Zurich.
[22] Schweizerischer Ingenieur- und Architektenverein (SIA). (2010) Merkblatt 2032: Graue Energie von Gebäuden. Zurich.
[23] Schweizerischer Ingenieur- und Architektenverein (SIA). (2011) Merkblatt 2039: Mobilität – Energiebedarf in Abhängigkeit vom Gebäudestandort. Zurich.
[24] Schweizerischer Ingenieur- und Architektenverein (SIA). (2011) Merkblatt 2040: SIA-Effizienzpfad Energie. Zurich.
[25] Frischknecht, R. (2015) Regeln für die Ökobilanzierung von Baustoffen und Bauprodukten in der Schweiz. Plattform "Ökobilanzdaten im Baubereich", KBOB, eco-bau, IPB, Bern, retrieved from: http://www.eco-bau.ch/resources/uploads/Oekobilanzdaten/Plattform_OeDB_Memo_Produktspezifische%20Regeln_v3%200.pdf. Accessed 30 May 2016.
[26] Frischknecht, R. (2015) Regeln für die Ökobilanzierung von Gebäuden in der Schweiz, Version 1.0. Plattform "Ökobilanzdaten im Baubereich", KBOB, eco-bau, IPB, Bern, retrieved from: http://www.eco-bau.ch/resources/uploads/Gebaeudespezifische_Regeln.pdf. Accessed 30 May 2016.
[27] KBOB, eco-bau and IPB. (2014) ecoinvent Datenbestand v2.2+; Grundlage für die KBOB-Empfehlung 2009/1:2014: Ökobilanzdaten im Baubereich, from April 2014. Bern: KBOB.
[28] Frischknecht, R., Jungbluth, N., Althaus, H.-J., Bauer, C., Doka, G., Dones, R., Hellweg, S., Hischier, R., Humbert, S., Margni, M. and Nemecek, T. (2007)

Implementation of Life Cycle Impact Assessment Methods. ecoinvent Report No. 3, v2.0. Swiss Centre for Life Cycle Inventories. Dübendorf.

[29] The IPCC Fifth Assessment Report. (2013) *Climate Change 2013: The Physical Science Basis*. Working Group I. Geneva: IPCC Secretariat.

[30] Frischknecht, R., Büsser Knöpfel, S. (2013) *Swiss Eco-Factors 2013 According to the Ecological Scarcity Method: Methodological Fundamentals and Their Application in Switzerland*. Environmental Studies no. 1330. Bern: Federal Office for the Environment.

[31] Webpage of the Swiss steel center. Available at www.szs.ch. Accessed 25 August 2014.

[32] Sansom, M., Meijer, J. (2002) *Life-Cycle Assessment (LCA) for Steel Construction*. European Commission technical steel research (EUR 20570 EN), Brussels.

[33] KBOB-recommendation. (2014) LCA data for the construction sector 2009/1:2014. Available at www.eco-bau.ch. Accessed 29 October 2015.

Chapter 4

Sustainable steel construction

4.1 ENVIRONMENTAL ASPECTS OF STEEL PRODUCTION

Raban Siebers and Diana Fischer

Iron ore, the main raw material used for producing steel, is one of the most abundant constituents of the earth's crust. Steel is completely recyclable, and a tried and tested recycling system has become established worldwide. Steel scrap is a much sought-after secondary raw material, and thanks to the ease of separation it can be recovered from every steel product, whether it be a razor blade, a refrigerator or a wrecked car. At the same time, steel recycling also results in a sustainable method of 'disposal' of worn-out everyday items and reduces the amount of waste produced. The cradle-to-cradle concept thus makes steel a resource-efficient material for construction.

More stringent quality requirements and increasing cost pressure have not only led to wide-ranging rationalisation measures being adopted in the European steel industry in the past but also to considerable technological advances. This is particularly apparent in the significant improvements in energy and material efficiency that have been achieved. Efficiency improvement means increasing the output-input relationship for all of the material resources used. Important factors affecting environmental performance have been reduced considerably, for example, by the German steel industry, since the 1960s: energy consumption is down by 38%, CO_2 emissions are down 44%, water consumption is 50% less and dust emissions are 90% lower. By-products produced during production of steel are used almost completely. For example, the slag is used in cement manufacturing, process gases are returned to the production process as energy sources, and other by-products like benzene, sulphur and tar are used by other industries.

Developments involving higher strength grades of steel have made a decisive contribution to resource efficiency in the construction sector in the past as well. The steels themselves have become more resource efficient. Thermomechanically

Sustainable Steel Buildings: A Practical Guide for Structures and Envelopes, First Edition.
Edited by Bernhard Hauke, Markus Kuhnhenne, Mark Lawson and Milan Veljkovic.

Siège FNEL

Location:	Luxembourg, G.D. Luxembourg
Architect:	hsa heisbourg strotz architectes
Building description:	The new federal headquarters of the FNEL (the National Scout and Guide Federation of Luxembourg) comprises administrative areas, meeting rooms, a library, a shop, a large multipurpose hall with a professional kitchen and storage areas as well as a small accommodation area.
Steel details:	The exterior skin is made up of wheatering steel panels and is a reference to the historical links between the FNEL and the steel industry in Luxembourg.
Sustainability:	As well as the usual amount of thought given to the importance of the insulation and the care given to the technical installations (low energy building), an appropriate choice of the building's constituent components – framework, walls and cladding – has been made.
Awards:	Steel Construction Award 2013 – Luxembourg.

Figure 1 View of the south façade. © Gilles Martin.

Figure 2 Detail of the ventilation system next to the window. © Gilles Martin.

Example provided by Infosteel.

(TM) rolled steels require significantly less alloying elements than normalised steels. Due to the weldability of TM-rolled steel, preheating during welding can be reduced significantly. The higher load-carrying capacity of the steels enables components with smaller design cross sections to be used, and the material consumption is correspondingly less, which is important in bridges and high rise buildings (see Section 4.5). The fact that the higher strength is achieved mainly by means of modern rolling processes and less by means of elaborate alloying means the energy and environmental balances for the production are of the same order as those for steels of normal strength [1], [22].

4.2 PLANNING AND CONSTRUCTING

4.2.1 Sustainability aspects of tender and contracting

Luis Braganca, Joana Andrade and Ricardo Mateus

The starting point to improve built environment toward sustainability is for the design team and suppliers to have a sustainable-conscious outlook. It is also essential to develop the building with an integrated design process (IDP), meaning a multidisciplinary approach where all the players work collaboratively to the same objective. The IDP differentiates a conventional building from a sustainable building. All stakeholders, including the clients, take a more active role than usual, and the architect may be the team leader. The team includes all the specialists needed for the building development, from the structural engineer to the services engineer.

A typical IDP consists of [2]:

- interdisciplinary work between architects, engineers, costing specialists, operations people and other relevant actors right from the beginning of the design process;
- discussion of the relative importance of various performance issues and the establishment of a consensus on this matter between client and designers;
- budget restrictions applied at the whole-building level, with no strict separation of budgets for individual building systems, such as HVAC or the building structure – this reflects the experience that extra expenditures for one system, for example, for sun shading devices, may reduce costs in another systems, for example, capital and operating costs for a cooling system;
- the addition of a specialist in the field of energy engineering and energy simulation;
- testing of various design assumptions through the use of energy simulations throughout the process, to provide relatively objective information on this key aspect of performance;
- the addition of subject specialists (e.g. for daylighting, thermal storage, comfort, materials selection, etc.) for short consultations with the design team;

- clear articulation of performance targets and strategies, to be updated throughout the process by the design team;
- in some cases, a design facilitator is added to the team to raise performance issues throughout the process and ensure specialist inputs as required.

In this sense, creating a sustainable building implies that all stakeholders are engaged in the whole process. The process of having all the stakeholders' input is referred to as 'charrette', the term used to refer to an effort to create a plan, being the basis for launching the design effort. Although there is not a standard charrette process for sustainable building, it can be summarized as follows [3]:

1. provide the stakeholders and participants with an overview of the goals of the client, the building program, the budget, the project schedule and other pertinent information – this should be done prior to the charrette and be reviewed by the charrette facilitator at the start of the charrette;
2. describe the building's sustainable objectives and goals of the owner with respect to sustainability;
3. conduct an open, uncritical brainstorming session with the goal of generating as many ideas as possible for making the proposed facility into a sustainable building – typically, the measures proposed by the participants are recorded on charts situated in the room where the brainstorming takes place;
4. organise the results of the brainstorming into the sustainability categories: land use, resources consumption, energy efficiency, health and comfort, functionality, costs, and post the measures on the brainstorming charts;
5. have the participants select their top measures from the brainstorming session – a typical method is to provide each participant with three to five starts that he or she can indicate as favorite green building measures;
6. sort the measures by order of interest by identifying those suggestions that receive more attention;
7. list the top measures by order of interest and begin a process of determining how to achieve these top measures;
8. determine the cost of the entire project based on the owner's program and direction and the results of the charrette;
9. if the process has failed to achieve the desired level of sustainability, an additional brainstorming should be performed;
10. if the process has failed to meet budgetary constraints, the owner has the choice of directing the project team to revisit the results to keep the project within the budget or allow higher capital costs.

The charrette, being the first stage in the process, brings together the design and the construction team with the client in an active, participatory process, offering stakeholders an opportunity to come away with a detailed understanding of the project.

Larger building companies have already developed codes of conduct in which they oblige themselves and other companies working with them to comply with

Rya Wastewater Treatment Plant

Location:	Gothenburg, Sweden
Architects:	KUB Arkitekter; PerEric Persson
Building description:	Transparent building that is a part of a wastewater treatment plant.
Steel details:	Steel frame. The building's upper part is composed mainly of steel.
Sustainability:	The world's largest discfilter plant with rotating screens for filtering wastewater.
Awards:	Winner of Stålbyggnadspriset (Steel building prize) in 2011.

Figure 1 Rya Wastewater Treatment Plant. © Kasper Dudzik.

Example provided by Swedish Institute of Steel Construction.

requirements for the protection of man and environment. The rise of voluntary agreements by contractors emphasizes the increasing importance of sustainability in the construction sector. The use of construction products with environmental product declarations (EPDs) is also increasing. Materials suppliers start to be selected according to their codes and principles when managing and producing their products and business. Products are selected according to their proved reduced environmental impact and good quality rather than by their cost.

Also, European manufacturers have already improved their resource efficiency because of generally high commodity prices and environmental regulations. Furthermore, a product's origin has to be kept in mind in the procurement of building materials. The FSC Certificate, for example, shows that the timber being used comes from sustainable forestry. SustSteel, a similar certificate for steel, was developed at EUROFER, the European steel association. It includes not just environmental but also social aspects, such as occupational safety and health in steel mills. If steel is obtained from one of these certified steel mills, the purchasers can be sure that the materials they bought comply with independently certified and internationally recognised social and quality standards.

Companies with certification labels, such as a product eco-label, environmental management system standards (ISO 14001), the Eco-Management and Audit Scheme (EMAS certification), quality certification (ISO 9000) or occupational health and safety (OHSAS 18001), are preferred to those with none.

4.3 SUSTAINABLE BUILDING QUALITY

4.3.1 Space efficiency

Luis Braganca, Joana Andrade and Ricardo Mateus

The space efficiency of any building relates to three factors: (1) the quantity of space, generally calculated in terms of floor area, though occasionally volume may also be relevant; (2).the number of users, potential and actual; and (3) the amount of time the space is used. High space efficiency is fundamental to achieving a sustainable built environment. Improving space efficiency means to meet floor area needs without compromising or even increasing land use. It brings benefits in the three fields of sustainability (see Chapter 1.2). On the environmental side, it enables a reduction on the environmental impacts per square metre of the construction, and during operation, by reducing heating and cooling, ventilation and lighting.

On the social side, in bring users satisfaction. If space use is efficient it means that it is adequate to its function. It influences positively the users' well-being and the ability to perform well the work and the tasks designed to be developed in the space. An efficient space also enables a pleasant indoor environment and a clearly arranged design.

Finally, space efficiency contributes to decreasing the construction and operational costs, and it contributes to increasing profitability of land and buildings, especially in areas where real estate prices and rents are high. In a high rise office or residential building, the net-to-gross floor area of a typical floor slab is of crucial economic interest to the developer, because it designates the space efficiency of the floors. At the same time, as the more efficient the typical floor slab is, the more usable area the developer gets and the more income is derived from the building [4].

Two types of space efficiency can be distinguished: (1) ratio of ground floor to usable floor space and (2) ratio of gross floor area to usable floor space. The first demonstrates that compact, multistorey buildings are more efficient than single-storey buildings, in terms of the relation between the ground and usable area. For instance, space efficiency of a high rise office building can be achieved by maximizing the gross floor area (GFA) and net (usable) floor area (NFA). Local codes and regulations must be considered, and in order to enable the developer and owner to get maximum returns from the high cost of land, the floors must have sufficient functional space [4].

TMVW Office

Location:	Ghent, Belgium
Architect:	Donald Desmet of Signum+ Architects
Building description:	Exceptional office from the Tussengemeentelijke Maatschappij Vlaanderen voor Waterbedeling (TMVW) with a large atrium and a very spacious look and feel.
Steel details:	16.2 m column-free spans.
Sustainability:	The large spans have created a very flexible grid; all interior walls can be moved, which allows a relatively easy modification of floor or office layout.

Figure 1 TMVW office Ghent. © Slimline Buildings.

Figure 2 16.2 m column-free spans. © Slimline Buildings.

Example provided by Slimline Buildings.

Residential and Office Building Lindenplatz

Location:	Baden, Switzerland
Architect:	Rolf Graf & Partner Architekten
Building description:	Cubic residential and office building in an inner-city environment.
Steel details:	Steel was chosen as a construction material because this meant a 60% reduction in construction weight compared to a conventional concrete building. Hence, no expensive reinforcing measures for the existing structures were required and a less massive foundation could be designed.
Sustainability:	The loadbearing structure is a steel frame construction with an innovative floor system. For good accessibility and ease of exchangeability, building services and loadbearing structure were strictly separated throughout the building. The construction was very fast due to the fact that the floor system is prefabricated industrially. The structural work of the eight-storey building was completed in only 8 weeks. By optimising the steel structure it was possible to minimise resource consumption. The slender steel structure consists entirely of recycled steel.

Figure 1 Residential and office building in Baden. © rgp Rolf Graf & Partner Architekten.

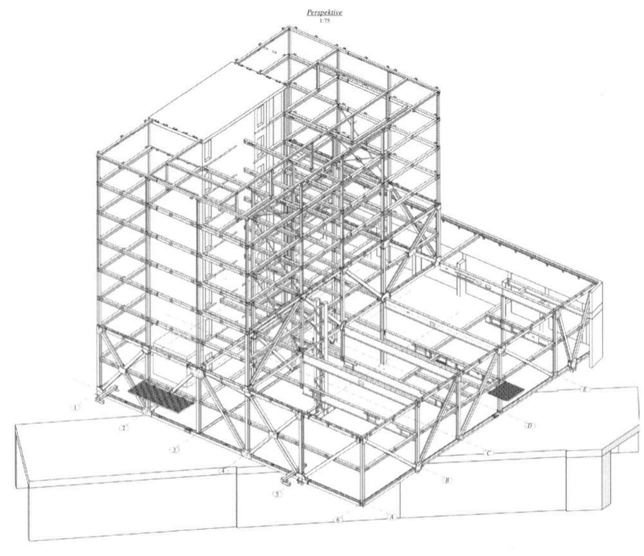

Figure 2 3D drawing of the residential and office building in Baden. © H. Wetter AG.

Example provided by Stahlbau Zentrum Schweiz.

As a good example, for the same gross floor area, if thick loadbearing walls are used, the usable space will be lower than if a lighter formed solution is adopted. Structural steel leads to slender efficient buildings that maximise usable floor space – with longer spans for open column-free spaces. Also it allows the integration of services systems into structural zones, leading to reduction of floor-to-floor heights. Minimising the number of columns makes it easier to subdivide and customize spaces. Steel-built buildings are often more adaptable, with greater potential for alterations to be made over time, extending the lifetime of the structure, as can be seen in the next section.

Extended steel spans can create large, open-plan, column-free internal spaces, with many clients now demanding column grid spacing over 15 m. In single-storey buildings, rolled beams provide clear spans of over 50 m. Trussed or lattice construction can extend this to 150 m.

A typical steel column occupies 75% less floor space than an equivalent concrete column, and steel beam depths are around half that of timber beams. Therefore, up to 12% of space can be gained when using steel construction instead of conventional methods. Wall thicknesses can be thinner because of excellent spanning capacity of modern lightweight cladding in comparison to thicker brick walls. This is particularly relevant for heavily constrained sites, where steel construction can be the key to overcoming spatial challenges.

Parking structures benefit from smaller structural steel columns and longer spans. Structural steel frames will typically span 17 m, allowing for a drive lane and two parking bays without any intervening columns. The use of the smaller steel columns at the front of the parking bays creates less intrusion into the parking space than larger concrete columns [5].

4.3.2 Flexibility and building conversion

Luis Braganca, Joana Andrade, Ricardo Mateus and Jan-Pieter den Hollander

The sustainability concept involves a long-term perspective: promote development that meets today's generations without compromising the ability of future ones to satisfy their own needs [6], [7]. Adaptable buildings can to be adjusted to new times, needs and expectations, and are becoming more important due to the following trends [8]:

- rising energy costs will force building owners and users to improve the building fabric to more cost-effective solutions;
- rising material costs may also promote reuse of materials and components economically more viable;
- rising temperatures may result in some buildings becoming uncomfortably hot and losing their market value unless they can be upgraded.

Adaptability can be seen as a design characteristic that embodies spatial, structural and service strategies, allowing the building or facility to respond to users' needs over time [7], enable improving the building's lifespan and hence reducing its environmental impact. Also, preserving existing buildings – built heritage – can represent social benefits as cultural identity and sense of belonging, or even economic benefits as tourism boosts local economies and is often driven by the cultural heritage of the city. The built environment works as a repository of cultural meanings [9]. This represents an extra aspect in promoting adaptability.

Beyond social concerns, energy demand, land and materials need also increase the need to maintain existing buildings to extend their life expectancy and use less energy [10]. Flexible and adaptable buildings prevent urban decay and allow upcycling of these buildings in urban regeneration projects. (See Figure 4.1 for current practice for inflexible office buildings.)

A building's flexibility brings also economic benefits. If flexibility is foreseen during design, maintenance and refurbishment activities, both time and costs are reduced [11], [12], and thus a building's adaptability can be identified as a cornerstone to sustainability (Figure 4.2).

Often buildings are designed in a way that it is not feasible to adjust the building to a new function. The focus of investors and owners is especially on the economic possibilities, and so the demolition is the only option. Demolition contractors must be paid, and therefore the value of the building drops below zero (Figure 4.1 red). The design of the building has to be focused on minimizing the 'drops' (Figure 4.1 green) in value. It is important to make the building easily adaptable for different variants on a function and even different functions. It is advantageous when the building can accommodate a variety of office layouts and when possibilities for apartments and/or shops are given.

The main characteristics of an adaptable building are that it is (1) available, (2) extendable, (3) flexible, (4) refitable, (5) moveable, (6) reusable and (7) recyclable.

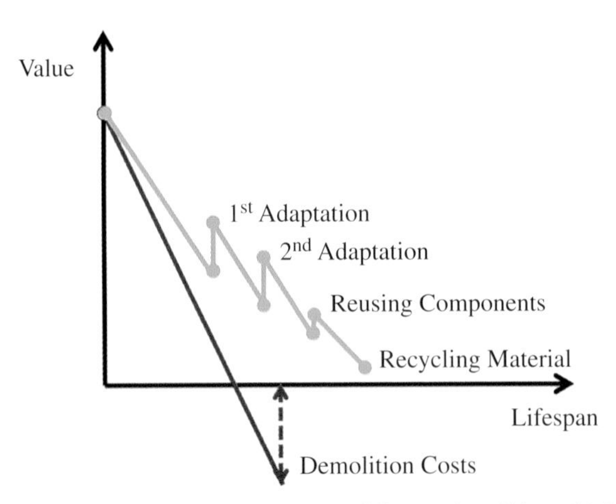

Figure 4.1 Current practice for inflexible office buildings – demolition with loss of value (red), and business model of the adaptive building – focus on high sustained value (green). Qualitative approach © bauforumstahl.

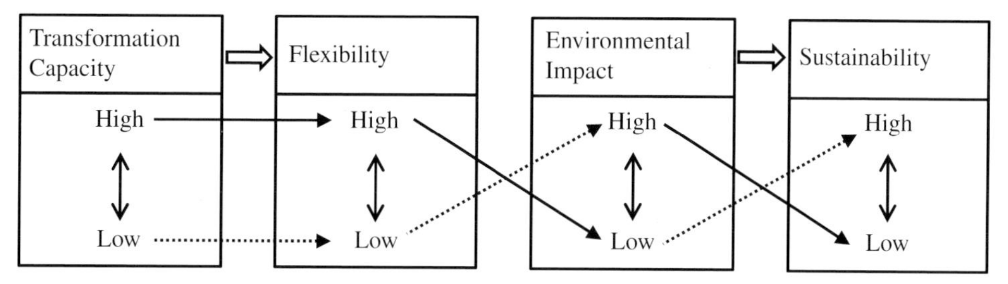

Figure 4.2 Relation between building conversion capacity and sustainability (after Durmisevic [13]).

Applying these abilities to buildings increases the number of loops between design and disassembly/demolition, thereby extending a building's life cycle and reducing its environmental impact. Nevertheless, flexibility can be achieved on various levels at different points in time in regard to special and technical aspects. Spatial flexibility regards organization changes as enlargement or reduction of usable spaces. It means extendibility, partitioning, multifunctionality or functional mutation. On the other hand, technical flexibility means the ability of building components and systems to be easily replaced, displaced, reconfigured, reused and recycled. Indicators are accessibility, replaceability, reconfiguration and separation. Also, while changes to utility and media supply and floor coverings might occur more often, when there are new users or new space requirements, a structural frame may only require slight modifications after several years. When flexibility is considered during design, the need to take preliminary measures for potential conversions and their implementation is more important. In fact, if a structure's flexibility is not considered in advance, the building owner will be faced with significant additional costs when modifications become necessary.

Flexibility is already being accounted for in sustainability assessment methods. It is being evaluated through (1) easiness of deconstruction and disassembly, (2) modularity of the building, (3) spatial structure, (4) indoor height clearance and (5) accessibility of utilities conducts and cables (see Chapter 5).

When looking at the characteristics of flexibility and ease of building conversion, the potential of spatial and technical flexibility in steel buildings is important. Steel buildings can be flexible in many ways [14], [15]. Steel has a high load capacity to weight ratio, achieving long-span buildings. This increases column-free spaces, enabling changes in the interior spaces without major interventions or material losses and leading to an optimization of the floor area. The floor plan can be easily adjusted to new requirements through installation/removal of internal lightweight partitions (see Section 4.4).

The position of utilities' cables and conducts (electricity, media, water, etc.) also influences the flexibility of the building. They should be easily accessible and positioned in a way that maintenance activities and possible changes are doable without losing wall-finishing materials. By means of composite floors, using steel decking, openings for utilities in the vicinity of columns are possible. Composite beams can already be designed with additional openings, or openings can also be created at a later point.

Modularity is presented as a major aspect to buildings' sustainability. It leads to faster construction, improved quality and reduced resources and waste. Components can be added, relocated or removed according to the users' requirements. At the same time it has an inherent share in steel buildings. Bolted joints are usually chosen for on-site assembly, and hence they can be unbolted to facilitate extension or dismantling. Moreover, conversion of steel structures does not lead to dust, vibrations or excessive noise, meaning that the building can partially be kept in use during conversion work.

At the building's end of life, steel members can be easily disassembled, allowing the reuse of its components in other buildings. This means that the lifespan of steel members can extend and enhance other buildings' flexibility. When there is no way of reuse, steel members are recycled.

4.3.3 Design for deconstruction, reuse and recycling

Annette Hillebrandt

4.3.3.1 *Being sustainably effective in architecture*

Thinking about the importance of steel in the building and construction industry – as an architect or a dedicated structural engineer – immediately brings about visions of great and important projects. On closer inspection, it becomes obvious that the secret of success of these projects is their combination of lightness and loadbearing capacity. This is the reason why – amongst many others – Philip Johnson´s 'Glass House' is an icon of transparency and Mies van der Rohe´s 'Crown Hall' for the College of Architecture on the Illinois Institute of Technology campus is the perfect embodiment of elegance. The great span of the Golden Gate Bridge and Frei Otto´s vast and complex tensile cable structure at Montreal Expo 1967 demonstrate not only steel´s great capability but also the enormous potential of this material to create symbols of building culture and to write architectural history.

Furthermore – as architect or engineer – thinking about steel will be influenced by factors like prefabrication, fast construction time a 'dry' building site. These aspects, in addition to its lightness and performance, make steel a sustainable building material with great potential for the future.

The relationship between steel and sustainability in the original sense of the word – namely, the ability of a system to meet the needs of present and future generations in equal measure [16] – is clearly apparent: 99% of the steel used in the building and construction industry is used again or is recycled (see Chapter 3.11).

Looking ahead, we may not have an energy problem – the sun provides more energy than we need – but we may have a resource problem. This means that steel is the perfect example for the circular economy of the future. But what does this mean in comparison with other building materials?

Recovery rates of over 95% have also been published for mineral building materials in the demolition waste, of which over 75% can be regarded as utilisation [17]. So where is there a need for action?

4.3.3.2 *Conservation of resources – target: 'A European recycling society'*

Generally speaking, more than 50% of all waste in Germany is attributable to the building and construction industry [18]. This means that there is not only a waste-disposal problem but an enormous waste of resources. In contrast, the European Council speaks of a 'European recycling society with a high level of resource efficiency'. The council emphasises that 'waste prevention should be the first priority of waste management, and that reuse and material recycling should be preferred to energy recovery from waste' and has published the following waste hierarchy [19] (see Chapter 2.3.1):

1 prevention;
2 preparing for reuse;
3 recycling;
4 other recovery, especially energy recovery and backfilling;
5 disposal.

'Waste prevention' in the building and construction industry doubtless includes the strategy for revitalising unused building stock as well as upgrading and conversion of existing sites. 'Reuse' means 'any operation by which products or components that are not waste are used again for the same purpose for which they were conceived'. 'Recycling' means 'any recovery operation by which waste materials are reprocessed into products, materials or substances whether for the original or other purposes'. It does not include 'energy recovery and the reprocessing into materials that are to be used as fuels or for backfilling operations'. 'Recovery' is defined as 'any operation the principal result of which is waste serving a useful purpose by replacing other materials which would otherwise have been used to fulfill a particular function'. 'Disposal' – used in the context of waste from the building and construction industry – means 'any operation which is not recovery', in other words 'deposit in or onto land', that is to say landfill, even when subjected to prior repackaging [19].

Concerning reuse and recycling, the guideline says: 'in order to…move towards a European recycling society, Member States shall take the necessary measures designed to achieve the following targets: by 2020 the preparing for reuse, recycling and other material recovery, including backfilling operations using waste to substitute other materials of non-hazardous construction and demolition waste… shall be increased to a minimum of 70% by weight' [19].

At first this appears to be ambitious, but it is nevertheless way removed from a true recycling economy. This target of recovery of 70% of all nonhazardous construction and demolition waste does not only mean recycling to the same level of quality or components able to be reused on a like-for-like basis, but also includes downcycling.

Regardless of whether it is considered globally and put in the context of the enormous quantities of raw materials needed by a growing world population in the future, or even considered from a national point of view, for countries with few natural resources, the demand for a rate of recovery of 70% is problematic. It is necessary to have a 'real recycling rate' as high as possible with the same level of quality as the starting product. So this does not include downcycling to a level of less quality (Figure 4.3).

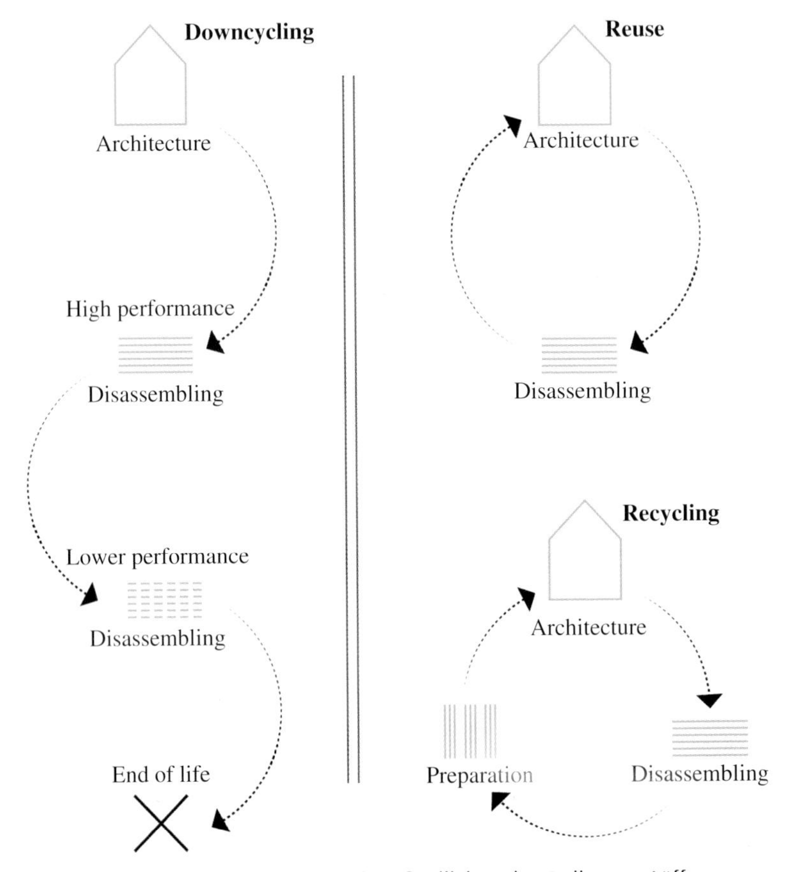

Figure 4.3 Downcycling, reuse and recycling. © Hillebrandt – Düllmann+Lüffe.

With steel it is relatively easy to achieve real recycling.

The situation is somewhat different with those materials that can only be recycled on levels of lower quality: for example, when a timber beam is recycled for the first time it can be processed into boards (e.g. oriented strand board or OSB), in the next recycling step, it can be processed into a board of even less quality (e.g. medium density fibreboard or MDF), and finally there is only the possibility of processing the board into pellets and thus utilising thermal recovery. The material loses part of its strength capability with every step, which is a clear downcycling process keeping in mind that wood is a renewable resource and therefore has other advantages in sustainability considerations.

The limited recyclability of mineral waste can easily be seen, for example, from the fact that only some 10%–20% of suitably sorted broken bricks can be used in the form of brick dust as a raw material in the production process without impairing the technical properties of the brick [20].

The content of recycled aggregate used in concrete is currently less than 1% [21]. There is a need for better information and political control instruments in order to increase the rate of recycled material.

The 'direct use of components' by reuse is also a good way to conserve resources. It means the renewed use of a component for the same intended purpose while maintaining the original shape of the product. However, reuse will remain a relatively small aspect in comparison to recycling. This is due to the fact that the capability of many components – and in particular the components like doors, windows or glazing – is continually subject to improvements (soundproofing, thermal insulation, fire protection) so that direct use in a new building to satisfy current regulations and requirements is unlikely.

Exchanges – trade markets – for used building components offer a good approach here, although the number of parts in stock is mostly limited, and the architectural planning process has to be reversed completely in order for these exchanges to turn into a successful model. The existing supply of used components has to be researched first as a precondition of planning decisions. Furthermore, advance financing is necessary on the part of the investor to secure the selected building material, and that has to be done before the planning process is completed. Perhaps this idea will gain more importance in times of future scarcity of raw materials.

'Further use' as a strategy for waste prevention/conservation of resources is understood to mean a used product being used again in its existing form but for another purpose, such as a former structural steel section used as simple lintel. Here, too, the range of applications is as limited as the market. Components that are subjected to reuse or further use have generally gone through their life cycle already. Reuse and further use therefore finally also lead to material recycling with relinquishing of the products shape (Figure 4.4).

4.3.3.3 *The planner's options and responsibility – designing for dismantling and recycling*

4.3.3.4 *The resource-conserving frame construction*

Possibilities for transformations that have to be made in the future may be difficult in a building stock based on a 'solid-construction' mentality that prevails in parts of Europe. Because of future scarcity of raw materials and the need for utilization of inherent buildings, the reuse and conversion of building stock will be given preference over new buildings. Therefore, in order to be able to transform one building typology into another at reasonable costs in the future – such as administrative buildings into residential (or vice versa) – it is necessary to plan for flexibility in floor plan and type of the structure from the very beginning. The conversion of solid wall-construction buildings is not generally possible in economical or ecological ways because the effort required for dismantling, conversion

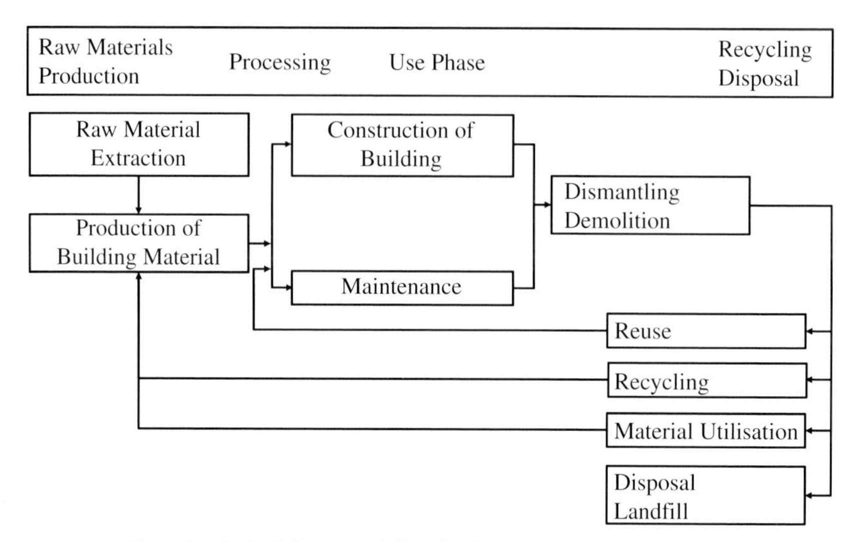

Figure 4.4 Life cycle of a building material. © bauforumstahl.

and upgrading outweighs any saving effects. However, frame construction must also be planned for future uses. Provision has to be made for sufficient heights between floors in buildings, including double floor construction and suspended ceilings, to satisfy all groups of subsequent users, for example, when allowing the transformation of an office building into a residential building.

Only a frame construction in combination with a flexible ducting system for services can provide the necessary floor-plan flexibility.

A further decisive factor is, of course, planning loadbearing reserve into the construction, by which a building typology can be transformed or building performance can be improved to future requirements.

With taller buildings, the question of fire protection arises, as a concrete frame seems to offer an advantage because of its fire protection being intrinsic to the design. Steel structures are often protected by gypsum plasterboards, for which there is currently no reuse potential but recycling activities have started.

4.3.3.5 *The resource-conserving choice of material*

As a basic principle to obtain sustainability, the planner should check all materials for their ability to be part of a biological or technical cycle. The biological cycle describes the return of organically based building materials and building products into decomposition or composting processes. This is the case with clay wall cladding boards, which have a reinforcement fabric made of natural fibres and are thus able to be returned to a biological cycle. Decomposition processes often proceed without any additional energy. In the technical cycle, resources such as metals, glass or plastics are separated and reprocessed, usually requiring a large amount of energy to produce new building materials or building products.

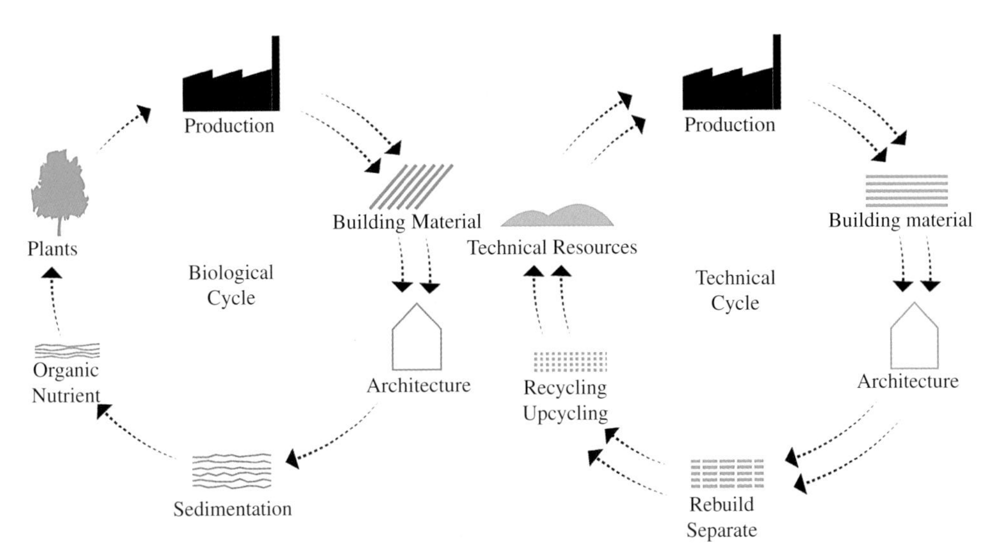

Figure 4.5 Biological and technical cycles according to the 'cradle-to-cradle' philosophy of Braungart and McDonough. © Hillebrandt – Düllmann-Lüffe.

These cycles only work successfully as sustainable, of course, if the additional effort needed for transport, processing and reforming is provided by renewable energies (see Figure 4.5).

The practice of coating or painting and the tendency to design using composite building materials often run contrary to the reintroduction into a real material loop. A paradigm shift will only succeed here if the 'polluter pays principle', stipulated by the European Parliament, for the costs of waste management is implemented in practice. This principle implies that the system that generates the waste is responsible for declaring the impacts of waste processing until the end-of-waste status is reached [19]. Focusing on steel, it shows that steel coating with zinc for protection against corrosion is not problematic. Zinc vaporises during the melting operation in the electric furnace, and over 90% is recovered. The combined rate of recycling and reuse of zinc-coated steel is even 99% [22]. Compared with coatings, hot-dip galvanising consumes even up to three times fewer resources [23]. As a basic principle, it is therefore advisable to give preference to the use of coatings that remain an intrinsic part of the material group, such as hot-dip galvanising of steel. The surface aesthetics of an untreated material determined by its chromaticity and surface structure alone impress due to a dignified and perceptible ageing process. In this respect, the potential for steel in building envelopes and interior design elements is very relevant (Figure 4.6).

Architects may elicit the whole range of different surfaces of steel. Nevertheless, there is a large potential in further development and wider distribution of this material. Various surface finishes may be discovered: golden shimmering corundum-blasted stainless steel, blue-black iridescent hot-rolled steel, polished stainless steel with a mirror-like finish and so forth.

Figure 4.6 'Metallwerkstück' (metal workpiece). Architects: m.schneider a.hillebrandt office building Bad Laasphe, Germany, 2010 © Cornelis Gollhardt.

Steel is impermeable to vapour and water and is thus suitable for shower trays or wall cladding without any additional waterproofing in areas exposed to water spray. It is suitable for stairs and built-in furniture without any additional coating.

4.3.3.6 *The separable assembling and demountable connecting*

Looking upon buildings as reservoirs from which building materials can be recovered, this has been a tradition going back to antiquity. Reuse has been a very pragmatic form of conservation of resources, mostly in the form of working time and manpower. The value of reused components can be seen by looking back to the period following the Second World War: without an easy-to-handle component in the form of a building brick and easily removable lime mortar, it would not have been possible to rebuild Europe within such a short period of time. Large-format slabs of natural stone can be reused in many more different ways than those that are slate sized; a longer-span roof girder will find a new field of application more readily than a shorter one.

For reuse, further use, recycling and recovery of building substances and components, it is essential to design for dismantling and recycling. Design for dismantling means connecting components by using demountable connections and forgoing the use of glued joints. Furthermore, it avoides the use of composite building materials and systems in which the most varied groups of materials are bonded together. A large assortment of separable connections has been proven over the centuries, such as screws, clamps, click connections and suspension clamping systems. Other than that there is an advantage of a material-intrinsic joint there is no need for dismantling and separating prior to recycling. When in doubt, the planner should always prefer a construction form that contains the smallest number of materials from different groups in order to achieve a reduction in the effort involved for dismantling, separating and sorting, as well as for the contamination of the building materials used by bonded component layers.

4.3.3.7 *The building as a planned store of raw materials – urban mining*

Building in a positive economic sense means creating and maintaining value and making a long-term investment into the future. By contrast, today a widespread attitude of pressure to succeed financially from buildings in the short term as return on investment has been established. However, anyone wishing to invest seriously in the future by not becoming dependent on making quick money will immediately understand the benefits gained when building in a recyclable and dismantleable manner: the design of the house is flexible, it can easily be repaired with a minimum of effort and it can be upgraded using existing parts.

In order to overcome the global scarcity of raw materials and significantly increase recycling in the building and construction industry, there is a need for the stimulating concept of 'urban mining' – the house as a raw material mine. As far as the building and construction industry is concerned, this would be the practical implementation of the 'promotion of eco-design' of the European Parliament as 'the systematic integration of environmental aspects into product design with the aim to improve the environmental performance of the product throughout its whole life cycle' [19].

Some voluntary building-certification schemes such as of the DGNB offer an approach that enables the recycling potential of buildings (see Chapter 5).

Poltical control tools, however, require clear, measurable factors and a reliable basis for calculation. Planners should be enabled to calculate clearly the amount of recoverable material streams within the whole life cycle of the building process, as well as the effort in time and finance to accomplish this. On this basis, it may be possible to decide whether it may become necessary to implement a security bond or residential property charge to cover the risk toward society. The true, sustainable value of a property will then become apparent to developers, purchasers or investors for the first time. The aim is first to protect society from loss of resources, and also to increase ′product responsibility′ for buildings. Despite the desire for measurable, sustainable methods of construction and government support, an innovation-oriented society must always be judged on the space it may create for intelligent exceptions and individual solutions.

4.3.3.8 *The future of building construction*

Three-dimensional printing using the most differing materials is on the rise. Nowadays 3D printing technology enables some of the most complex objects to be fabricated. It therefore seems logical to use this technology in building and construction. The promise of being able to make the art of construction superfluous and to design a house using a single material without all the necessary details appears to be quite appealing. Smaller examples have already been produced, for example, in concrete reinforced with steel fibres (unfortunately not to dismantle or recycle). At the moment, though, most houses are still larger than available 3D printers.

A particular challenge, however, is the complexity of house building: the structure needs to be loadbearing, wind- and rainproof, thermally insulated and soundproof, plus protective against fire and secure. Putting all of these parameters into practice with a 3D printed object using one material seems to be quite far off. Last but not least, the house of the future needs to be completely recyclable. There is perhaps even more potential for steel construction here: the house of the future might be 3D printed with closed-pore steel foam and thus satisfy all desires.

Planners and developers or investors have to make decisions considering the conservation of resources and waste prevention now:

- at an urban development level, conversions should be given priority over new buildings, the reuse of 'brownfield' (post industrial) land priority over greenfield development;
- at the level of buildings and its structures, avoiding creating basements and therefore save excavating soil and bonded seals;
- a compact-shaped building results in a good energy- and resource-efficient envelope surface area to volume;
- a flexible frame construction instead of a highly predefined solid construction simplifies the transformation at a later date;
- loadbearing reserve in the structure facilitates the desired flexibility of use;
- a modular building construction improves the chances to reuse or further use of the components at a later date;
- the aim for forgoing the use of coatings will improve the accuracy of sorting materials according to material type to prevent contamination of material;
- as a basic principle, renewable raw materials or materials that can be recycled to the same level of quality – for example, steel – are to be preferred to materials undergoing downcycling at the end of their useful life or those that are destined for combustion or landfill;
- large formats, whether for loadbearing components or as panels in the building envelope, also encourage subsequent reuse or further use of the components;
- at the level of responsibility toward society, investors and planners should take account of future demands by considering buildings as being four-dimensional, which includes an end-of-life cycle.

All these aspects are meant to support planners, investors and politicians in rethinking planning and building strategies in order to be sustainably effective in architecture.

4.4 MULTISTOREY BUILDINGS

Johann Eisele, Richard Stroetmann and Benjamin Trautmann

4.4.1 Introduction

Throughout European countries exists a high vacancy rate of office buildings, and partial these have become a visible problem in large towns. The increase in the average vacancy rate, for example in the 125 largest German towns has risen from approximately 1% in about 1990 to 7.5% (and over 10% in some large towns, with the maximum rate in Frankfurt of 18.5%) (Figure 4.7).

Increasing vacancy figures point to a structural problem, especially when at the same time, new buildings that are intended to satisfy current needs are being planned and completed. New forms of working like home office, desk sharing and nonterritorial office arrangements have led to changed technical and spatial requirements in the construction of offices. These have a markedly negative

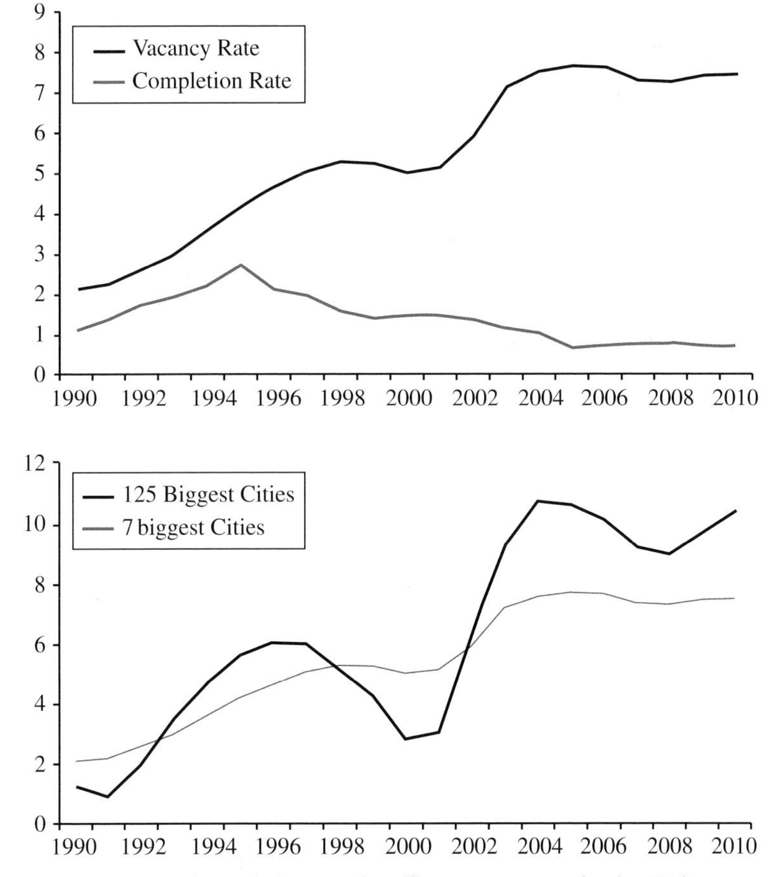

Figure 4.7 Vacancy and completion rates for office space: average for the 125 largest towns in Germany as a percentage of the entire existing stock of office space [24].

D2 Tower

Location:	Paris – La Défense, France
Architect:	Anthony Bechu and Tom Sheehan
Building description:	High rise building (171 m) of 37 floors with Lozenge-pattern external steel grid structure.
Steel details:	Total of 5300 t of steel. Including 3000 t of HISTAR Jumbo sections, mainly HEA 600. As the worksite was difficult to access and included a portion partly over the ring road, subtle organization and new implementation methods and processes were required for the work to go forward. This could only be achieved in steel construction.
Sustainability:	The D2 Tower is the first tower in La Défense to enable 30% of material saving through its sophisticated steel exo-structure. All technical installations comply with the highest criteria of energy efficiency.
Awards:	NF-HQE, energy performances Cref –30%. Conformity to 2012 Thermal Rules (RT2012). BREEAM. The project is a nominee for the Archi Design Club Awards 2015. The project is the winner of 2015 steel architecture Eiffel Trophy (Working category).

Figure 1 View from below the arched steel structure. © ConstruirAcier.

Figure 2 General view of the D2 Tower in Paris – La Défense. © ConstruirAcier.

Figure 3 External diagrid structure with its diamond-shaped windows creates a unique lighting effect. © ConstruirAcier.

Example provided by ConstruirAcier.

impact on the marketability of existing buildings because of their lack of flexibility in the building structure. Users always expect modern concepts and furnishings, which can often be better achieved in new buildings than in renovated or modernised existing buildings [25].

Using the typology of office and administration buildings as an example, it is apparent that multistorey building design is independent of the user (and the use) in order to satisfy the adaptations of the working world, and also demographic changes. Whereas up until the 1970s, buildings were characterised by mainly rigid structures that were designed for the same use into the future, cost-optimised investment properties for special uses and short-term amortisation of the investment are now being constructed. Negative consequences of this type of development are an inability to adapt to other requirements, declining marketability and eventually vacancy and demolition of the building. This conflicts with the central idea of sustainability, which is a long, useful life and resource efficiency – that companies use today to emphasise their corporate social responsibility and that in view of the maintenance costs of property are becoming increasingly important [26]. The aim of project development should be to construct buildings that offer optimal flexibility/versatility at acceptable investment costs. This ensures a relatively long, useful life and thus a high level of sustainability.

4.4.2 Building forms

The planning of multistorey buildings often begins long before architects become involved and prior to the 'charrette' (Section 4.2.1). The developer's representatives or company advisors discuss location issues, prepare planning needs, carry out feasibility studies and specify the desired building concepts, furnishing and the finishes. At this stage, the architecture is often limited to planning the building's shell and core [27]. The diversity of building forms is constrained during the design process by the shape and size of the site, the ideas of the developer and the eventual user regarding the interior arrangement. During the iterative development process, guidelines relating to building and planning legislation, regulations covering fire safety, acoustics and building physics, and the planning of the building services are taken into consideration in order to perform the detailed design of the building form and structure (Figure 4.8).

The diversity of possible designs and the resultant dependencies show how difficult it is to specify definite parameters for the final form of the building. However, basic design decisions regarding the use of materials and energy strategy have to be made, and these are being increasingly subjected to stricter regulations. Structures made up of single-storey enclosures, building shells with a large number of penetrations and material-intensive structures whose level of 'grey energy' is unreasonably high are not acceptable from a sustainability point of view. Similarly, a land-intensive design approach for a site with high land prices cannot be regarded as sustainable from the point of view of investment costs. An economic and ecological design approach therefore has to be chosen based on the total of all of the provisions and concepts of the specific design and that additionally ensures the flexibility and versatility of the interior arrangement to current and future uses.

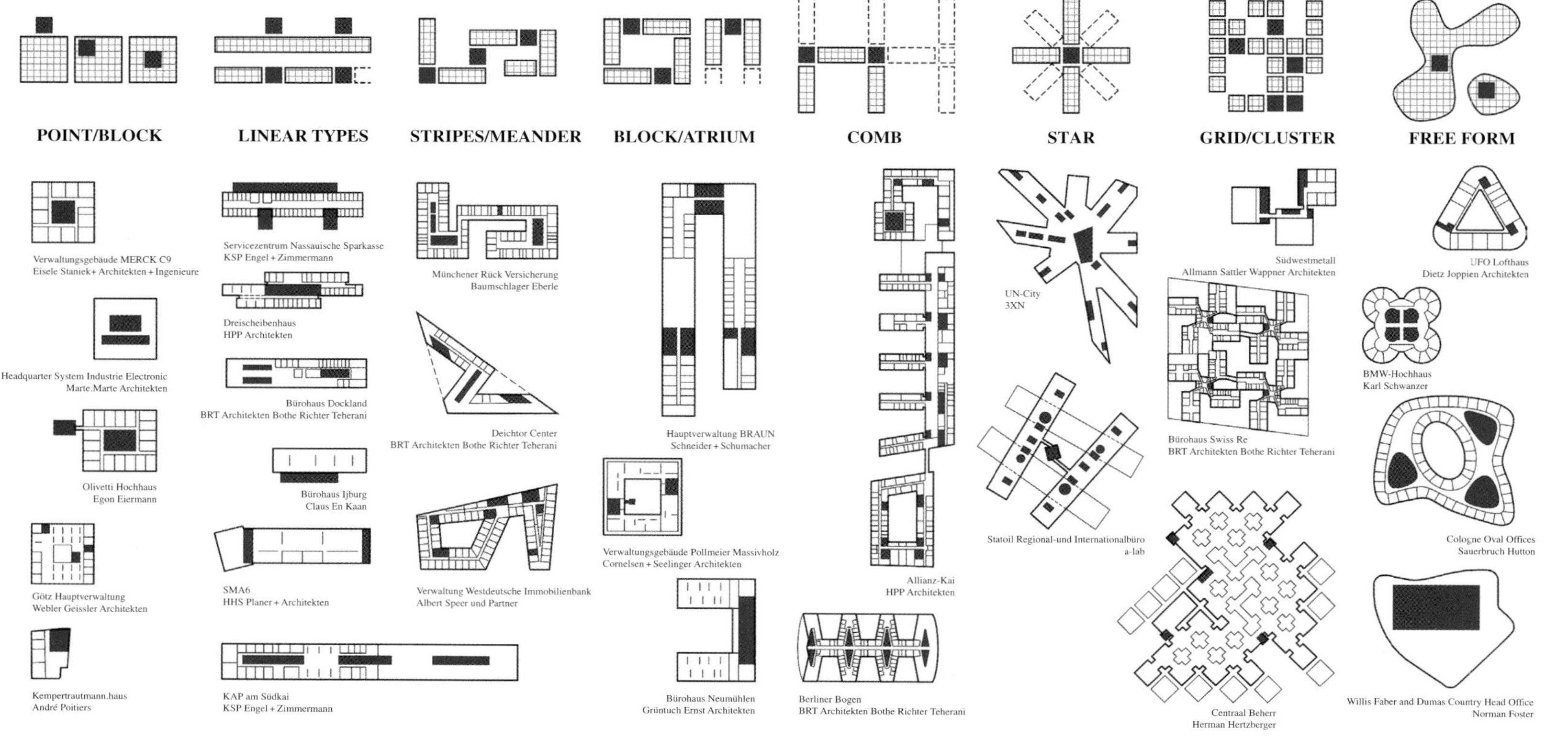

Figure 4.8 Typological concepts: point/block – linear types – stripes/meander – block/atrium – comb – star – grid/cluster – free form.

4.4.3 Floor plan design

The reconcilability of the structural requirements for the different forms of standard floors plays a decisive role in the planning process in order to be able to achieve flexibility or versatility of the building structure. At the same time, without considerable additional expense the prerequisites for upper floorlevels have to be combined with the structural requirements of the ground floor, where special uses are usually required. This is often the case for basement floors, where underground car parks often prescribe a rigid grid pattern dictated by the size of the parking bays (Figure 4.9).

Using a simple model, it is possible to show the basic interdependencies of the individual parameters of a multistorey building – use, building depth, height between floors and typical sizes of construction and facade grids. Such a parameter model should not be regarded as a final building design but more as an abstract model. It describes the building types examined using the lowest common denominator and thus represents an abstract set of rules that facilitate the planning of multistorey buildings (Figure 4.10).

It is not usually possible to consider the facade and structural grids separately because it is useful for the structural grid to be a multiple of the facade grid. The number of units of the facade grid in the construction depends among other things on the properties of the selected floor system.

Long span construction without internal columns or downstand beams is often most costly in construction terms but provides for greater freedom of the internal space for different uses; therefore column-free space is regarded as being

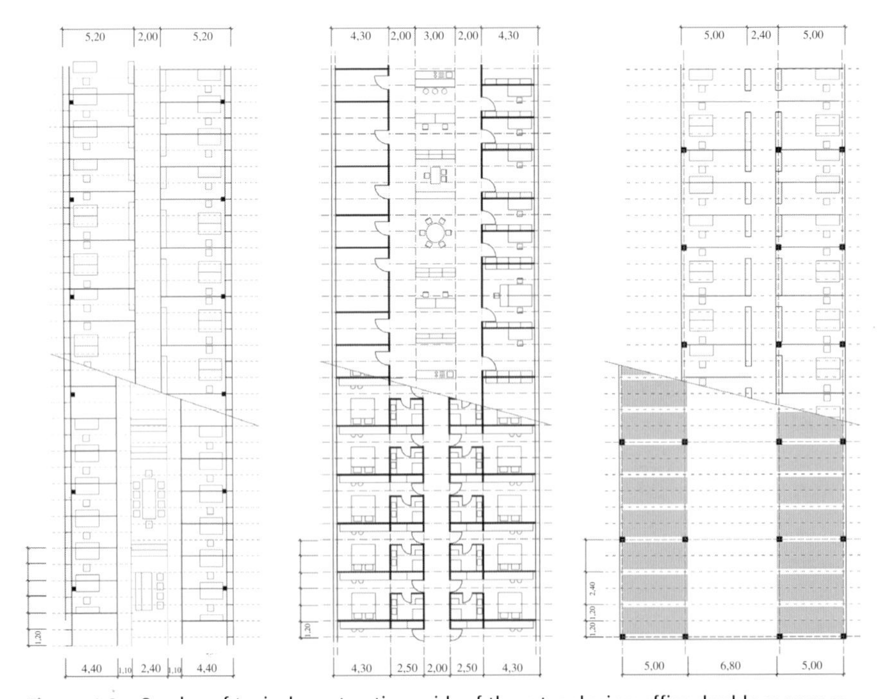

Figure 4.9 Overlap of typical construction grids of three typologies: office double sequence with office triple sequence, office with hotel and office with parking.

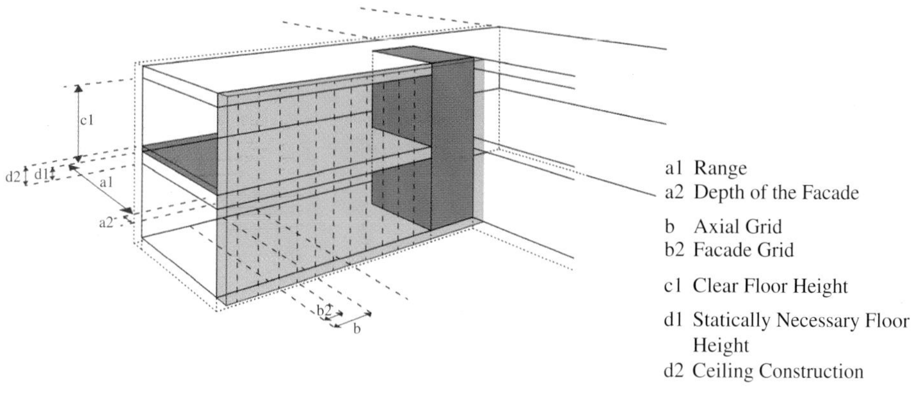

Figure 4.10 Model of spatial parameters.

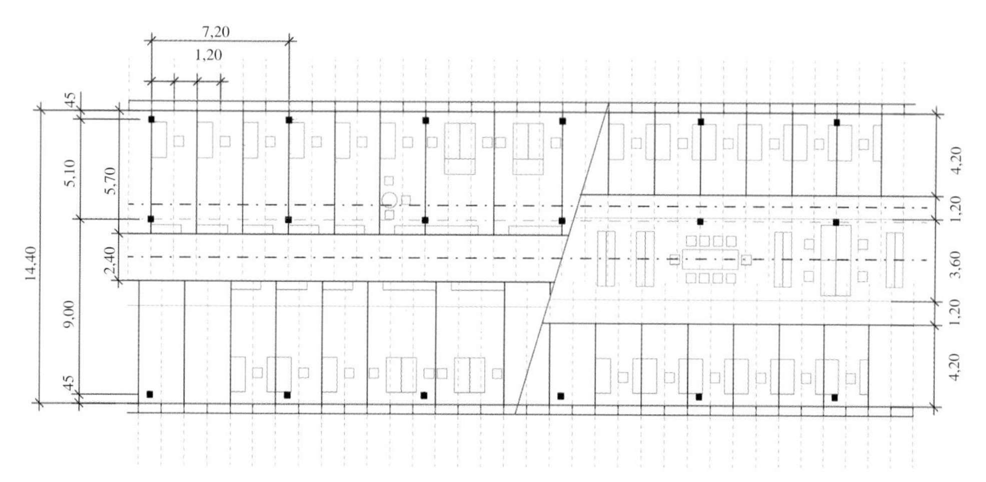

Figure 4.11 Position of centre columns that either interfere or do not interfere with the use of the space on other floors (right- or left-hand side, respectively).

highly flexible. Special areas, such as the reception and car parks in the basement also benefit from the freedom that the absence of columns offers. The layout of the upper floors can be arranged as required. Shorter span structures are supported by one or two rows of internal columns and often require downstand edge beams, and their design restricts the flexibility of the internal planning significantly. The aim therefore has to be to maintain flexibility and versatility for the interior space for the different uses (Figure 4.11).

Also linking the necessary circulation areas vertically and horizontally is essential for a functioning versatility in multistorey buildings. For example, in office buildings, several 400 m^2 units can be linked to the necessary stairwells, whereas in residential properties only a limited number of apartments can be linked directly via stairwells. Alternatives such as incorporating an access gallery have to be considered with regard to the resultant quality of the affected dwellings. The future conversion of buildings may be taken into consideration directly when the building is designed or, alternatively, provisions may be made via extensions such as stairwells at a later date (Figure 4.12).

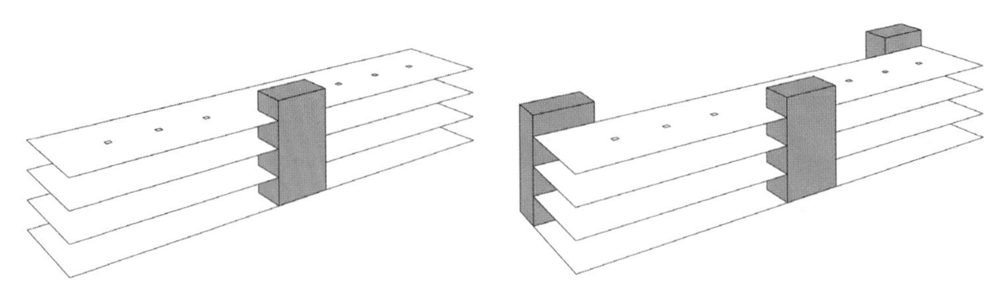

Figure 4.12 Transformation of offices into dwellings – addition of more closely spaced vertical core accesses.

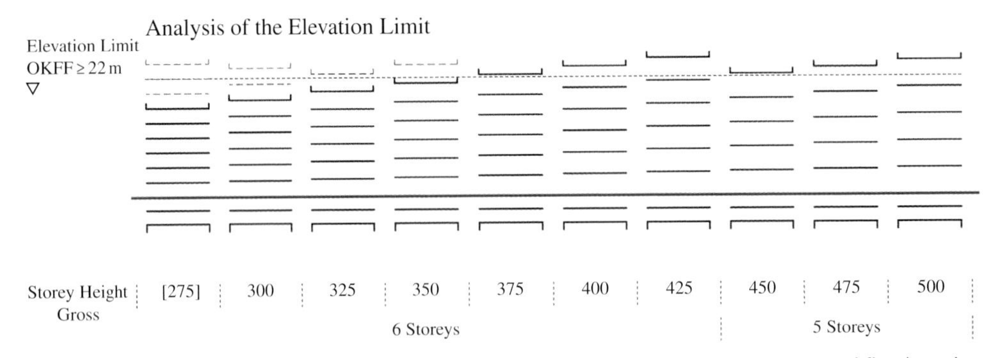

Figure 4.13 The effect of the height limit (here 22 m above the upper surface of the ground floor) on the number of possible storeys in a high rise building.

4.4.4 Building height and height between floors

The building height of multi-storey buildings can differ significantly despite them having the same number of storeys because of the requirements for the available room height and the height of deck constructions depending on the user/use. The building height, which includes the sum of the heights of all internal space, the floor structure and services, has to be given careful consideration with regard to possible uses. Space that is too low can restrict the flexibility for future uses because different room sizes have to have a minimum height for a comfortable room atmosphere. Oversized heights between the floors result in considerably higher costs due to the cost of facades, partitions and building services. Where planning imposes a height restriction (specification of the height of the eaves in the development plan or a high rise building limit), special attention should be given to control of the building height. From the investor's point of view, omitting a storey could make the development unprofitable, therefore decisions relating to the storey height should achieve the desired measure of spatial flexibility/versatility without affecting the investment costs (Figure 4.13).

4.4.5 Flexibility and variability

As mentioned in Section 4.4.1, the vacancy rate of office buildings is a serious and increasing problem. Rapid technical development and changing processes in the working

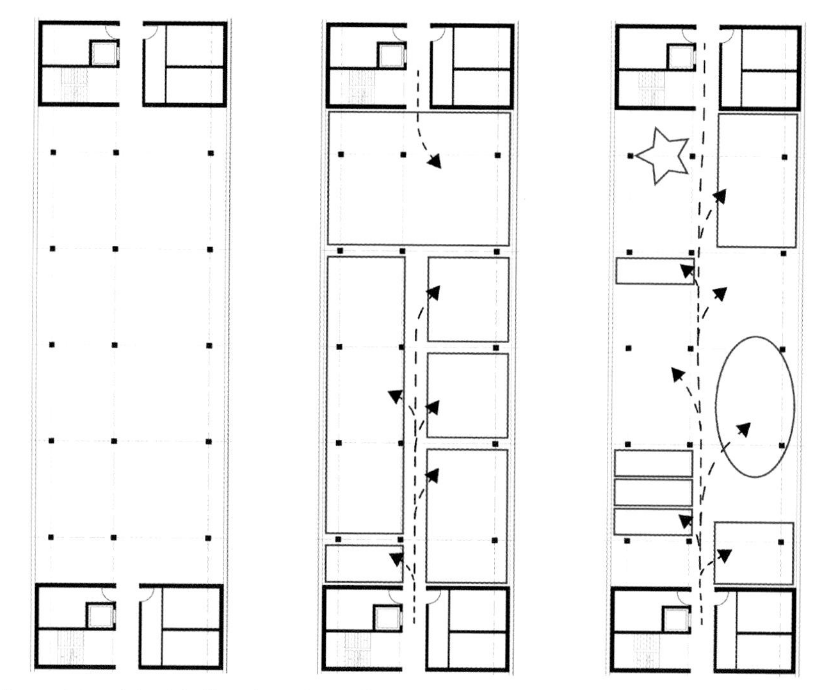

Figure 4.14 'Elastic' allocations of an office unit.

world cannot be implemented in many office and administration buildings today. Inadequate ducts and cables of the building services or inflexible layout mean that considerable financial effort is needed to make office buildings suitable to meet the demands of today's user. Comparing the costs for modernisation with those for a new building that meets current requirements, including demolition of the existing building, is a common process and often results in a decision being made to construct a new building. It is not only the investors of speculative buildings who are looking to achieve a high degree of flexibility to satisfy the specific requirements of users who were not usually identified at the planning stage. Even developers who intend to use a property themselves extend their requirements by including the possibility to rent out part or the whole of the building to a third party later. Therefore, it is important to design a building structure that can be used flexibly in various office layouts (Figure 4.14).

The mixture of different uses of buildings poses major organisational, technical and economic challenges for the partners involved in design and construction, because the differing types of structure have to be superimposed. This is especially the case for different uses on different levels, which begins with the choice of building plan form and the structural grid and continues with the development of the interior layout [28]. For the versatility of multistorey buildings, the market viability of different uses at the selected location should be investigated in detail in order to assess and coordinate all building parameters for the different uses within a viable framework.

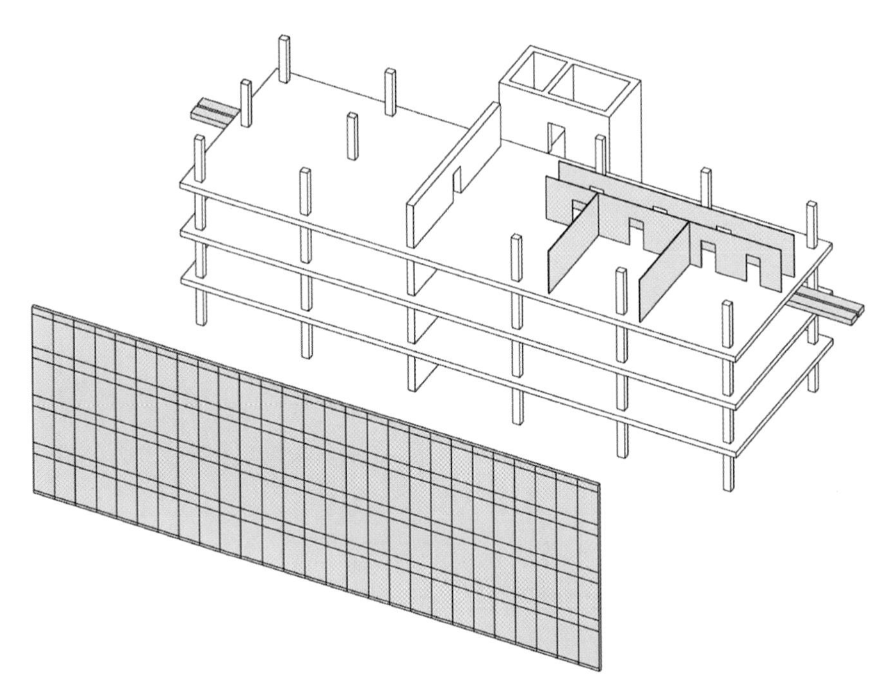

Figure 4.15 Model of parameters. Whether the blue elements are suitable and should be retained has to be checked prior to the transformation process.

Multistorey buildings of the future have to be planned in such a way that they can be adapted with little financial effort to changing social and sociological demands within their life cycle in order to achieve profitability and avoid the property becoming unusable. In addition to flexibility for the same use, this might require the building or parts of the building being suitable for different uses. (For examples of steel composite solutions and their life-cycle assessment (LCA) comparisons, see Chapter 6.3.) This necessitates a 'robust' primary supporting structure that enables the requirements of the widest possible range of uses to be met without major constructional changes. It has to fulfil the spatial, functional and building-legislation-related requirements for the different uses and at the same time satisfy the client's design requirements. The availability of secondary and tertiary systems, building services elements or the facade construction should be investigated when there is a change of use from both economic and ecological points of view. This does not limit the sustainability merely to energy-related values and the durability of all components but particularly places value on spatial flexibility/versatility and recyclability (Figure 4.15).

4.4.6 Demands placed on the structural system

The competitiveness of the areas for rent and use depends amongst other things on whether the space defined by the structure offers the necessary flexibility to changes in use. The structural system of the building is of particular importance

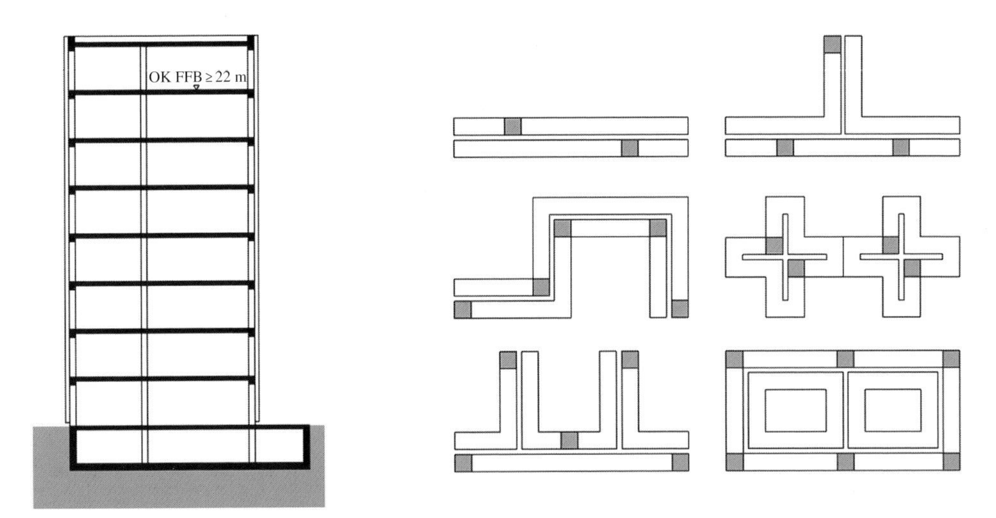

Figure 4.16 Typical variants of floor plan designs for office buildings with centre corridors and building cores; the section shows an office building with underground parking [29], [30].

from the start. The positions of columns and shear walls determine the possible uses of the space and either allow or restrict the flexibility of the floor layout (Figure 4.16).

The building's geometry and different functional areas place different technical and design demands on the structural system. For example, the column grids, building services and design requirements differ between underground car parks, conference areas and offices. In the reception and conference areas, the shape and size of the columns and the floor soffits is important.

In underground car parks, the column grid is predetermined by the parking spaces, but only a small storey height is required. By contrast, the arrangement of the columns in the area of the office is usually determined by the size of the offices and the facade grid in the longitudinal direction of the building. Typical grid dimensions are 1.20 m, 1.25 m, 1.35 m and 1.50 m. The width of an individual office can be achieved using double the grid dimensions. For a workstation with standard furniture, a grid of at least 2 × 1.20 m is needed. For the arrangement of an underground car park, this grid covers the minimum required parking space width of 2.30 m, for example, in Germany; 2.5 m is often required in other countries. In the transverse direction of the building, the position of the columns depends on the office layout. The room depths for the widely used cellular office are 4.50–7.20 m. By contrast, the Business club layout requires large areas, and these also require a building depth of 12 m [31]. To avoid elaborate support structures, the column grids should be the same on all floors or a transfer structure has to be used (often at ground floor).

The requirements with respect to floor plan design and the heights between floors can change over the life cycle of buildings. This is especially true for commercially used office properties with changing tenants. The adaptability to meet changing user demands should be taken into consideration at the building's planning stage in order to maintain its marketability. Appropriate preliminary work involving the design of the structural system and the nonstructural elements (e.g. the use of flexible partition wall systems) should be conducted taking the relevant use scenarios into account in order to allow changes in use with minimal modification (see also [32]).

4.4.7 Floor systems

Steel-concrete composite structures are used for multistorey buildings in technologically advanced countries and have differing market shares. The advantages of this method of construction are the high degree of prefabrication, its light weight, the rational and fast method of construction and the reduced effect of adverse weather conditions. With regard to the structural design of the floor systems, two basic construction principles may be differentiated: floor systems with downstand beams as composite beams and slim floor systems with integrated beams (Figure 4.17; see also [30]).

In short to medium spans (grids up to approximately 11 × 8 m), it is possible to use flat slabs with integrated steel beams (slim floors). The advantages of this method of construction are the low construction height and the installation flexibility. This reduces the storey heights and therefore the facade areas, and also the building volume that has to be heated. The steel beams integrated in the floor slab are often in the form of welded sections with a wider bottom flange to support the slabs. Prestressed concrete hollow-core slabs are often used to span between the beams because they are economical, manufactured in transportable dimensions and have long spans. The usual spans of the prestressed slabs are 1 to 1.5 times the span of the supporting beams (Figure 4.18).

For longer spans, slabs with downstand beams are suitable and create a free open floor plan design (Figure 4.19). The usual spans are 6–15 m for the secondary beams and 9–12 m for the main beams. This means it is possible to span common widths of office buildings without using internal columns. The spacing of the secondary beams determines the span of the slabs, which is often between 2.5 and 4 m. Rolled or welded I-sections are used as the long-span floor beams. The slab

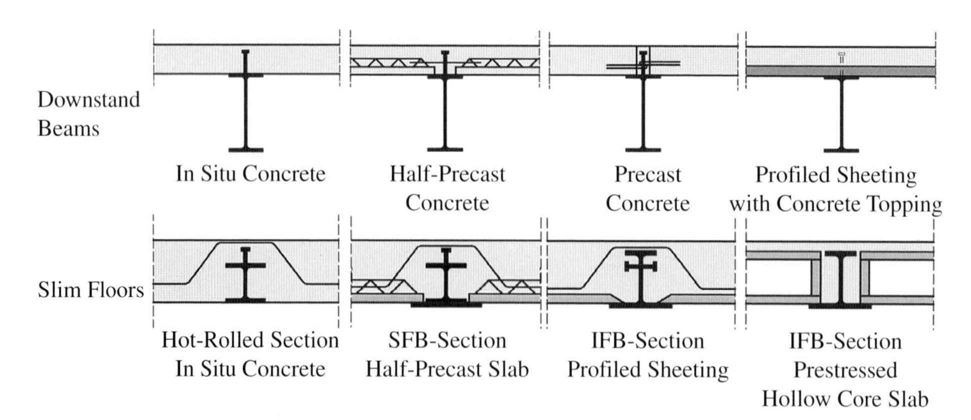

Figure 4.17 Downstand and slim floor beams with different slab types.

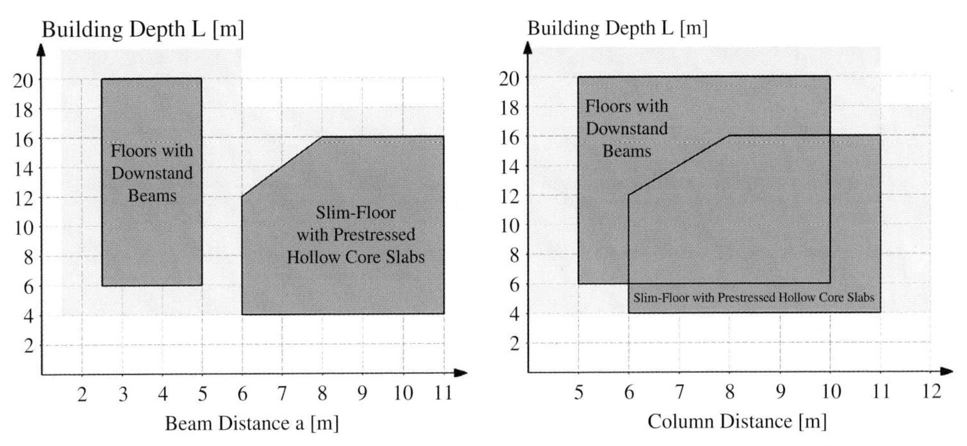

Figure 4.18 Possible beam and column spans with use of flat slabs and downstand beam floors.

Figure 4.19 Spanning a building without columns using a TOPfloor INTEGRAL composite deck – view of underside of deck and installation of element. © H. Wetter AG, CH-Stetten.

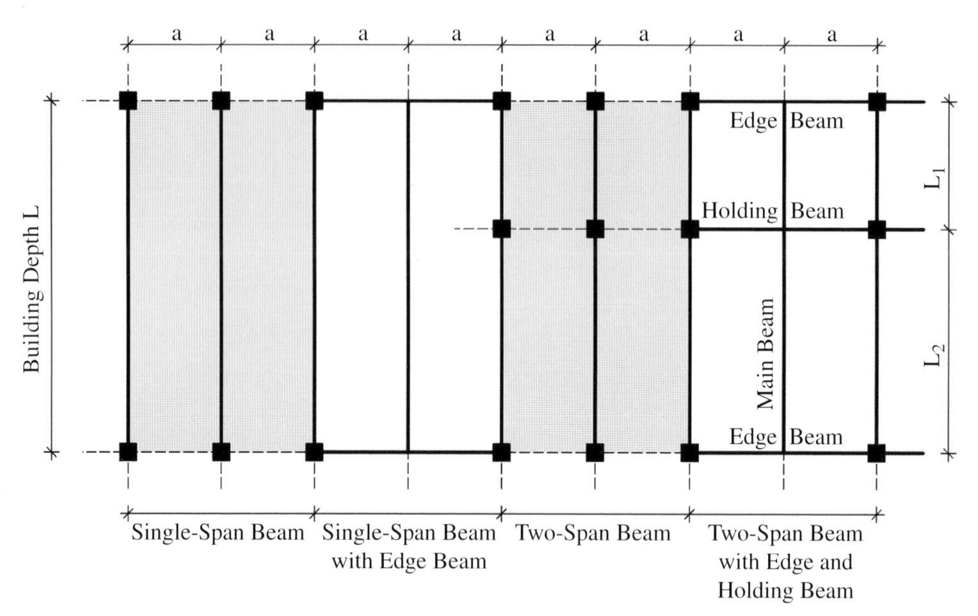

Figure 4.20 Configuration of downstand beams as single and two-span beams.

can be made from cast-in-place concrete, precast concrete elements with grouted joints and semiprecast elements or profiled steel sheeting with cast-in-place concrete. The latter variant has the advantage that the sheeting replaces the formwork, and if the geometry and end anchorage are appropriate, it can be taken into account as reinforcement. Furthermore, due to its lightness, profiled sheeting can be placed manually and used to support suspended ceilings and also the cable and pipe systems below the floor slab.

Based on the decision as to whether or not internal columns are required the preferred variants for floor systems can then be identified. In addition to the free span, other criteria influencing the choice of system are the depth of the floor construction (on which the storey height depends), requirements for building services, and wether the underside of the floor is exposed or covered by a suspended ceiling (Figure 4.20).

Figure 4.20 shows different variants for arranging columns and beams in a rectangular floor plan. The first arrangement (left) shows the use of long span beams without columns using composite beams, each of which is connected directly to the facade columns. If the columns are arranged with a larger spacing, edge downstand beams are necessary to support the floor beams (second and fourth part of the square). The same is true if an inner column is used (fourth field). The positioning of these columns is generally performed off-centre, for example, on the axis of partition walls in the central corridor if conventional cellular offices are arranged eather side of the corridor. Flat slabs are more economic for shorter [30].

Beams may also be designed as continuous beams to limit deformation and crack widths in concrete slabs, to increase the modal mass when excited dynamically (due to human-induced vibrations) and also to optimise the use of materials.

This is possible by arranging continuous reinforcement in the concrete slab and using contact pieces to transmit the compressive forces in the bottom flanges of the composite beams. Plasticity theory-based design models take the redistribution of moments from the column area to the span area into account to obtain an economical design.

Imposed loads for office areas are given in EN 1991-1-1 [33] as 3 kN/m^2 for regular use and 5 kN/m^2 in areas with heavy equipment, storage etc. (including an allowance for movable partitions). The superimposed dead load is determined by the particular floor system. Suspended ceilings and raised floors are particularly flexible because installations can be carried out easily, if required. In addition, the construction has a small self-weight. In contrast, screeded floors are heavier but can be produced with low construction heights.

For the floor systems and columns, the ultimate and serviceability limit states have to be verified on the basis of the Eurocodes EN 1990 to EN 1994. For the ultimate limit state analysis, verifications are carried out of the cross-section resistance, stability, shear connection and the transfer of shear forces. The verification of the serviceability limit states is carried out by defining the deformation criteria, controlling the crack widths in the slab (if required) and limiting human-induced vibrations [34].

To increase material efficiency, it is recommended to use higher strength steels that lead to weight savings. The importance of deformation and vibrations increases with higher slenderness of the components. Eurocodes 3 and 4 do not present any quantified limits for the serviceability limit state. In Eurocode 2, guide values are given for the deflection limits of beams, slabs and cantilevers. It is recommended to use a limit of span/250 as a combination of all deflections due to load. To avoid damage to components that are connected to the supporting members (e.g. partition walls or facades), a limit of span/500 is suggested. No explicit limiting values are available from the manufacturers of the different partition wall and facade systems. Therefore, one often has to rely on recommended guide values.

Composite beams are usually precambered, taking the permanent actions and the effects of creep and shrinkage into account so that a large part of the resultant deformation has already been compensated. Possible precamber to compensate for the deflection from imposed loads has to be specified for the individual case. When determining deflections, the effects of the method of construction should be taken into consideration. This includes, for example, whether the beams are propped or unpropped during construction. When considering the deformation of partition walls and facade elements, the time when installation occurs and the stage of construction when props are removed should be considered.

In addition to the deflections, the vibration behaviour of floor systems has to be taken into account. Vibration in office and administration buildings is caused by people walking. The vibration behaviour can be decisive for the design if the ratio of stiffness to mass is small and the system is only lightly damped [35]. The one-step root mean square (OS-RMS) method [30] enables the vibration behaviour to be designed so that it is acceptable for the comfort of the user and the human-induced floor vibrations can be predicted based on the building use.

4.4.8 Columns

In multistorey buildings in steel and composite construction, a range of steel columns and steel-concrete composite columns are used (Figure 4.21). A combination of composite slabs with reinforced concrete columns is also possible, but it requires suitable solutions for the joining technique and installation. For building heights up to 22 m, the loads supported per column are moderate, so that common I-shaped and hollow steel sections can be used. Other forms of sections are possible if required for design or structural reasons.

When using steel columns, additional measures are required to comply with fire protection requirements. Fire protection systems in the form of boarding, plasterboard and intumescent coating systems are available. The choice of the cross-section shape depends on the design loads, structural aspects and detailing. Rolled open sections are now produced with yield stresses up to 500 N/mm^2. Hollow steel sections are also available in higher strengths. The decision on the steel grade depends on the loading, the slenderness of the column, the availability and general sustainability criteria. With columns in the low to intermediate slenderness, higher steel strengths can be beneficial and minimise materials use [36].

Steel-concrete composite columns offer a higher load-carrying capacity than steel columns with the same external dimensions of the cross-section. The method of construction is different for concrete-filled hollow steel sections and for H-sections partially encased in concrete and sections that are fully encased in concrete. The concrete may be reinforced by longitudinal and stirrup reinforcement and possibly spiral reinforcement for concrete-filled circular hollow sections. For composite columns, the increase in the load-carrying capacity compared with steel columns is typically up to 80%. Due to the additional use of concrete and reinforcing steel, the fire resistance classifications for office buildings required by regulations can be complied without any additional protection measures when suitable designed.

Concrete-filled hollow steel sections do not require any formwork, and they are resistant to abrasion, wear and impact and can be given an aesthetically appealing colouring by the coating system. They can be designed to be particularly slender in axial compression, and square or circular cross sections exhibit the same bending resistance in both planes of bending. Vent holes are provided to release the vapour pressure in the case of fire. As the steel tube makes little contribution to the resistance when the column is in fire conditions, the concrete and the reinforcement

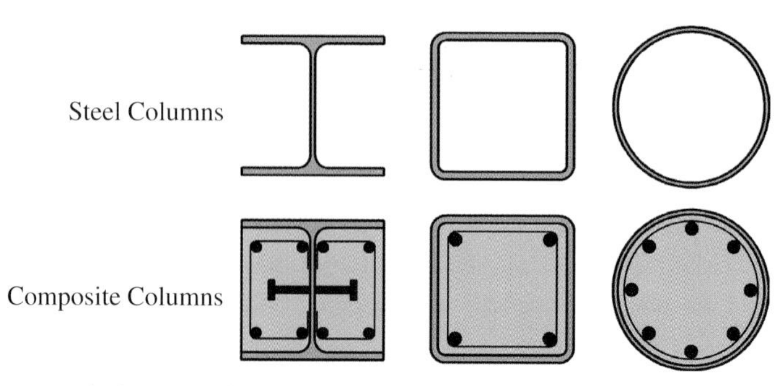

Figure 4.21 Typical cross-section of steel and composite columns.

provide the main resistance to loading. Additional structural steel sections can be introduced into the concrete core within the hollow sections in order to increase the resistance further. Fin plates are often used for beam-to-column connections and are welded onto, and if necessary, into the hollow section. The use of seating cleats can be attached to square or rectangular hollow section columns.

I-sections that are encased partially in concrete are generally constructed more economically than concrete-filled hollow steel sections. Joining is simpler when connecting downstand beams and splicing the columns. A corrosion protection coating of appropriate colour is applied to the exposed steel flanges. The column flanges prevent abrasion and spalling and act as formwork when concreting. Reinforcement cages and shear connectors (e.g. headed stud connectors, stirrups through or welded to the web) are incorporated in the concrete between the flanges. The orientation of the steel section leads to considerably different bending resistances about both principal axes.

The design rules for composite columns covered by Eurocode 4 apply to steel grades up to S460 and normal weight concrete of strength classes up to C50/60. The use of both high strength steel and concrete reduces column cross-sections for the same load-carrying capacities. In the event of fire, the concrete encasement protects the column web but the flanges are much less effective. Adequate longitudinal reinforcement is placed in the concrete encasement. For a given fire resistance, the benefits from using higher steel yield strengths is often small without additional measures. Although completely concrete encased steel sections are protected in the event of fire, they have to be fully shuttered for concreting.

4.4.9 Innovative joint systems

In steel and composite structures, the use of building units that are preassembled in the factory can be beneficial when they are installed using bolted, or other joints at the construction site. Boltless joining techniques are occasionally used, such as for the Deutsche Post building in Saarbrucken, where the cast-in-place concrete of the composite slab ensured the integrity of the composite beams and columns installed using cleating and contact and plug-in joints (Figure 4.22; [37]).

Figure 4.22 Left: composite deck system with profiled sheeting and downstand beam partially encased in concrete at Postamt 1 in Saarbrucken; centre and right: installation-friendly cleated and plug-in joints. © stahl + verbundbau gmbh.

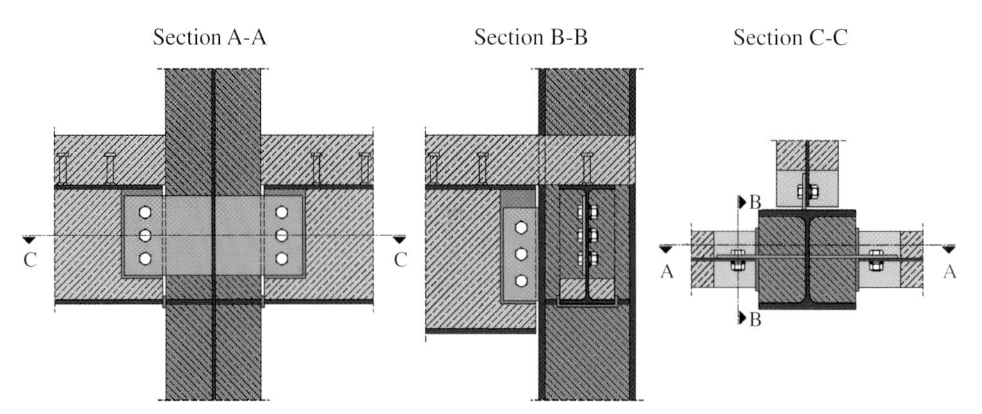

Figure 4.23 Types of joint for the column and beam carried out using I-sections and partial concrete encasement.

When properly used, the manufacturing and installation principle leads to ease of dismantling in the future. Loosening the bolted joints and dismantling mechanically allows the components to be reused in another similar structure. Concrete slabs of composite structures can be separated from the steel components using saw cuts and high pressure water jet technology. Cutting torches are often employed to quickly separate the steel components into transportable units.

With respect to the design of the joints, it is first necessary to decide between the use of downstand beam and slim floor systems, which depends only on the required construction depth. The shape of the steel beam and its position relative to the concrete slab offers various opportunities for designing pinned or moment-resisting joints. Furthermore, different types of connections are used for beam-beam and beam-column joints and splices between columns.

Figure 4.23 shows as an example of the connection of an I-beam that is partially encased in concrete to an H-column that is partially encased in concrete. The transfer of the shear force is via a fin plate in each case. To connect the edge beams, the fin plate projects from the column so that easier installation is possible. The concrete encasement has appropriate recesses in order to be able to create the bolted connection on site. These can be filled with mineral wool after the beam has been installed and enclosed by fire protection boards. This avoids direct heat impact in the event of a fire. Continuity of the edge beams can be achieved by welding in stiffeners in the cross section of the columns at the height of the lower flange, using an arrangement of contact pieces after installation of the beams, and by installing reinforcing bars parallel to the edge downstand beam, which are placed within the effective width of the slab.

4.5 HIGH STRENGTH STEEL

Richard Stroetmann

The development of the mechanical and engineering properties of steel has seen considerable progress in the past decades. Figure 4.24 shows how the strength classes available have changed over the past 70 years, together with the respective

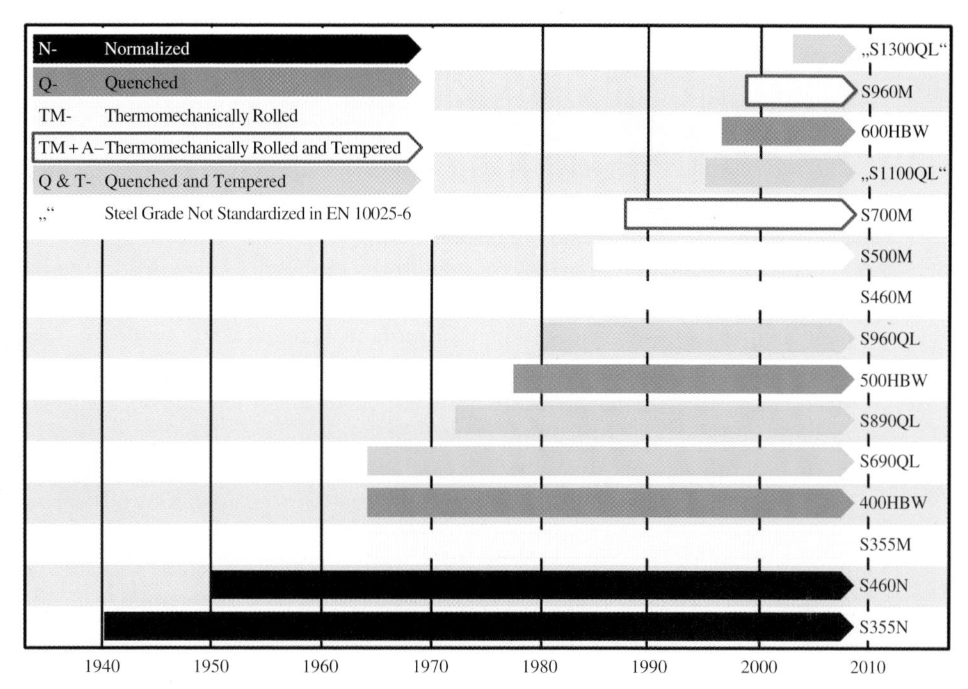

Figure 4.24 Development of the yield strength of various steel grades (see [38]).

treatments involved [38], [39]. Steels St 48, St Si and St 52 with a yield strength (f_y) up to 360 N/mm^2 were regarded as high strength when they were developed in the 1920s [40], but weldable rolled steels are now available with yield strengths up to 1300 N/mm^2.

Steel is used in many products and the properties (such as strength, hardness, toughness or workability) are matched to the respective requirements in a product-specific manner. In high tech sectors, such as the car industry, innovative materials are an important part of product development, and they find a quick route to the marketplace. This is particularly true for the higher and high strength steels that are being used increasingly in the automotive industry as so-called 'tailored blanks' and 'tailored products'. The construction industry is also benefiting from these developments and the resultant opportunities they offer.

There are no clear limits to the designations 'normal strength', 'higher strength', 'high strength' and 'ultra high strength'. The classification depends on the field of application. In the building and construction industry, steels with yield strengths above 355 N/mm^2 are regarded as higher strength (see also [44]). In the following, this classification will be used in view of the scope of Eurocode 3 for steels up to S700.

The use of higher strength steels is beneficial where the increased strength leads to a reduction in member weight but where the necessary structural and functional characteristics are still maintained ([41]–[43]). This is particularly the case with highly stressed steel construction (e.g. power stations or bridges) but might also be the case with multistorey construction, factory buildings or hangar-type structures. In structures where the weight has a considerable influence on the structural solution, a reduction in the mass is doubly beneficial: the effects of actions on the components are reduced, which in turn makes additional materials savings possible. Examples are long-span structures, such as bridges, exhibition halls, storage and production facilities, and stadium roofs. Further efficiency can be achieved by use of lightweight composite and hybrid methods of construction. The use of higher strength steels often also offers advantages from a functional point of view, as beams with longer spans allow the number of columns to be reduced and thus provide a more flexible design of the floor plan.

The use of higher strength steels improves resource efficiency and the Life Cycle Assesment (LCA). The strength of steel increases at a rate lower than cost compared to steels of normal strength. Economic benefits can be achieved in use of materials and in manufacturing and installation. Thermomechanically rolled steels are particularly beneficial because of their relatively low alloy contents and carbon equivalents, and they have good weldability and high toughness.

In order that higher strength steels can be used as effectively as possible, material-specific aspects must be taken into consideration when planning and installing steel and composite structures. The effects of stability, fatigue and limits on deflections and vibration control have to be observed more carefully with increasing member slenderness and stress level. Higher strength steels are beneficial if yield strength and tensile strength control the design solution. This is particularly the case with constructions that are under predominantly static and tensile loading, and loading under compression or bending in the compact and middle slenderness regions.

The importance of higher strength steel grades in the building and construction industry will continue to grow, as it will with the expansion and adaptation of technical specifications, such as Eurocode 3. Technical support for the planning and execution coupled with greater availability of the rolled products will help boost the introduction of these steels.

4.5.1 Metallurgical background

The methods that are available for increasing the strength of steels are alloying, heat treating, thermomechanical processing and cold working (stretching), and these are often used in combination. They change the grain size and microstructure and cause pinning of the grain boundaries, which produces steels of higher yield and tensile strengths. In physical metallurgy, plastic flow of steel means there is migration of dislocations in the lattice structure. Measures that lead to increased strength make these movements more difficult or block them completely (see Figure 4.25).

In the iron-carbon system, iron changes its face-centred cubic lattice structure (austenite, γ-iron) into a body-centred cubic structure (ferrite, α-iron) on cooling. The lattice structure can accommodate up to a maximum of 0.02% carbon in the form of foreign atoms. Any carbon in excess of this amount combines with iron during the cooling process to form iron carbide, Fe_3C or cementite (Figure 4.26). Depending on the cooling rate, differing amounts of the different phases form, which leads to steels with different properties (Figure 4.27). High cooling rates force carbon into solution at the interstices in the tetragonal martensite lattice.

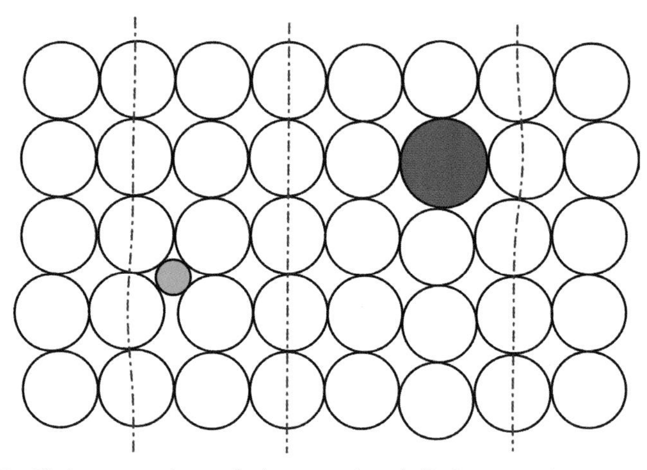

Figure 4.25 Displacements due to the incorporation of alloying atoms in the lattice structure.

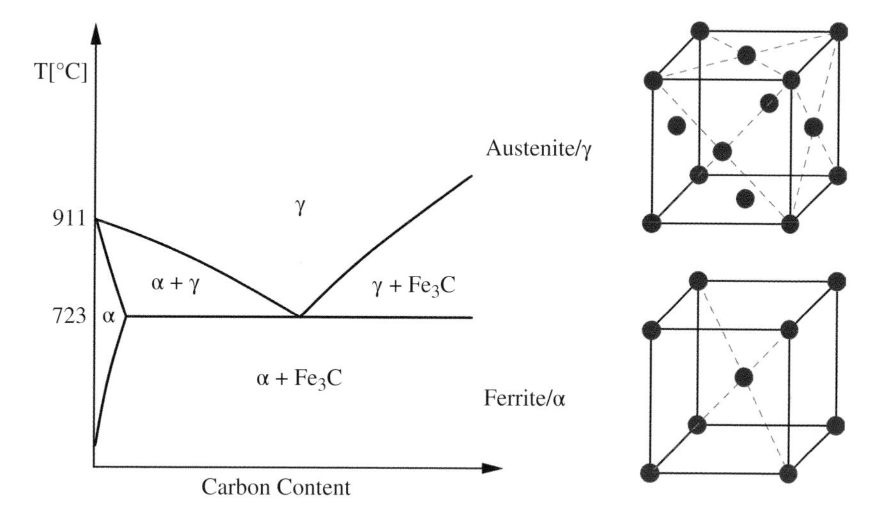

Figure 4.26 Schematic iron-carbon diagram.

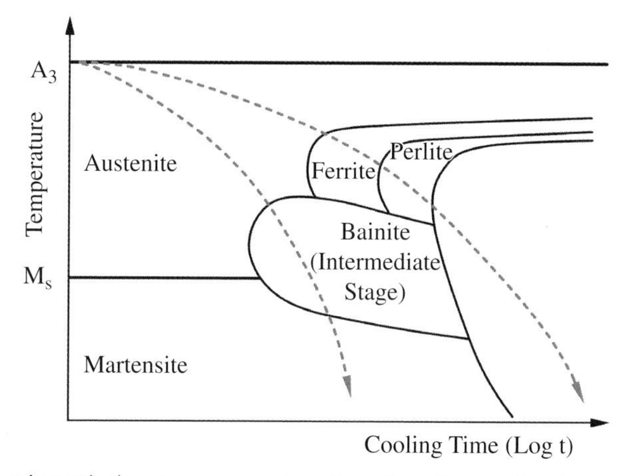

Figure 4.27 Schematic time-temperature-transformation diagram showing a continuous transformation.

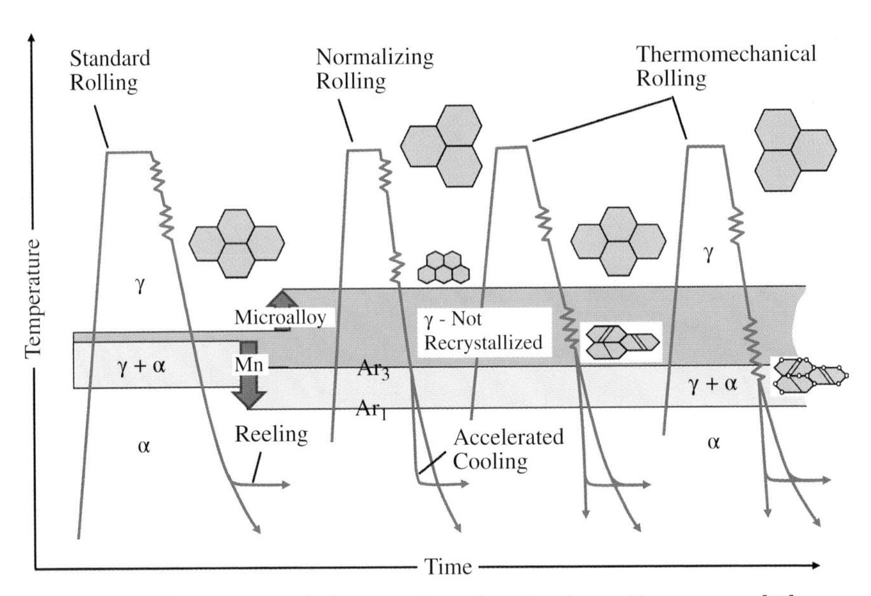

Figure 4.28 Roll sequences and phase transformation of various rolling processes [43].

Platelet-like crystals form, which also appear as an acicular structure in the micrograph. The high lattice strains result in high strength and embrittlement of the steel. Reheating (tempering at a temperature up to about 700°C) improves the diffusivity. Extremely fine carbide particles then precipitate out from the martensite lattice. This reduces the hardness and increases the toughness. The structure is transformed into a more equilibrium-like state.

The carbon content is limited in order to maintain the weldability of the steels. It is a maximum of 0.27% for steels specified by EN 10025. Together with the economical alloying due to carbon, substitution elements for iron, such as manganese, chromium and nickel, are added. 'Microalloying' elements, for example, niobium, vanadium and titanium, are also used in combination with suitable heat treatment processes, and these result in a fine grain microstructure with higher strength and toughness. The term 'microalloying' is used because of the small quantities of these elements used; they form carbides and nitrides and are necessary for grain refinement and precipitate hardening.

Normalised steels are produced by conventional hot rolling and subsequent normalising or by normalising rolling. Normalising is carried out in a continuous process (walking beam furnace) or stationary (batch furnace). Normalising rolling at a temperature above the recrystallization temperature involves carrying out the rolling and normalising operations together (Figure 4.28), which is possible for certain plate thicknesses and steel grades. Normalising produces a regular ferritic-perlitic microstructure. Higher yields and tensile strengths are obtained by use of alloying elements, such as carbon, silicon, manganese and vanadium. This results in higher carbon equivalents with corresponding constraints for welding. EN 10025-3 regulates the technical delivery conditions for

Table 4.1 Standards including delivery conditions for steels.

Standard	Version	Title	Range of Yield Strength [N/mm^2]
prEN 10025	2014–05	Hot-rolled products of structural steels	
prEN 10025-1	2014–05	General technical delivery conditions	
prEN 10025-2	2014–05	Technical delivery conditions for nonalloy structural steels	235–500
prEN 10025-3	2014–05	Technical delivery conditions for normalized/ normalized rolled weldable fine grain structural steels	275–460
prEN 10025-4	2014–05	Technical delivery conditions for thermomechanical rolled weldable fine grain structural steels	275–500
prEN 10025-5	2014–05	Technical delivery conditions for structural steels with improved atmospheric corrosion resistance	235–460
prEN 10025-6	2014–05	Technical delivery conditions for flat products of high yield strength structural steels in the quenched and tempered condition	460–960
EN 10149	2013–12	Hot finished structural hollow sections of nonalloy and fine grain steels	315–960
EN 10210	2006–05	Hot finished structural hollow sections of nonalloy and fine grain steels	235–460
EN 10219	2006–05	Cold formed welded structural hollow sections of nonalloy and fine grain steels	235–460
EN 10293	2015–01	Steel castings for general engineering uses	200–1000
EN 10340	2007–10	Steel castings for structural uses	200–700

normalised and normalised rolled weldable fine grain structural steel up to a yield strength of 460 N/mm^2 (see Table 4.1).

Quenched and tempered steels are heated to the recrystallization temperature after rolling, then quenched with water at a high cooling rate (Q = quenching) and subsequently tempered at about 600°C. Direct hardening after rolling (DQ = direct quenching) with subsequent tempering is equivalent metallurgically to conventional quenching and tempering. Quenching results in the formation of a hard martensitic-bainitic microstructure with higher strength, which is relaxed on tempering and results in improved toughness properties. The technical delivery conditions of EN 10025-6 [45] regulate quenched and tempered steels up to a yield strength of S960 in the quality groups Q, QL and QL1. These differ essentially in the permitted levels of the associated impurities, phosphorus and sulphur, and in their impact energy values and the corresponding ductility at low temperatures.

The high strengths of the quenched and tempered steels are achieved in combination with increased amounts of alloying constituents. This is reflected

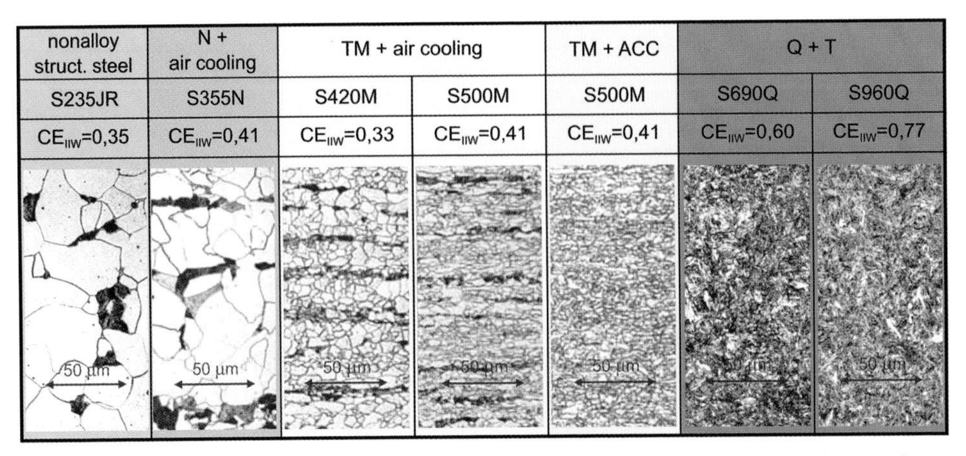

Figure 4.29 Microstructure of various steels (see also [46]). The carbon contents are given by way of example. They are dependent on the nominal thickness of the elements and vary according to various producers considering the limit values in EN 10025.

in the carbon equivalent and the welding of the steels. With increasing plate thickness, through hardening the material becomes more complex because the cooling rate, which is decisive for the formation of martensite, decreases from the surface to the core.

Thermomechanical rolled steels are made using different processes, possibly combined with heat treatment by quenching and tempering. The aim when coordinating the alloy components and the manufacturing process is to combine high yield and tensile strengths with high ductility and good weldability. Microalloying elements like niobium delay the recrystallization of the austenite between the individual roll passes and promote the formation of extremely fine grains in the microstructure. Unlike the conventional process, the rolling process is not a pure forming process but is used specifically to achieve the steel properties. As with cold forming, the yield strength is increased due to the lower final rolling temperature. There is an extremely fine elongated transformation microstructure in the direction of rolling, with high strength and toughness values.

After the rolling process, heat treatment can follow by accelerated cooling, and if necessary, tempering. With regard to cooling, it is necessary to differentiate between ACC (accelerated cooling), DQ (direct quenching) and QST (quenching and self-tempering) processes. These are associated with different cooling rates at the surface and in the core, which lead to different microstructure formation, strength and homogeneity (see Figure 4.29).

The technical delivery conditions for thermomechanically rolled weldable fine grain structural steels are currently regulated up to a yield strength of 500 N/mm^2 in EN 10025-4 [47] for thicknesses up to 120 mm for flat products and 150 mm for long products. The standard also covers steels that are heat treated by accelerated cooling after rolling and possibly self-hardened (ACC or QST process) but not direct hardening (DQ) and liquid annealing (NQ). EN 10149 [48] contains technical delivery conditions for steels up to grade S960MC that are suitable for cold working.

Table 4.1 shows a selection of relevant standards for the construction industry with the strength classes and quality groups in each case.

4.5.2 Designing in accordance with Eurocodes

With the introduction of EN 1993-1-12 [49], the use of higher strength steels in the building and construction industry is regulated not only as part of general building approvals but also in connection with product standards and the EN 1090-2 standard for execution. The different parts of Eurocode 3 are valid initially for steels up to S460 grade. Part 1-12 contains additional rules to extend EN 1993 to steel grades up to S700. The necessary deviations and supplementary design rules are given, which are valid in connection with the national annexes of the respective countries. The working groups of CEN TC 250 SC 3 are currently working on an integration of the rules of Part 1-12 in the remaining parts of the EN 1993 standard so that the use for steels up to S700 will be generally applicable with the new edition. A general overview of the current status of standardisation is given below. A detailed presentation of the technical rules is presented in [50] and [51].

In EN 1993-1-12, quenched and tempered steels S500 to S690 of the Q/QL/QL1 qualities in accordance with EN 10025-6 [45] and thermomechanically rolled steels for cold forming S500MC to S700MC in accordance with EN 10149-2 [48] are permitted as higher strength steels. The plastic deformability of steels decreases with increasing yield strength, the difference between the tensile strength and the yield strength reduces.

The design rules for steel and composite construction presuppose a minimum ductility, and Table 4.2 shows a comparison of the requirements for normal and higher strength steels. Redistributions of moments of continuous beams and the development of the plastic resistance of the supporting structure are generally not allowed.

Table 4.2 Ductility requirements for steels according to EN 1993-1-1 and EN 1993-1-12.

Steel Grade	up to S460	Above S460 up to S700
Ratio f_u/f_y	≥ 1.10	≥ 1.05
Elongation at failure	$\geq 15\%$	$\geq 10\%$
Ultimate strain ε_u	$\geq 15\, f_y/E$	$\geq 15\, f_y/E$

4.6 BATCH HOT-DIP GALVANIZING

Murray Cook and Holger Glinde

4.6.1 Introduction

Hot-dip galvanizing is a process by which iron and steel can be treated to prevent rusting. An article with a chemically clean surface is dipped into a bath of molten zinc, which reacts with the iron and forms a uniquely protective coating. Zinc is a

Car Park 'Silo 2' – Toulouse-Blagnac Airport

Location:	Toulouse (France)
Architect:	SCAU architects, with Azéma Architects, Tsuba
Building description:	The biggest multistorey car park in Europe. Steel construction with seven storeys and 3200 parking spaces.
Steel details:	4000 tonnes of steel for the structure, 70,000 m^2 metal deck floor. Mostly galvanized for optimized corrosion protection. Facade in stainless steel.
Sustainability:	For the building design, the special recyclability of steel was taken into account. The 50% open facade provides a natural ventilation.

Figure 1 Inside view of the access route in the carpark. © ConstruirAcier.

Figure 2 Stainless steel façade of the carpark. © Castel & Fromaget.

Example provided by ConstruirAcier.

Hengrove Park Leisure Centre

Location: Bristol, Great Britain

Architect: LA Architects and Ramboll

Building description: The Hengrove Park Leisure Centre has been built on the site of a former airport. Its facilities include a 10-lane, 50 m international-standard swimming pool, a 20 m teaching pool with a moveable floor, sports hall, a dedicated spin studio, climbing wall, a 150-station fitness gym, a healthy living centre, café and crèche.

Steel details: The pool has a galvanized structural steel frame with the roof supported on a network of cellular beams spanning 37.5 m. The steelwork in this area is partially hidden by a suspended ceiling formed from a series of acoustic baffles that filter light and sound.

Sustainability: The Local Authority requirements called for the main structure to have a guaranteed life of 60 years with minimal maintenance. Traditional painting methods, utilising chlorinated rubber, were quickly discounted and the benefits of a factory-applied, robust, homogenous finish led to hot-dip galvanizing to EN ISO 1461 being selected. In addition to the main frame, galvanized components were used extensively throughout the project including structural steel frames around plant equipment and galvanized angle framework that supported the extensive suspended ceilings throughout the facility.

Awards: BREEAM 'excellent' rating.

Figure 1 Hengrove Park Leisure Centre. © Liz Eve.

Figure 2 The roof of the pool is supported on a series of cellular beams spanning 37.5 m. © Liz Eve.

Example provided by Industrieverband Feuerverzinken.

metal that is widely used in corrosion protection but is also recognised for its essentiality for human health and ecosystems and for its ease of recycling.

Galvanized steel plays a vital role in our everyday lives. It is used in construction, transport, agriculture, renewable energy and everywhere that good corrosion protection and long life are essential. For example, galvanized steel lighting columns light our roads and galvanized steel structures support solar panels that provide renewable energy.

There are many other important industries that make use of galvanizing, and a large proportion of galvanized steel in Europe is used in construction. However, it is a very versatile process and articles ranging in size from nuts and bolts to large structural sections can be protected. When the galvanized coating is applied in the batch process (to EN ISO 1461), the coating is thick, tough and gives complete coverage of the steel article. This combination cannot be achieved by other zinc coatings.

There are many hundreds of batch galvanizing plants across Europe, serving their local steel fabrication and manufacturing industries and keeping transport distances to a minimum. Steel fabricators normally work closely with their local galvanizing plant, but it is easy to identify a local galvanizer through the network of national or regional galvanizers associations that operate across Europe and globally. See also www.stahlbauverbindet.de.

4.6.2 The galvanizing process

Galvanizing is a corrosion protection process for steel that involves dipping cleaned iron or steel components into molten zinc (at a temperature of around 450°C). A series of zinc-iron alloy layers are formed by a metallurgical reaction between the iron and zinc, creating a strong bond between steel and the coating. A typical time of immersion is about 4 or 5 minutes, but it can be longer for heavy elements that have high thermal inertia or where the zinc is required to penetrate internal voids (Figure 4.30).

On withdrawal from the galvanizing bath, a layer of molten zinc is deposited on top of the alloy layer. Often on cooling this exhibits the bright shiny appearance associated with galvanized products. In reality, there is no demarcation between steel and zinc but instead a gradual transition through the series of alloy layers that provide the metallurgical bond. Conditions in the galvanizing plant, such as temperature, humidity and air quality do not affect the quality of the galvanized coating (Figure 4.31).

4.6.3 Batch galvanized coatings

Hot-dip galvanized coatings provide a thick protective layer that is also tough and abrasion resistant. One of zinc's most important characteristics is its ability to protect steel against corrosion. The life and durability of steel are greatly improved when coated with zinc. No other material can provide such efficient and

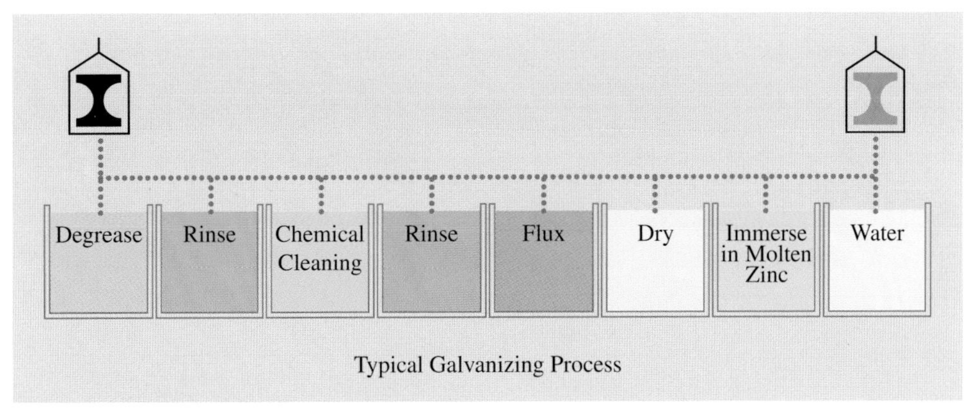

Figure 4.30 A typical galvanizing process. © EGGA.

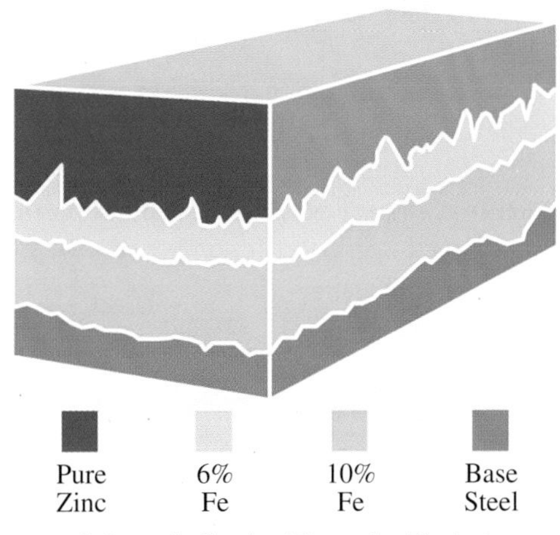

Figure 4.31 Schematic of a typical galvanized coating. © EGGA.

cost-effective protection for steel. In addition, the tough metallurgical bond between the coating and steel ensures its resistance to abrasion and impact. When left unprotected, steel may corrode when exposed to the environment. Zinc coatings prevent corrosion of steel in two ways – a physical barrier and electrochemical protection. Zinc coatings provide a continuous, impervious metallic barrier that does not allow moisture and oxygen to reach the steel. The metallic zinc surface reacts with the atmosphere to form a compact, adherent patina that is insoluble in rainwater. Typical coating thicknesses range from 45 μm to over 200 μm.

With zinc corrosion rates normally less than 1 μm per year in most European countries, a typical 85 μm coating can provide many decades of maintenance-free life (in rural and urban environments), and the thick coatings provided by batch galvanizing give added reassurance in more aggressive conditions. Research over many years has shown that the life of this barrier protection is proportional to the

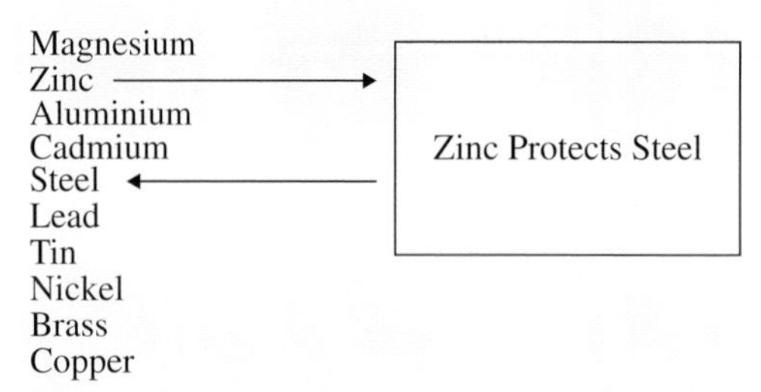

Figure 4.32 Zinc's position in the electrochemical series gives it a unique ability to protect steel. © EGGA.

zinc coating thickness, so doubling the coating thickness will double the life of the coating.

Zinc also has the ability to galvanically protect steel (Figure 4.32). When bare steel is exposed to moisture, such as at a cut edge or damaged area, a galvanic cell is formed. The zinc around the point of damage corrodes in preference to the steel and forms corrosion products that precipitate on the steel surface and protect it. There is no sideways corrosion at points of damage.

4.6.4 Sustainability

The sustainability credentials of the use of galvanizing are simple. As the use of steel that has been batch galvanized to EN ISO 1461 leads to lower economic and environmental costs of maintenance of steel structures.

The long-term durability provided by galvanizing is achieved at relatively low environmental burden in terms of energy and other globally relevant impacts.

The overall economic cost of corrosion has been studied in several countries, and it is estimated at up to 4% of gross domestic product. Several studies have demonstrated the economic and environmental costs associated with the repeated maintenance and painting of steel structures. These burdens can be significantly reduced by an initial investment in long-term protection systems. In social housing projects, future maintenance costs will be borne by the local authorities. In public infrastructure projects, use of galvanized steel leads to lower maintenance costs, releasing public funds for other purposes.

4.6.4.1 *Life-cycle environmental information*

In 2005, a pan-European life cycle inventory (LCI) study of hot-dip galvanized products lead to an average result for typical galvanized products. The objective of the work was to deliver LCI datasets for the galvanizing process, sometimes known as 'the service', using data submitted by European national galvanisers associations from their members' operations. This involved quantifying the average energy, resource consumption and emission of substances to the environment, resulting in an LCI of a sample of processes operating in several plants in Europe, according to the defined system boundaries. The sample covered about 937,000 tonnes of steel galvanized by 46 plants.

The systems under consideration have the purpose of processing steel pieces and steel products to protect the surface of the steel from the environment. The functional unit was thus expressed in terms of 1 tonne of averaged zinc-coated steel product.

Energy and environmental results are expressed by reference to the functional unit, but an extension of the analysis provides data about the system, independently of the steel product, in order to focus attention on 'the service'. Such results were expressed in terms of '1 kg of zinc alloy ready for coating purposes'. This represents a useful measure of the energy and environmental costs of 'the service'. This LCI data is available, on request from the European General Galvanizers Association (EGGA) to LCA professionals and customers who wish to generate an environmental product declaration for a galvanized steel construction product.

EGGA has published an EPD based on this European LCI study. An EPD for batch galvanized structural steelwork has also been published in Germany according to the IBU system for building products (see Chapter 3.11.2). The galvanizing industry is committed to this open publication of environmental data on its processes and products and is actively engaged in partnership with the steel industry on the numerous methodological challenges that are faced to properly establish the benefits of steel construction within a life-cycle approach to sustainability (Figure 4.33).

4.6.4.2 *Environmental performance of the galvanizing process*

Galvanizing is always carried out in an industrial works that contains all stages of the process. There are many galvanizing plants in most countries and steel does not have to travel great distances to a nearby plant, keeping transport costs and environmental impacts as low as possible.

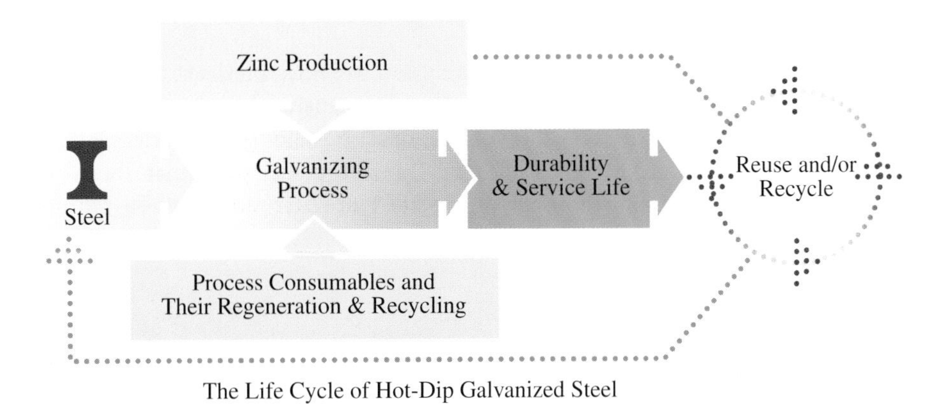

Figure 4.33 The life cycle of hot-dip galvanized steel. © EGGA.

The main consumable, zinc, is used very efficiently in the process. The hot-dip operation ensures that any zinc that is not deposited on the steel is returned to the galvanizing bath. Zinc that oxidizes on the surface is removed as an ash and is readily recycled (sometimes on site). Dross formed at the bottom of the bath is removed periodically and has a high market value for recycling.

Energy is required to heat the hot-dip galvanizing bath, and this is usually supplied by natural gas. In some countries, baths are heated by electricity or fuel oil. Although the galvanizing industry has made great efforts to manage its energy use efficiently. In some countries, the galvanizing industry has set targets for energy efficiency and encouraged improved energy management and has introduced new technologies to achieve these targets. Examples of these advances are

- improved burner technology for greater energy efficiency;
- more efficient bath lids (used during maintenance and/or downtime);
- greater use of waste heat for heating of pretreatment tanks.

Emissions within the plant are carefully controlled to avoid effects on the surrounding neighbourhood. Galvanizing plants are regulated under the EU Industrial Emissions Directive ([38]; see also [52]). The industry has cooperated in the publication of an EU Best Practice Reference Note (BREF) for hot-dip galvanizing. The principal requirement of the BREF is to capture the nonhazardous particulates during dipping. These particulates are then filtered using either scrubbers or bag filters.

Pre-treatment steps in the process are mainly aimed at cleaning the steel articles. Process consumables, such as hydrochloric acid and flux solutions, all have important recycling and/or regeneration routes. For example:

- Spent hydrochloric acid solutions are used to produce iron chloride for use in treating municipal waste water. Many plants remove iron and zinc and recycle regenerated acid to the pretreatment tanks.
- Improved monitoring and maintenance of flux tanks means that these are rarely discarded to waste, and only small volumes of sludge require periodic disposal. Closed-loop flux recycling is used in many plants.
- Ambient temperature acidic and biological degreasers have been developed.

Galvanizing plants use relatively low volumes of water compared to other coating technologies. Any wastewater that is generated can be treated and returned to the process, with only low volumes of stable solids sent for external disposal. In some cases, it has been possible for galvanizing plants to eliminate the use of mains water by harvesting rainwater on the site. Rainwater can be collected through gutters and stored for later use.

4.6.4.3 *Recycling*

Use of recycled zinc: There are two important sources of zinc used in the galvanizing process. Refined zinc is produced from a mix of both mined ores and recycled feedstocks. It is estimated that, on average, refined zinc contains about 25% of

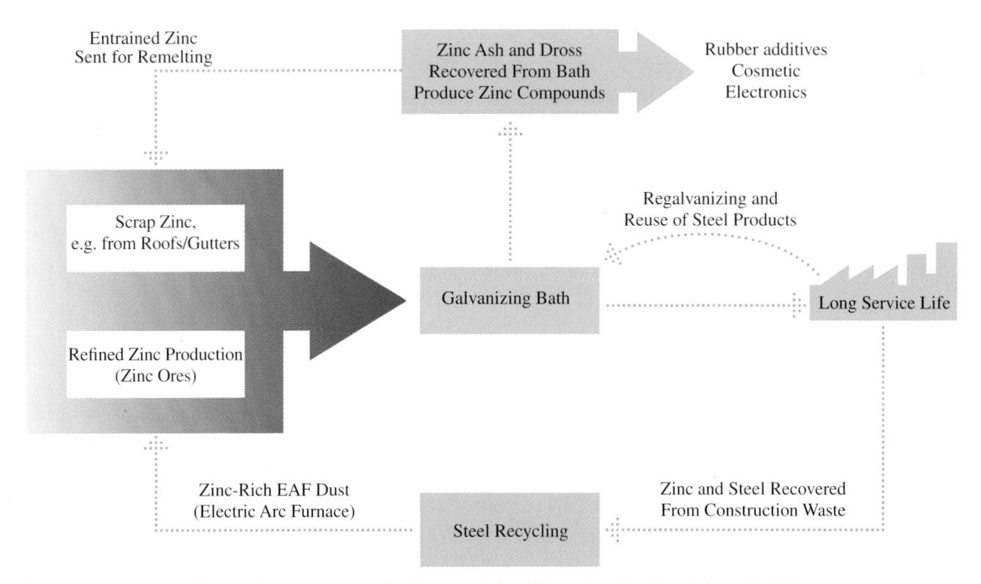

Figure 4.34 Recycling and reuse are at the heart of the life cycle of galvanizing. © EGGA.

recycled feedstock. In 2012, a major Japanese zinc producer announced it operates on 100% recycled feedstock derived from steel industry flue dusts. Galvanizers are also important purchasers of remelt zinc – scrap zinc from, for example, old zinc roofs that have been cleaned and remelted into ingot form. The refined zinc purchased by galvanizing plants contains a high proportion of recycled zinc, and fully recycled zinc is often purchased to supplement use of refined zinc (Figure 4.34).

The production of one kilogram of refined zinc (from ore) requires gross energy of 50 MJ, although only 20 MJ of this energy is used directly in zinc production. Secondary (remelted) zinc used by general galvanizers requires just 2.5 MJ to produce.

Recycling of process residues: During the galvanizing process, any zinc that does not form a coating on the steel remains in the bath for further reuse. There is no loss of materials that may occur during spray application of other coating types. Zinc ash (from surface oxidation of the galvanizing bath) and dross (a mix of zinc and iron that accumulates at the bottom of the galvanizing bath) are fully recovered. Any zinc metal within the crude ash is directly recycled for further use, often in the same galvanizing process. The fine ash and dross are then sold to make zinc dust and compounds for a variety of applications such as rubber additives, cosmetics and electronic components.

Reuse of galvanized steelwork: Many galvanized steel products can be removed, re-galvanized and returned to use. For example, highway guardrails are often removed and replaced during routine highway maintenance and resurfacing. The redundant barriers are returned to the galvanizing plant for re-galvanizing and are then used again in similar applications. The zinc-rich acid that is produced by stripping the remaining coating is used for production of zinc compounds for the chemical industry.

Recycling galvanized steelwork: Galvanized steel can be recycled easily with other steel scrap in the electric arc furnace (EAF) steel production process. Zinc evaporates early in the process and is collected in the EAF dust that is then recycled in specialist facilities and often returns to refined zinc production. In a study from the GDMB Society for Mining, Metallurgy, Resource and Environmental Technology conducted in 2006, it was found that the European steel industry (EU27) produced 1,290,750 tonnes of EAF dust, which contained 296,872 tonnes of zinc (i.e., 23% by weight). Ninety-three per cent of this zinc (276,920 tonnes) was recycled [38], [53].

4.6.4.4 *Essentiality of zinc for the environment*

Although zinc is well recognised for its positive effects for humans and ecosystems, it is also important to avoid high concentrations in the environment. Industrial emissions of zinc have been steadily falling over the past decades. Where locally high zinc concentrations may occur, for example, in highly mineralised areas, nature has a remarkable ability to adapt. Nature also has mechanisms to bind zinc to reduce its so-called bioavailability. Bioavailability has been defined as 'the amount or concentration of a chemical (metal) that can be absorbed by an organism thereby creating the potential for toxicity or the necessary concentration for survival' [54]. It is, however, not simply a function of the chemical form of the substance. Rather, it is largely influenced by the characteristics of the receiving environment. Hence, factors such as water hardness and pH have to be taken into account. It is these bioavailability effects that explain why the apparently high soil zinc concentrations around large galvanized structures, such as electricity transmission towers, do not produce the toxic effects that may be predicted in the laboratory. These factors have long been recognized as important, but there was insufficient scientific knowledge to allow a quantitative prediction of zinc's bioavailability in a given set of conditions. To address this, the galvanizing industry has contributed to extensive research to develop clear predictive models to quantify zinc bioavailability in waters, sediments and soils. This has provided the necessary reassurance for the continued widespread use of zinc for corrosion protection of outdoor structures [55].

4.6.5 Example: 72 years young – the Lydlinch Bridge

In 1942, the UK Ministry of Defence was considering outline plans for the D-Day invasion. Where and when the landings would take place was top secret, but the speedy movement of the invasion force to the south coast ports was a common factor for all alternatives. One such route, the A357 through Dorset to Poole Harbour, needed to be improved. At Lydlinch, the picturesque, narrow stone bridge over the River Lydden would not withstand the weight of heavy tanks. To solve this problem, Canadian army engineers erected a temporary galvanized steel Callender-Hamilton bridge alongside the older structure. The tanks and heavy equipment were diverted over the galvanized bridge (Figure 4.35).

Figure 4.35 This hot-dip galvanized Callender-Hamilton bridge was built 1942 at Lydlinch. © Iqbal Johal.

Temporary structure: The bridge was not intended to be a permanent structure but has stayed in service, having been passed into Dorset County Council's control after the war. It has carried the road's eastbound traffic ever since. The bridge has seen only minor changes to its original design since it was erected. Timber deck repairs were carried out in 1985 and 2009. The only work of any structural note was to strengthen the bridge in 1996 to enable it to conform to new standards in order to carry 40 t lorries. At the time, Dorset's chief bridge engineer, Ted Taylor, said, 'We have had no real trouble ensuring that this 'temporary bridge' is brought up to the new standard and the bridge was in remarkably good shape'. In fact, the strengthening consisted of bolting 'T' sections to the existing transverse deck beams and the addition of some longitudinal beams. The two main trusses were left as they were in 1942. Engineering forethought meant that on a few sections where a lot of cutting and readjustment of design had taken place, the sections were re-galvanized.

Inspection: As a follow-up to an inspection carried out by Galvanizers Association (GA) in conjunction with Dorset County Engineers in 1999, the bridge was reinspected in October 2014 by GA staff with assistance from Dorset County Engineers. The bridge still looked in very good condition, and on first sight the areas that look to have a slightly weary appearance are mostly due to dirt deposits and growth of moss on the steel surfaces. Coating thickness measurements were taken on steel members chosen at random on both sides of the bridge after wire brushing to remove surface contaminants or build-up of corrosion products. The members inspected included the main truss diagonals, joining plates and some bolt heads. Average coating thicknesses on the diagonal trusses ranged from 126 μm to 167 μm. On the plate sections the average thicknesses were 131 μm to 136 μm. On bolt heads, the average coating thicknesses ranged from 55 μm to 91 μm. Les Lock, project engineer, said, 'Despite all that has been thrown at it, floodwater, mud and grit salt, after 72 years the galvanized exposed members are still in very good condition.'

72 years young: Having started life as a temporary structure, the Callender-Hamiliton bridge at Two Fords, Lydlinch, is still in fine condition 72 years after it was first installed. The galvanized coating has stood the test of time exceptionally well, and taking the remaining coating thicknesses into account alongside zinc corrosion data, the coating can be expected to provide a life well in excess of 100 years.

4.7 UPE CHANNELS

Raban Siebers

In U-profiled rolled steels, it is necessary to differentiate between round-edged U (or UNP) channels with an inclined inner flange and sharp-edged UPE channels with parallel inner flange surfaces. In both cases, the height of the U profile is 80–400 mm and the width is 45 or 50 mm to 110 or 115 mm. The channels can be ordered in lengths of 8–16 m for a channel height up to 300 m or of 16 m for channel heights of 300–400 mm.

The design of the UPE channel, which is standardised in the German DIN 1026-2, is largely the same as that of the UNP channel, which is standardised in DIN 1026-1, both with tolerances in accordance with EN 10279. However, the UPE channel has a thinner web and a parallel flange that is a little thicker and wider. The inner surfaces of the flange are parallel, and the ends of the flange are not rounded. The UPE range thus also complements the IPE in the range of the nominal heights as presented in Figure 4.36.

UPE channels have higher resistances in both directions than the old U channels. Along the stronger axis (y-y), the UPE channel exhibits a slightly higher stiffness, and along the weaker axis (z-z) the stiffness is some 20%–27% higher than the corresponding U section. In some cases, the U channel can be replaced by a smaller UPE channel – resulting in cost and weight savings. With a comparable loadbearing capacity around the strong axis a weight saving of up to 11% (W_y/G) can be achieved as a result of the thinner webs in the UPE channels. With comparable loadbearing capacities around the weak axis, weight savings (W_z/G) of up to 27% can be achieved (Figure 4.37).

UPE channels are also superior to conventional U channels in installation. Wedge-shaped washers are no longer necessary because the inner surfaces of the flanges are parallel and not angled (Figure 4.38). Elimination of the washers and the shorter installation time also reduce costs.

Furthermore, it is no longer necessary to chamber the connecting plate because the internal chamber of the UPE channels is of a constant depth up to the radii (Figure 4.39). This eliminates a production operation, namely, flame cutting of the bevels, and thus reduces production costs.

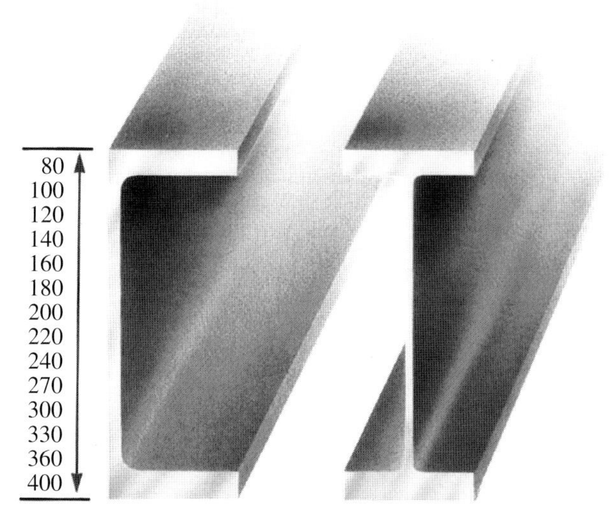

Figure 4.36 Cross section of a UPE channel; nominal height adapted to match IPE range.

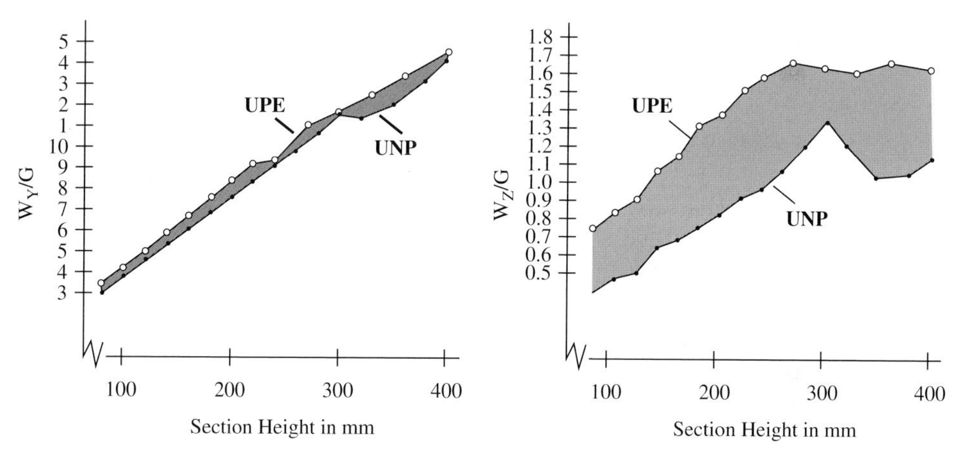

Figure 4.37 Weight savings of UPE channels compared with UNP channels. © Salzgitter Mannesmann Stahlhandel.

Figure 4.38 Benefits during erection as a result of eliminating washers. © Salzgitter Mannesmann Stahlhandel.

These channels are often used as lightweight beams in peripheral areas in order to form a flush slab edge. Other applications are for frame members subjected to low tension or compression, as supporting structures for wall panels (cladding rails) or as a trimmer beam for steel staircases (Figure 4.40) [56].

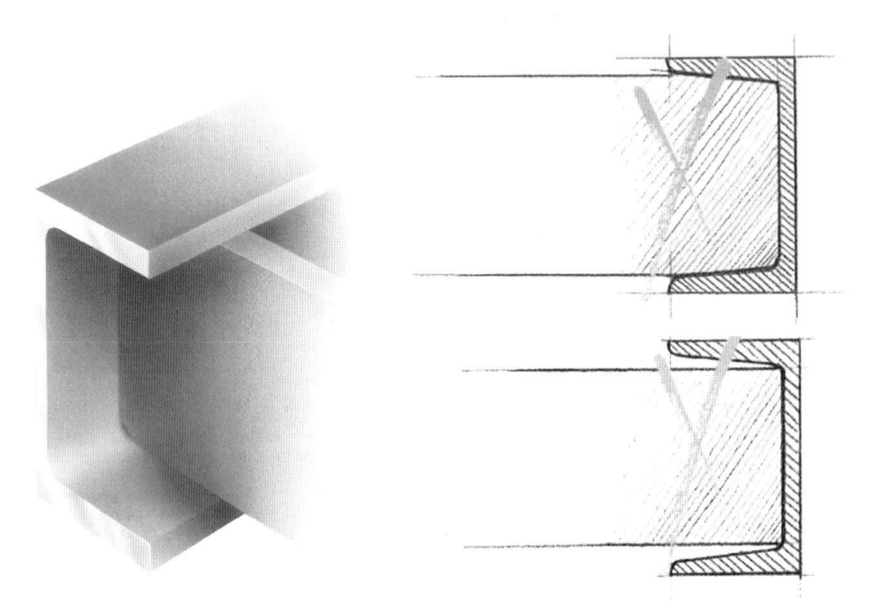

Figure 4.39 Benefits during production due to fixed separation of the flanges. © Salzgitter Mannesmann Stahlhandel.

Figure 4.40 External steel staircase; an area of application for UPE channels.

4.8 OPTIMISATION OF MATERIAL CONSUMPTION IN STEEL COLUMNS

Marc May

The traditionally used steel grades S235 and S275 may be considered as increasingly uneconomical, due to their higher steel consumption compared to the steel grade S355 and high strength steel S460. For columns in multistorey buildings, hot-rolled column sections in grade S355 and S460 are more efficient and sustainable. For the higher yield strength of the latter grades, it is possible to reduce the required cross section for a given loadbearing capacity. Further savings arise in fabrication, such as easier welding and handling, less surface protection and less transport.

Figure 4.41 shows a comparison of buckling resistances for various buckling lengths of some popular rolled section sizes for two steel grades. The smaller the buckling length, the greater is the influence of high strength steels on the buckling resistance. In multistorey buildings, the buckling length is equal to the typical storey height of 3–4 m. In this range of typical buckling length, the optimisation potential in material of S460 steel is of the order of 20%–40% relative to S355 steel.

Figure 4.42 shows the buckling resistances of the popular size ranges HE and HD sections for a buckling length of 4 m in grades S355 and S460. The discrete points are connected with lines for rolled sections up to 400 kg/m weight.

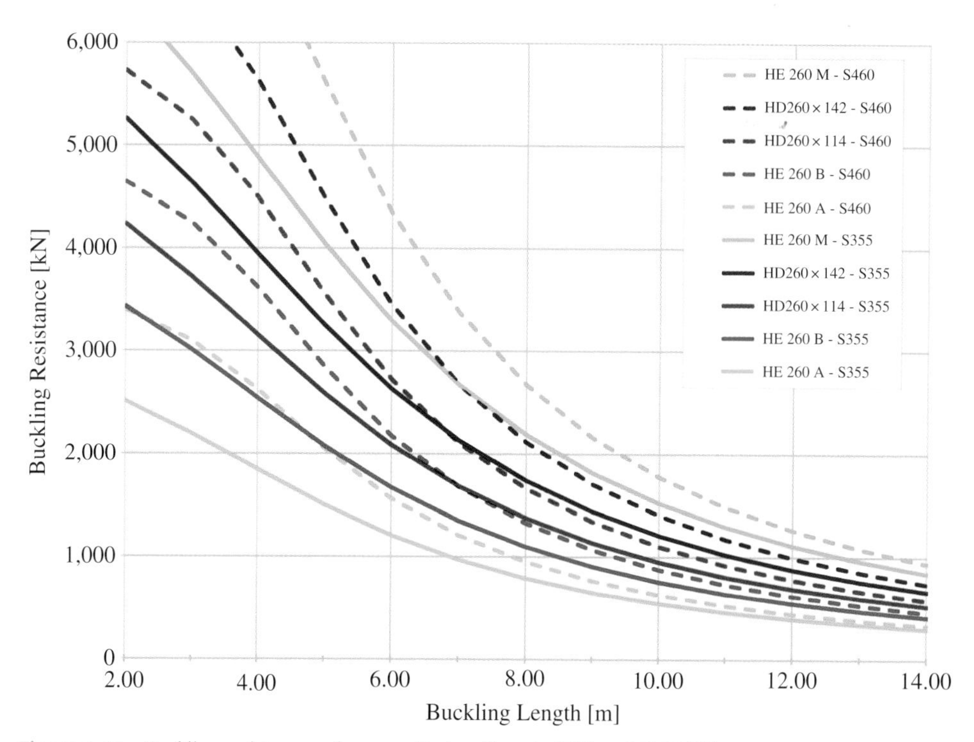

Figure 4.41 Buckling resistances of some rolled sections in S355 and S460 [57].

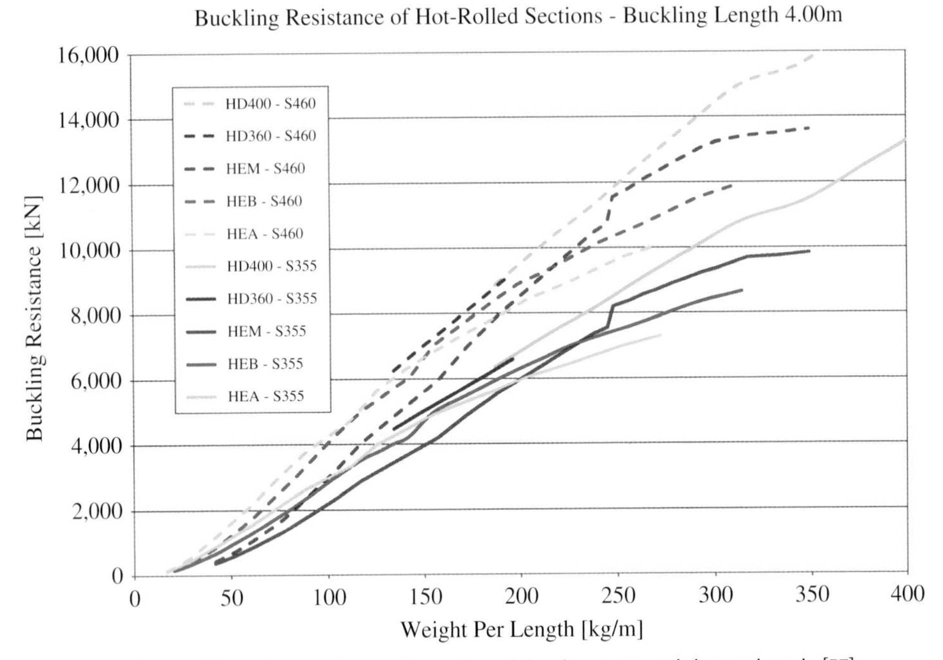

Figure 4.42 Curves of buckling resistances depending of the size range and the steel grade [57].

Discontinuities in the curves are caused by the buckling curve specified in EN1993-1-1, which changes when the height-to-width ratio h/b exceeds 1.2 and the yield strength is reduced with increasing steel thickness according to EN10025.

Regarding the reduction of material consumption independent of the steel grade, it can be observed that HEA sections up to 125 kg/m (i.e. HEA 400) are the most appropriate size range. HD sections are the most appropriate for weights exceeding 134 kg/m (i.e. HD360 × 134).

The optimization potential for a given buckling resistance and a buckling length of 4.0 m can be illustrated as follows: Consider a HEM320 in S355 steel with G = 245 kg/m, an area of A = 312 cm² and a radius of gyration i_z = 7.95 cm, which leads to a buckling resistance of $N_{b,Rd}$ = 7553 kN (see Figure 4.42). A similar buckling resistance of $N_{b,Rd}$ = 7583 kN is achieved with the section HD360 × 162 in S460 steel for G = 162 kg/m, A = 206,3 cm² and i_z = 9,49 cm (see Figure 4.42). Consequently, for this example, a reduction in material consumption of 34% can be reached through the use of an appropriate section in high strength steel S460.

Figure 4.43 shows the comparison of buckling resistances of rolled section sizes with similar section area of approximately 200 cm² (HEA500, HEB400, HEM240 and HD360 × 162).

The following parameters are relevant for the determination of buckling resistances:

Buckling length L_{cr} of the compression element.

Radius of gyration i: The bigger the radius of gyration, the smaller is the resulting slenderness, which further results in a higher buckling resistance. Figures 4.41–4.43 show buckling resistances for weak axis buckling, that is, i_z is

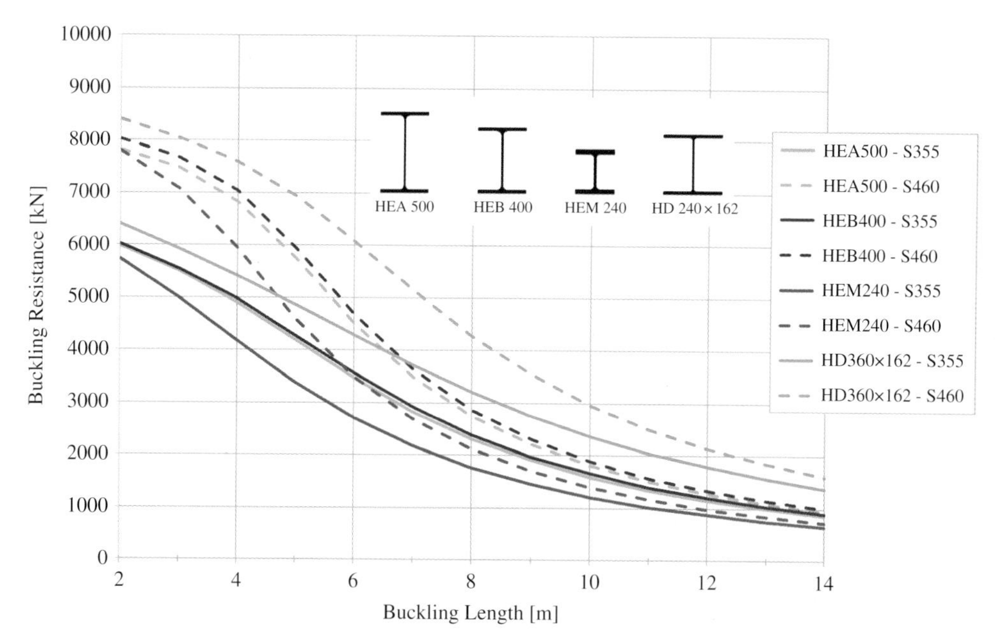

Figure 4.43 Comparison of buckling resistances of columns with similar section area and steel consumption [57].

governing the design. Obviously HD sections have the best possible material distribution for weak axis buckling of the available rolled section ranges that results in higher buckling resistances of HD sections compared to those of rolled sections with comparable section areas.

Yield strength f_y of the chosen steel grade and the relevant buckling curve according to EN1993-1-1, Table 6.2.

The gain in buckling resistance for a given buckling length of 4.0 m can be illustrated as follows:

Taking as an example the section HEM240 in S355 with G = 157 kg/m, A = 199.6 cm^2 and i_z = 9.49 cm gives $N_{b,Rd}$ = 4188 kN (Figure 4.43).

With a similar section area of A = 206.3 cm^2, the section HD360 × 162 in S460 with G = 162 kg/m and $i_z = \sqrt{(Iz/A)}$ = 9,49 cm achieves $N_{b,Rd}$ = 7583 kN (Figure 4.43).

Consequently, for this example a gain in buckling resistance of 81% for a similar material consumption can be reached through the use of an appropriate section in high strength steel S460.

4.9 COMPOSITE BEAMS

Bernhard Hauke

The development of new steel grades with higher strength and good working properties has lead to increased applications especially for pure steel structures. High performance steel grades are much less employed for steel/concrete composite

Briand Headquarters

Location:	Les Herbiers, France
Architect:	Bodreau Architecture
Building description:	A steel fabricator has made its headquarters an impressive showroom of its skills and know-how. Three levels of open-space offices have been set up around a planted inside patio. On one side, the building shows a monolithic 11 m high glass facade; on the opposite side, the green roof is slowly lowering up to the ground level.
Steel details:	200 t of steel. Cellular beams: laminated profiles and welded plate girders – for some of them, the cellular shapes are the logotype of the company. Circular hollow sections columns, cables bracing, mixed floors.
Sustainability:	The use of 100% recyclable materials. Thermal inertia optimization by half-burying the ground level and covering roofs with greenery.
Awards:	BBC building (very low consumption building). Winner of the Vendée's department Architectural prize in 2014: Public award and Architecture award for office building.

Figure 1 The glass facade. © Nautilus.

Figure 2 Inside view with working spaces. © ConstruirAcier.

Example provided by ConstruirAcier.

members because most codes and design recommendations are based on conventional grades. Moreover, the overall economic advantage of composite members with high performance materials – which are naturally slightly more expensive per unit – is often not evident at first glance.

In this section technical and economic conditions for the applications of composite beams utilising increased strength of high performance steel and concrete are discussed. Market possibilities for those high performance beams are seen for long spans and higher loads as well as for members that may be subjected to very high loads. In addition, because of the smaller dimensions and lesser installation weights, advantageous conditions for prefabrication for building applications are evident. Several examples illustrate this, as follows.

4.9.1 Composite beams with moderate high strength materials

Composite members consisting of concrete slabs connected to steel beams through shear connectors are increasingly used in modern buildings. Due to the composite action, a significant increase in resistance and stiffness of the beams is gained, resulting in savings not only in construction depth but also in significantly lower steel and concrete consumption, which means a better environmental performance, such as carbon footprint or energy consumption. Therefore, composite beams are the first choice for long-span structures, providing the opportunity to achieve a flexible floor design for the adoption of changing demands. With the advent of high performance materials for composite members, these advantages are increasing.

In order to make optimum use of high strength steel in composite bending members, higher concrete strengths may also be considered. The increase of the resistance of the tension and compression chord of a composite member leads to an increasing demand concerning the capacity of shear connections. However, in order to make optimum use of higher concrete strengths in a composite member, sufficient ductility of the shear connectors is also required. Shear studs have proved their value as ductile shear connectors when embedded in concrete with conventional strengths. For higher strength concrete, however, the lower flexural deformation of the shear stud changes its loadbearing behaviour, with the result that a ductile deformation cannot always be achieved. An alternative is offered by so-called concrete dowels, also known as perfobond connectors [58]. These are sheet-metal strips with openings, which are welded, for example, onto the upper flange of a steel section and tie into the concrete slab. The shear connection between steel beam and concrete slab is established through the apertures in the steel plates, which are filled with concrete – the concrete dowels. Technical approvals for concrete dowel connectors are available.

For composite beams with conventional materials and with compact sections, the moment resistance can be well estimated with the plastic moment capacity Mpl based on stress blocks. The application of mid to high strength materials for composite members leads to higher member resistance. For high strength steel, for example, grade S460, higher strain values must be achieved reaching the yield point, which also requires a higher rotation capacity. Until reaching the plastic moment capacity, strain values far beyond the ultimate concrete strain would be necessary. Hence, for mid to high strength materials, Mpl overestimates the actual ultimate moment capacity Mu [59]. Then, either a reduction factor β on Mpl can be employed [59], without changing the familiar design concept, or the moment capacity is determined based on elasto-plastic material behaviour (Mep) obeying the ultimate concrete compression strain, according to EN1992.

4.9.2 Examples for high strength composite beams

4.9.2.1 Exhibition building Moscow

An exhibition building in Moscow has a long-spanning roof made with composite beams. The span of a standard beams is 10 m, the distance between them is 6 m. The governing loading conditions are a basic snow load of 180 kg/m^2 for Moscow, augmented by a factor of 2.6 for snow accumulation. The composite beams consist of welded steel sections in grade S235 and a 30 cm slab in concrete C30/37 (see Figure 4.44 and Table 4.3). If moderate high strength materials such as steel grade S460 and concrete C50/60 had been used a slab thickness of 20 cm and a steel section IPE450 would have been sufficient (see Figure 4.44 and Table 4.4), whilst still obeying the deflection limit of l/300. Herewith a cost advantage of about 10% could have been achieved compared to the built composite beams with conventional material strength. Furthermore, the floor system height could have been reduced from 79.5 cm for the conventional composite beam to 65 cm for the high strength composite beam.

4.9.2.2 Power plant heavy-duty floor

For a power plant a heavy-duty floor (1.7 t/m^2 dead load, 1.4 t/m^2 live load + 60 t lorry) has been built with two-span beams (U159) and a slab on top (D70) in reinforced concrete (Figure 4.45 and Table 4.5). The same floor has been redesigned

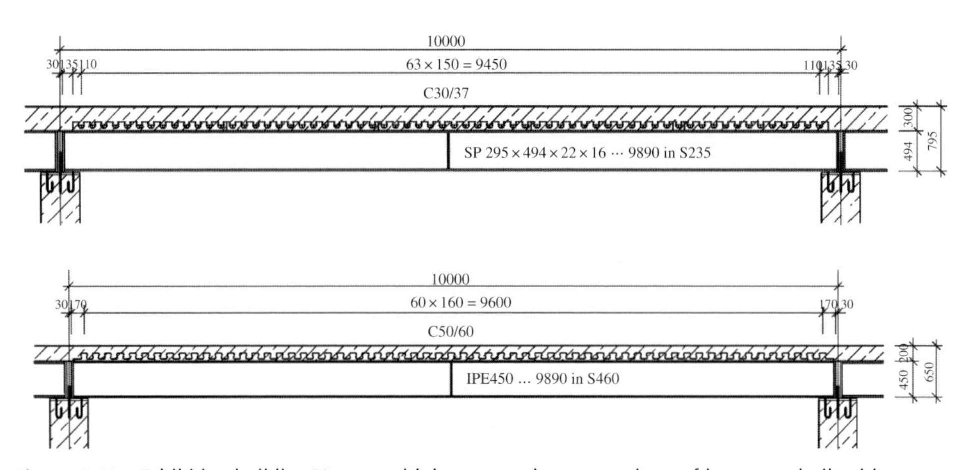

Figure 4.44 Exhibition building Moscow with long-spanning composite roof beams, as built with conventional materials and alternative with high strength materials.

Table 4.3 Exhibition building in Moscow: costs per conventional composite beam [60].

	Quantity	Unit Costs	Costs
Formwork, h = 30 cm	57 m²	70 €/m²	3990 €
Reinforcement	1.6 t	550 €/t	880 €
Concrete C30/37	18 m³	85 €/m³	1530 €
Steel I-beam, welded section in S235	1.6 t	1275 €/t	2275 €
Shear connector: concrete dowels S235	64 connectors	3.23 €/conn.	207 €
			ca. 9000 €

Table 4.4 Exhibition building Moscow: costs per high strength composite beam [60].

	Quantity	Unit Price	Price
Formwork, h = 20 cm	58 m²	70 €/m²	4060 €
Reinforcement	1.8 t	550 €/t	990 €
Concrete C50/60	12 m³	120 €/m³	1440 €
Steel I-beam IPE450 in S460	0.8 t	1320 €/t	1056 €
Shear connector: concrete dowels S460	61 pcs.	2.99 €/pc.	182 €
			ca. 8000 €

with single-span composite beams. On the one hand conventional materials (steel S235, concrete C30/37, shear studs) and on the other hand moderate high strength materials (steel S460, concrete C50/60, concrete dowels) have been assumed. The slab is presumed as conventional reinforced concrete.

For composite beams with conventional materials, costs are nearly the same (Table 4.6) compared to the RC floor (Table 4.5), but the height of the floor structure increases from originally 110 cm (RC solution) to 114 cm (Figure 4.45). If, on the other hand, moderate high strength materials are used (Table 4.7), costs cannot only be reduced by about 3% but the height of the floor system also cuts back to 90 cm (Figure 4.46).

Further savings can be achieved if instead of the classical cast-in-place concrete slab with formwork and temporary supports, easy to erect main and secondary composite beams are used on which either profiled steel decking (e.g. Holorib or Cofrasta) or semi-precast RC elements with lattice girders (e.g. Filigran) can be placed. Even more economical are high strength composite beams for larger spans of the main beams. The deflection limit of span l/300 has also been observed.

4.9.3 Economic application of composite beams

With a parametric study [60], the economic application of composite beams has been systematically looked at to obtain information for favourable material selection. The study [60] focused on floor systems with beams and slabs. The

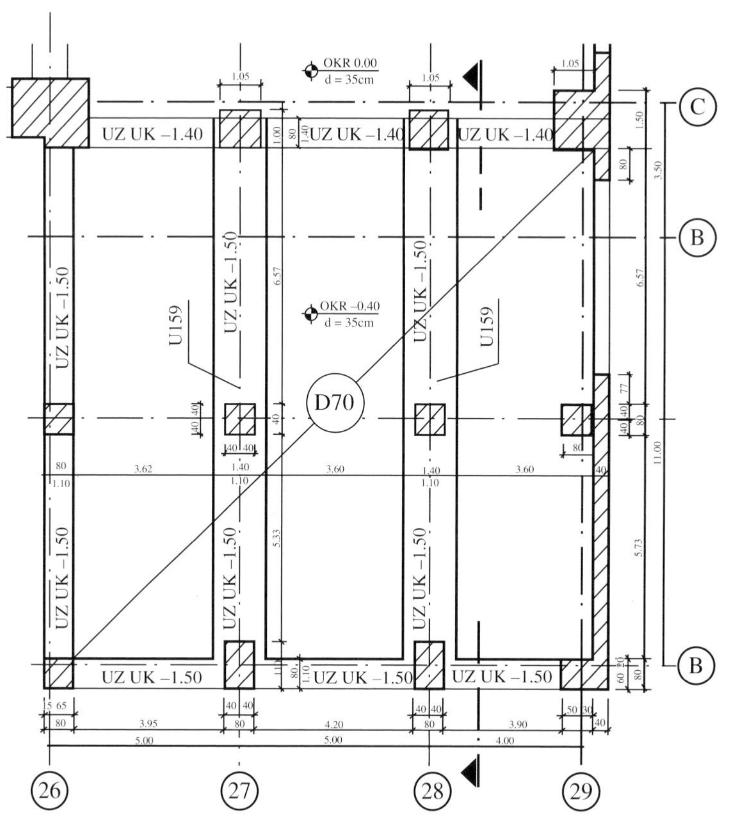

Figure 4.45 Power plant heavy-duty floor with slab D70 and beams U159.

Table 4.5 Power plant heavy-duty floor: costs per floor with RC-beams [60].

	Quantity	Unit Price	Price
Formwork, h = 35 cm	218 m²	70 €/m²	15,260 €
Reinforcement	9.6 t	550 €/t	5280 €
Concrete C30/37	90 m³	85 €/m³	7650 €
			ca. 28,000 €

Table 4.6 Power plant heavy-duty floor: costs per floor with conventional composite beams [60].

	Quantity	Unit Price	Price
Formwork, h = 35 cm	171 m²	70 €/m²	11,900 €
Reinforcement	6.3 t	550 €/t	3465 €
Concrete C30/37	62.6 m³	85 €/m³	5321 €
Steel beams HEA 800, HEB 700 in S235	5.1 t	1275 €/t	6503 €
Shear connectors: shear studs 22/125	184 pcs.	1.55 €/pc.	285 €
			ca. 27,500 €

Table 4.7 Power plant heavy-duty floor: costs per floor with high strength composite beams [60].

	Quantity	Unit Price	Price
Formwork, h = 30 cm	173 m²	70 €/m²	12,110 €
Reinforcement	7.5 t	550 €/t	4125 €
Concrete C50/60	53.6 m³	120 €/m³	6432 €
Steel beams HEB450, HEB360 in S460	2.7 t	1320 €/t	3564 €
Shear connectors: concrete dowels S460	144 pcs.	2.99 €/pc.	431 €
			ca. 27,000 €

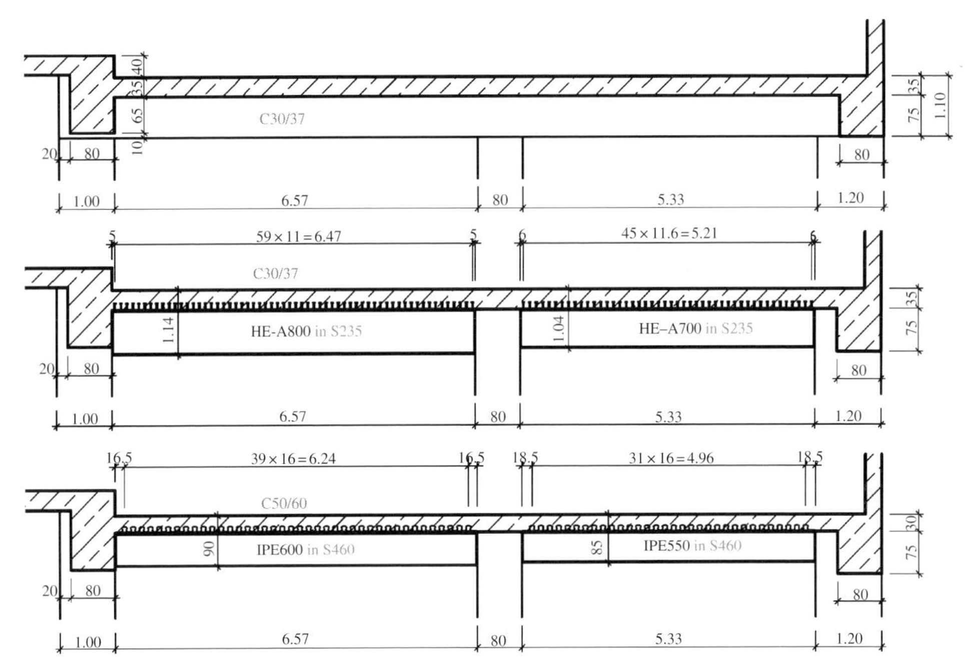

Figure 4.46 Power plant heavy-duty floor: as-built RC beam, alternative composite beams with conventional and high strength materials.

composite beams are modelled as simple beams. The slabs are cast-in-place reinforced concrete. Variable parameters are the steel grade (S235, S355 and S460), the concrete strength (C20/25–C70/85), the distance between the composite beams (b = 3 m, b = 6 m and b = 9 m), the span of the composite beams (l = 5 m–l = 20 m), the additional dead weight (g_2 = 200 kg/m² and g_2 = 500 kg/m²) as well as the live load (p = 500 kg/m² and p = 1000 kg/m²). The structural design is made according to Eurocode 4 for the composite beam and Eurocode 2 for the RC slab. For composite beams with steel grade S460 and concrete strength C50/60 or above, the moment resistance was determined based on ultimate strain, otherwise with the usual stress blocks. As shear connectors conventional headed studs 22 × 125

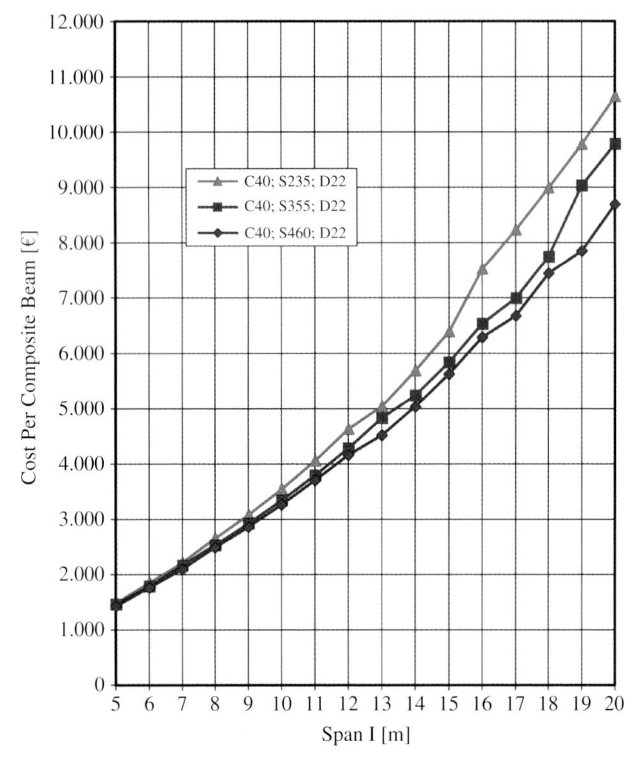

Figure 4.47 Cost comparison for composite beams with concrete C40/50, b = 3 m, g_2 = 200 kg/m^2 and p = 500 kg/m^2 [60].

were used, and from concrete strength C50/60 on concrete dowels to ensure sufficient ductility. The costs assumed [60] include material, fabrication and installation on site, for example, 1275 Euro/t for S235, 1295 Euro/t for S355 and 1320 Euro/t S460 steel.

Nearly independent of all other parameters, concrete strength C30/37 and C40/50 are the most economical whilst C60/75 and C70/85 are the least economic solutions. The reason is that for the concrete strength classes above C40/50, the strength gain is less than the associated additional costs. The results are slightly different for the structural steel: the highest grade S460 is nearly always the most economical, the lowest grade S235 is the least economical (Figures 4.47–4.49). In other words, the strength gain is more than the associated costs for high strength steels in composite beams. Hence it can be said that composite beams with high strength steel beams are advantageous compared to the ordinary steel S235. Especially for long spans and high load levels, the benefit is substantial [61]. For sustainability considerations, environmental benefits also arise. The resulting lower component masses lead to better LCA performances, for example, in carbon footprint or energy consumption.

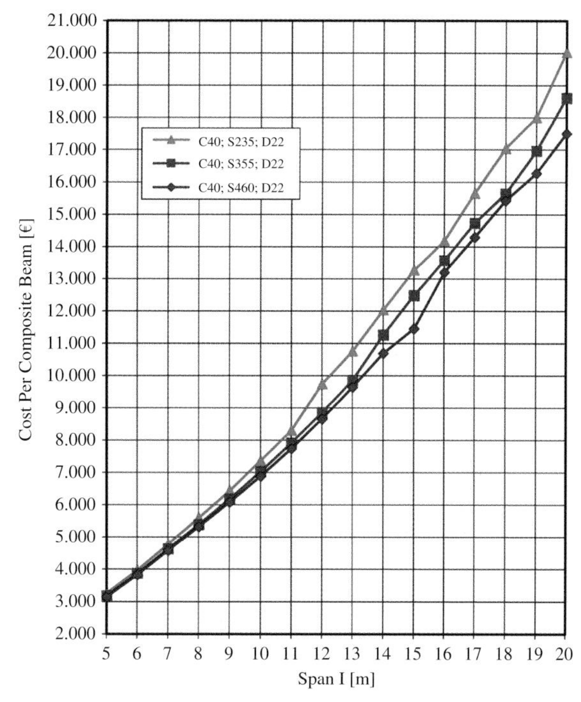

Figure 4.48 Cost comparison for composite beams with concrete C40/50, b = 6 m, g_2 = 200 kg/m² and p = 500 kg/m² [60].

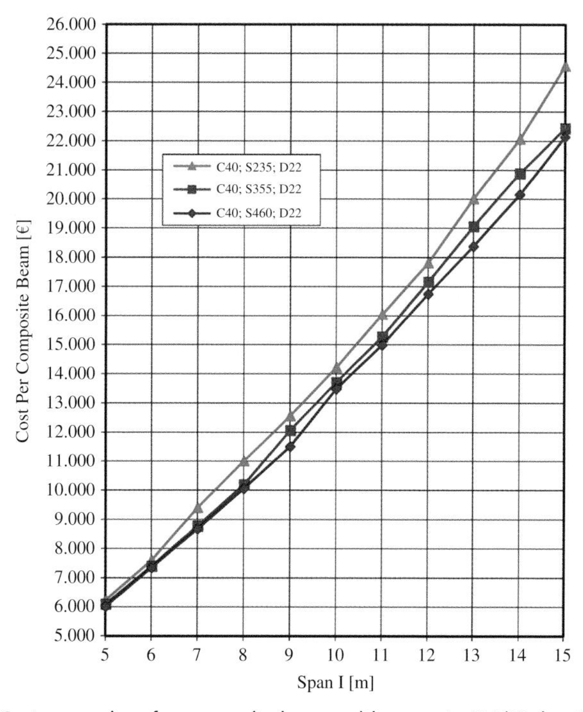

Figure 4.49 Cost comparison for composite beams with concrete C40/50, b = 9 m, g_2 = 500 kg/m² and p = 500 kg/m² [60].

4.10 FIRE-PROTECTIVE COATINGS IN STEEL CONSTRUCTION

Michael Overs and Diana Fischer

4.10.1 Possible ways of designing the fire protection system

Fire protection is very important in steel construction because steel loses strength and thus stability with increasing temperature. At a core temperature of 550°C, steel may lose about 40% of its strength. Therefore, modern steel construction often requires some form of fire protection. Fire-protection systems are usually divided into fire-resistance classes. In Europe, abbreviations such as R30, R60 or R90 are used. R30 is equated to a time period of 30 minutes that is required to rescue people from a burning building.

Various methods may be used to fire protect steel construction. These are divided into active systems, such as sprinklers, and passive fire-protection systems (Figure 4.50). The choice of a suitable system is basically dependent on aesthetic demands and economic aspects. Some passive fire-protection systems, for example, mineral or cement-based sprayed materials, have drawbacks, as steel loses its character. This makes harmonizing with modern architecture difficult. Further, in exposed applications, steel can corrode underneath these protection systems without careful choice of the protection system. One way to fire protect steel construction without losing specific character is to use intumescent coatings.

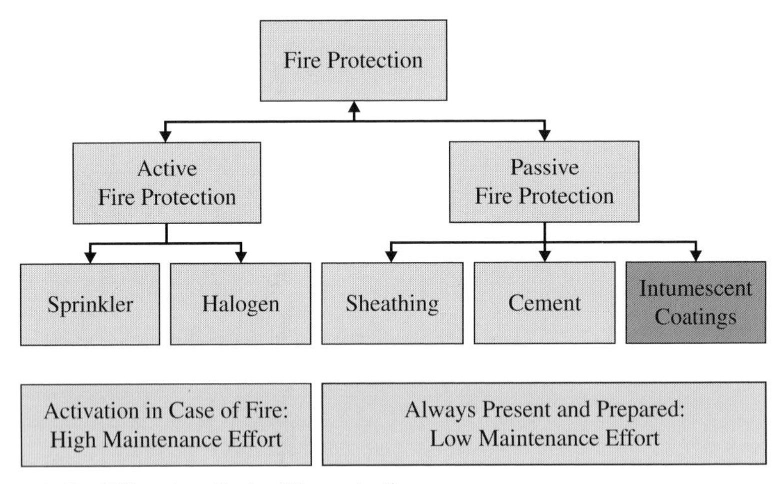

Figure 4.50 Different methods of fire protection.

4.10.2 Fire protection of steel using intumescent coatings

Intumescent coatings, also referred to as 'insulating layer formers', have been used successfully for about 50 years (Figure 4.51). The function of these systems is based on a chemical reaction that starts at about 185°C. The active components

Figure 4.51 Intumescent coatings are used, for example, in the World Finance Center and Jin Mao Tower in Shanghai (left) and in the Mode Gakuen Cocoon Tower in Tokyo (right). © AkzoNobel.

then undergo change, and above 300°C gases are released that cause the coatings to swell. A previously less than 1 mm thin coating then forms a foam that can expand to a hundred times its original thickness. This results in a spongy insulating layer that encases the steel and insulates it so that its loss of strength is slowed down. The rapidly drying, shockproof coating systems provide long periods of fire resistance of up to 3 hours (R180).

4.10.3 The structure of fire-protective coating systems

A fire-protective coating system usually consists of a primer, the intumescent layer, and a topcoat [Figure 4.52a]. The primer constitutes protection against corrosion and serves as a tie coat. This is essential if exposure to corrosion is expected. The topcoat provides the colour and protects the intumescent layer against moisture and other environmental influences. In addition, it can also facilitate cleaning. To accommodate the demands of owners and builders, in recent years products have also been developed that require only one or two layers. These include fire-protection systems for dry, internal rooms that dispense with the need for a topcoat, and systems in which all three of the above-stated functions can be performed by one single layer.

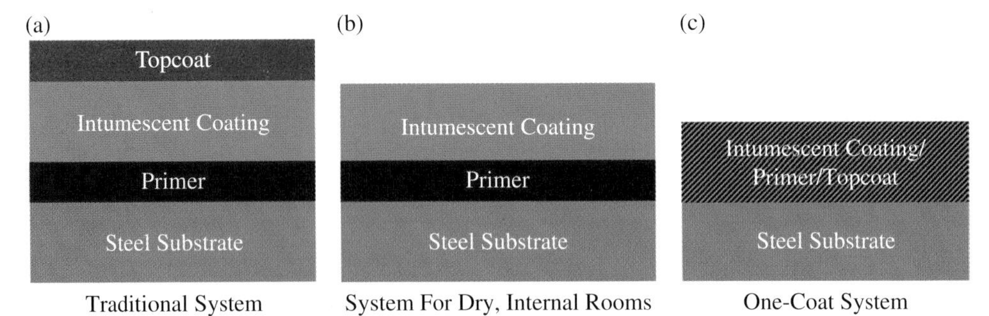

Figure 4.52 Fire-protective coating schemes in steel construction.

4.10.4 Sustainability of fire-protection coating systems

Even if the coating material fulfils the required fire-protection properties, its sustainability is still dependent on other factors, which can sometimes be of crucial importance. These include, inter alia, aesthetical, economic and environmental aspects (Figure 4.53). The importance of these criteria can differ for different applications. Therefore, it is vital that all parties involved agree from the start on which aspects are the most important.

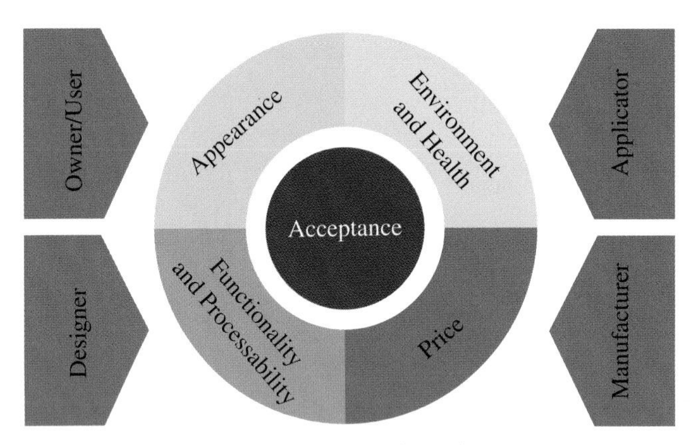

Figure 4.53 The sustainability of a coating system depends on many aspects.

4.10.4.1 *Visual appearance*

When it comes to design requirements, intumescent coatings offer many advantages compared to other protection systems. Fire-protective coatings, that are only a few millimetres thick and applied in line with the profile, emphasise the filigree nature of the structural steel design.

The visual appearance of the coating itself may also be of high priority for the building owner or user, in particular when the building has a representative function. In these cases, the choice of the colour for the topcoat and the careful application of each layer are important. Modern fire-protection systems can hardly be differentiated from conventional corrosion-protection systems. New materials and extensive work by the manufacturers mean that the products are thinner and more efficient. Previously, a 2–3 mm layer was required for R30 coatings. The same effect can now be achieved using a layer of less than 300 μm (Figure 4.54). This reduction in the coating thickness results in several benefits:

- the intumescent coating can be applied more easily, usually in a single operation, and it hardens more rapidly;
- thin layers can be applied more evenly and are thus more visually appealing;
- the cost of fire-protective coatings can be reduced significantly.

Unfortunately, there is rarely a clearly defined requirement with respect to the visual appearance. This is mostly limited solely to the colour shade. In order that building owners, planners and the company carrying out the work use the same prerequisites, these requirements should also be contained in the specifications. One possibility is to classify the visual requirements into different categories as defined by the IGSB the association for steel fire protection coatings (Interessengemeinschaft Stahl-Brandschutzbeschichtungen), a competence centre in all matters concerning fire protection of structural steelwork and related areas. The members of IGSB defined three finish standards with the specific designations Q1, Q2 and Q3 (Figure 4.55) to describe the visual appearance of intumescent coatings:

Q1: Technical finish – finish standard Q1 is suitable for surfaces that are not the subject of any requirements in respect of their appearance. An example of suitable uses would be steel profiles in industrial environments.

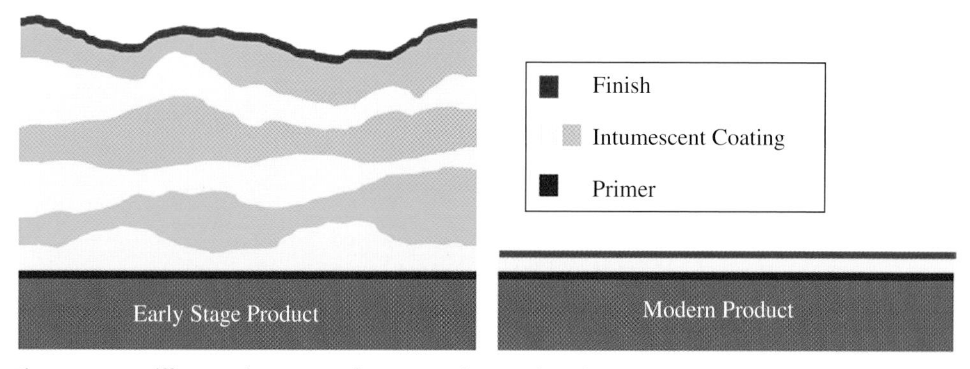

Figure 4.54 Differences between early stage products and modern products.

Figure 4.55 Where no or only minor requirements are made on the visual appearance of a fire-protective coating, the most suitable finish standards are Q1 (left). Decorative finishes (Q3) meet highest demands on their surfaces (right). © AkzoNobel.

Q2: Standard finish – finish standard Q2 is used to describe finishes that are subject to minor requirements in respect of their appearance. With Q2, brush traces, runs, inclusions and sags should no longer be visually noticeable at a distance of 5 m.

Q3: Decorative finish – finish standard Q3 is mainly requested for steel components that are directly accessible or visible, for example, supports in the entrance hall of representative buildings. At a distance of 3 m they should present a decorative appearance. This means that brush traces, runs, inclusions and sags should not be visually noticeable from that distance.

In addition, bespoke surfaces, including colour effects as well as any special textures, are possible. (See also [63]).

4.10.4.2 *Economic efficiency*

Fire-protective coatings are maintenance free and can protect steel construction for many decades, provided they are properly applied. Topcoats of UV-stable paint retain their original colour even if subject to high insolation. Should contamination occur, for instance, in highly frequented areas, the coatings may be cleaned by blasting, vacuuming or lightly brushing, or – subject to the manufacturer's recommendation – by using water and suitable cleaning agents. Thus, the life-cycle costs of such fire-protective coating systems are merely limited to the initial costs for their application.

If the visual appearance of the coating system is of high importance to the owner and user, the application has to be carried out with the appropriate care. The uniform application of a coating system requires sufficient time, particularly with regard to decorative finishes. The required drying time between applications of the various coats is also vital for the final appearance. This should be considered when estimating the costs. On the other hand, pressure on price can lead to the work being carried out by inadequately trained personnel with poor quality equipment and lack of diligence.

It is therefore helpful to involve the manufacturers of fire-protective coatings in the planning at an early stage, as with their specialist knowledge they can advise on how to make improvements and reduce costs. They may also help in formulating the tender specifications with sufficient precision to ease precise cost calculations.

4.10.4.3 *Environment and health*

Sustainable construction places high demands on the environmental quality of building products. Ensuring a high indoor air quality is an important criterion in building certification schemes such as DGNB, LEED and BREEAM (see Chapter 5). Thus, the emission of harmful substances such as volatile organic compounds (VOCs) has to be reduced to a minimum.

The numerous fire-protective coatings available enable a targeted selection to be made on the particular application and on the basis of health-related and ecological criteria. Solvent-based coatings offer benefits compared with aqueous coating materials when it comes to application, appearance and stability. However, they can have adverse health effects. For indoor applications, it is therefore recommended to choose low emitting, VOC- and halogen-free coatings.

The environmental quality of fire-protective coatings systems can be quantitatively verified via EPDs. Due to their small quantities applied, the LCA of a building is hardly influenced by such systems. To the contrary, thanks to the minor film thicknesses, other material- and resource-intensive protective measures can be avoided. (See also [62]).

4.11 BUILDING ENVELOPES IN STEEL

Markus Kuhnhenne, Dominik Pyschny and Matthias Brieden

4.11.1 Energy-efficient building envelope design

Sustainability is an important topic in the building and construction industry. It serves to maintain value in combination with protecting the environment and taking social needs and economics into account. Energy consumption plays a dominant role here because it has a very marked effect on the evaluation of a building with respect to sustainability. In order to be able to construct sustainable buildings, it is necessary to consider all aspects affecting the energy requirements of a building and evaluate them with respect to potential energy savings. In assessing the sustainability of buildings, the energy performance of building envelopes is of central importance (Figure 4.56).

Building envelopes of industrial buildings have to be verified according to minimum requirements regarding thermal insulation. Innovative solutions to reduce thermal bridges and irregularities could lead to new markets for envelope systems in steel.

Architects and engineers choose steel and composite materials according to requirements for load-bearing resistance, serviceability and fire protection. Cost effectiveness and aesthetical design are also important reasons for their choice, but

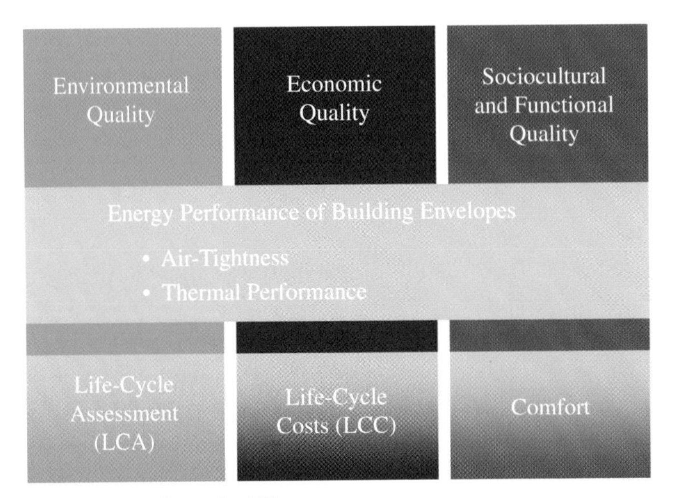

Figure 4.56 Assessment of sustainability.

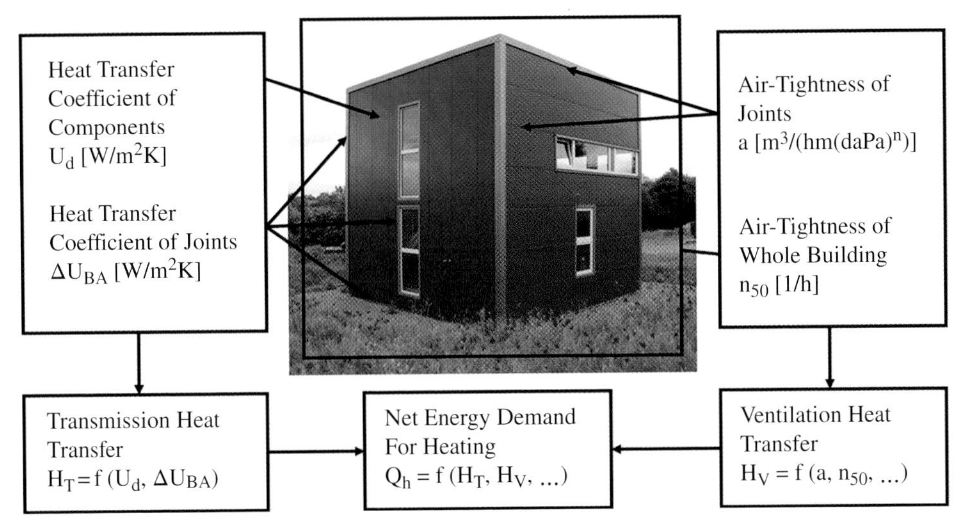

Figure 4.57 Factors of energy-efficient building envelope design.

energy efficiency is one of the dominant design aspect when planning a building. Figure 4.57 shows all factors of the energy performance of building envelopes that affect the energy efficiency of a building.

The energy-related performance of building envelopes is determined by their heat transmission and heat convection properties. Heat transmission takes place as one-dimensional heat flow in the thermally undisturbed control zone of elements of the building envelope; in addition, there are two- and three-dimensional heat flows within linear and point thermal bridges (Figure 4.58). According to EN ISO 13789 [64], the transmission thermal transmittance between the heated zone and the outside is calculated as follows:

The basic methodology for measuring the air-tightness of a whole building is to either pressurise or depressurise the entire building with respect to the

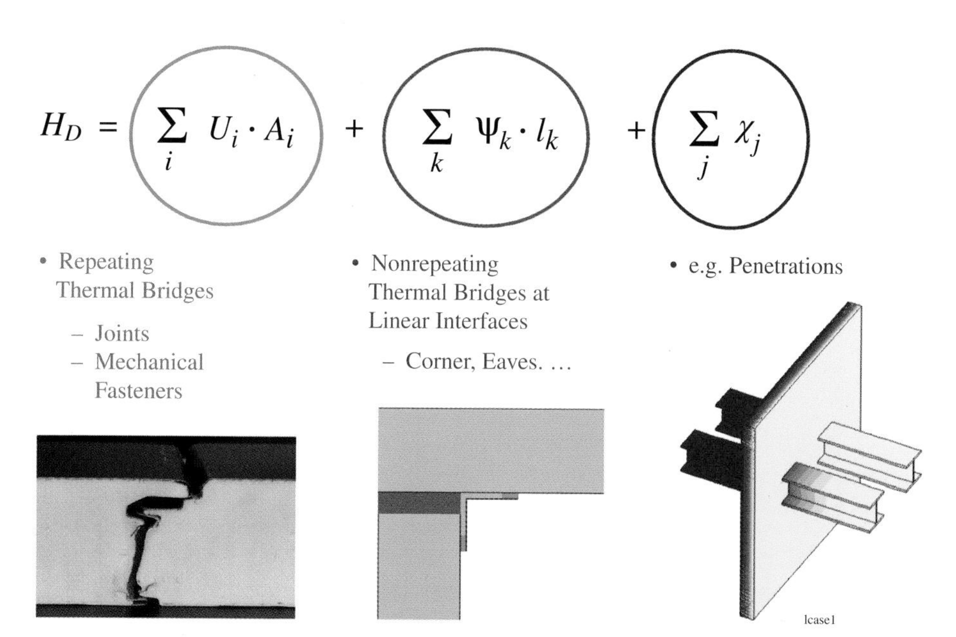

Figure 4.58 Thermal transmittance according to EN ISO 13789 [57].

ambient air pressure. Air egress or ingress can then be measured, giving an indication of the permeability of the building fabric.

Different countries use different metrics when defining levels of air-tightness, n_{50}, API and ALI, which makes comparisons difficult, particularly since some countries also quote their requirements at different pressures. For example, in France a pressure of 4 Pa is used as the test regime compared with the majority of countries, in which 50 Pa is used. In order to compare these different requirements they must be converted to a single metric. In order to do that a relationship needs to be derived to account for the different pressures, and an example geometry must be considered to allow conversion of units.

Lightweight metal construction is used primarily in industrial and commercial buildings. Two common forms of construction are double-skin designs and sandwich construction [65]. These two construction methods are examined in more detail in the following sections.

4.11.2 Thermal performance and air-tightness of sandwich constructions

4.11.2.1 *General*

Sandwich construction is made from individual, industrially manufactured insulated panels. These are ready-to-install roof and wall elements that consist of two thin metal layers and are available with linear-, trapezoidal- or wave-shaped profiles that are joined via core insulation (see Figure 4.59).

In Europe, approximately 150 million m^2 of cladding using steel sandwich systems are installed per year.

Media Library Marsan Mediatheque

Location:	Mont de Marsan, France
Architect:	archi5, Associate Architects: B. Huidobro
Building description:	The building has been designed as a covered cultural square of 58 × 58 m with alternating transparent and reflecting glass facades, double skin of 2 m wide. The interior space on the ground floor is completely open and centres around a patio, the design of which has been inspired by Matisse's paintings of acanthus leaves.
Steel details:	Very thin inclined steel columns – circular hollow sections – supporting the glass facade of the patio. Truss of the exterior facade realized with rectangular hollow steel sections. Suspended inside footbridge.
Sustainability:	Slender structural components, transparency and sunlight input, energy efficient climate system.
Awards:	International Architecture Award of Chicago Athenaeum Museum of Architecture and Design and Leaf awards 2013: Overall Winner and winner of Public Building of the Year – Culture.

Figure 1 Aerial view of the building. © Marsan Agglomeration.

Figure 2 Suspended footbridge inside the library building. © ConstruirAcier.

Figure 3 The reflecting and transparent façade. © ConstruirAcier.

Example provided by ConstruirAcier.

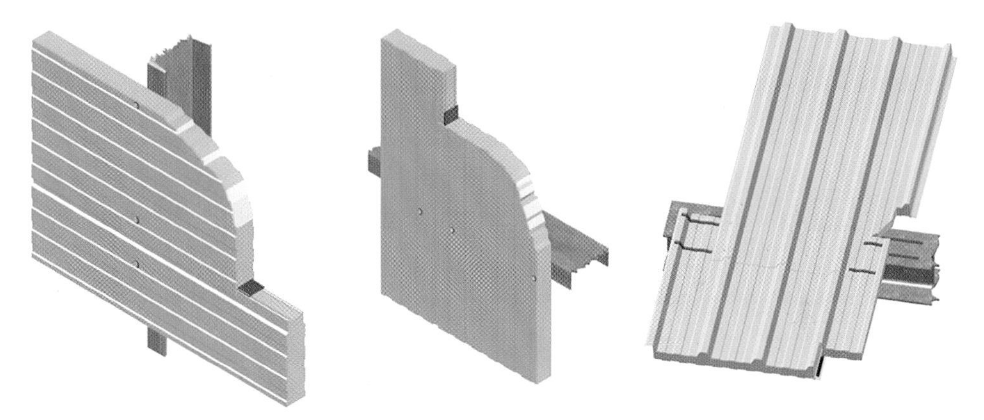

Figure 4.59 Example of sandwichpanel construction.

4.11.2.2 *Calculation of the thermal transmittance of a sandwich panel according to EN 14509 [66]*

Simplified method for the calculation of the thermal transmittance of a panel ($U_{d,S}$): The thermal transmittance of a panel $U_{d,S}$ can be calculated with a simplified method by using Equation (1) neglecting the influence of the profiled faces and using the linear thermal transmittance contribution factor of the joints (f_{joint}) for steel faces according to the generic type of joint (see Table 4.9).

$$U_{d,S} = \frac{1}{R_{si} + \frac{d_c}{\lambda_c} + R_{se}} \cdot \left(1 + f_{\text{joint}} \cdot \frac{1{,}0}{B}\right), \qquad (1)$$

where f_{joint} is the linear thermal transmittance contribution factor of the joints calculated for a joint spacing of 1.0 m (see Table 4.8).

Residential Complex 'La Corte del Futuro'

Location: Torre Boldone, Bergamo, Italy

Client & contractor: Vanoncini spa

Architect: Atelier2 – Gallotti e Imperadori Associati; Gian Pietro Imperadori (steel structures design)

Building description: The building is a residential complex that was designed considering the characteristics of the urban context and the use of innovative techniques. This led to the completion of a building matching the highest energy performance.

Steel details: The structural system, mainly achieved by means of dry construction, is composed of steel frames embedded in the building outer skin and floors. It is made of corrugated steel sheet filled with reinforced concrete. The floor plays the role of horizontal diaphragm connected to a vertical core acting as wind bracing, all of it providing a substantial thermal inertia to the building.

Sustainability: 'Hybrid' system responds better to the thermal fluctuations between winter and summer. Suitable integration of innovative constructive methods proposed by local companies. Reduced energy requirements due to increased insulation of the closings, a total elimination of the thermal bridges and the use of renewable energy sources. Thermal solar panels and photovoltaics for warm water supply and electricity.

Awards: The project has been awarded the 'Sustainability Prize' in the IX Edition of the IQU Competition, organized by the University of Ferrara.

The project is certified 'A+' CENED (Certificazione ENergetica degli EDifici) class with an energy consumption of 10 KWh/m^2 per year and Casaclima class 'A' with an energy consumption of 15 KWh/m^2 per year.

Figure 1 Residential complex 'La Corte del Futuro'. © Atelier2 – Gallotti e Imperadori Associati.

Figure 2 Construction site – the steel structure is still visible. © Atelier2 – Gallotti e Imperadori Associati.

Example provided by Fondazione Promozione Acciaio.

Table 4.8 Thermal transmittance contribution factor (f_{joint}) for steel faces with different joint types.

d_d [mm]	f_{joint} Type I	Type II No clip ($f_{joint,nc}$)	Type II Clip ($f_{joint,c}$)	Type III	Type IV	Type V	Type VI	Type VII	Type VIII
30	–	–	–	–	0.057	–	–	–	0.061
40	0.160	–	–	–	0.045	0.144	–	0.098	0.044
60	0.083	0.111	0.818	0.244	0.031	0.072	0.227	0.049	0.030
80	0.052	0.063	1.016	0.105	0.024	0.044	0.094	0.036	0.024
120	0.032	0.039	1.325	0.057	0.019	0.026	0.049	–	–
160	0.025	0.030	1.555	0.041	0.015	0.019	0.034	–	–
200	0.020	0.025	1.733	0.033	0.013	0.015	0.026	–	–

Accurate calculation of the thermal transmittance of a sandwich panel: The thermal transmittance of the panel should be determined by using a computer programme in accordance with EN ISO 10211 [67]. With the help of numerical methods, it is possible to determine the heat transfer of two- and three-dimensional situations reliably. The finite element method (FEM) is used for complex

Friem Headquarters

Location:	Segrate, Milan, Italy
Client:	FRIEM spa
Architect:	OnsiteStudio (Angelo Lunati, Luca Varesi); CeAS srl (structures)
Contractors:	Stahlbau Pichler srl (steel structures and facades), Edildam
Building description:	The building separates the production site and the circulation roads. The separation and frontiers are defined by an L shape as an architectonic element.
Steel details:	The bearing structure of the buildings is composed of steel profiles, HEA 300 for the columns and HEB 200 for the beams. More than 2000 unique perforated stainless steel sheets form the outer shell.
Sustainability:	Sandwich panels and windows are characterized by different material properties depending on their exposure to the sun. The water is collected and further used to irrigate the gardens. A photovoltaic installation of 200 kW and a heat pump working at low temperature improve the energy performance of the building. An automatic system controls the lighting of the internal spaces, taking into account the magnitude of incoming natural light and the presence of workers.
Awards:	The office building is certified 'A' CENED (Certificazione ENergetica degli EDifici).

Figure 1 Friem Headquarters, Milan. © Filippo Romano/Stahlbau Pichler.

Figure 2 The stainless steel facade provides natural lighting. © Filippo Romano/Stahlbau Pichler.

Figure 3 View from the inside of the production side. © Filippo Romano/Stahlbau Pichler.

Example provided by Fondazione Promozione Acciaio.

Table 4.9 Generic types of longitudinal joints.

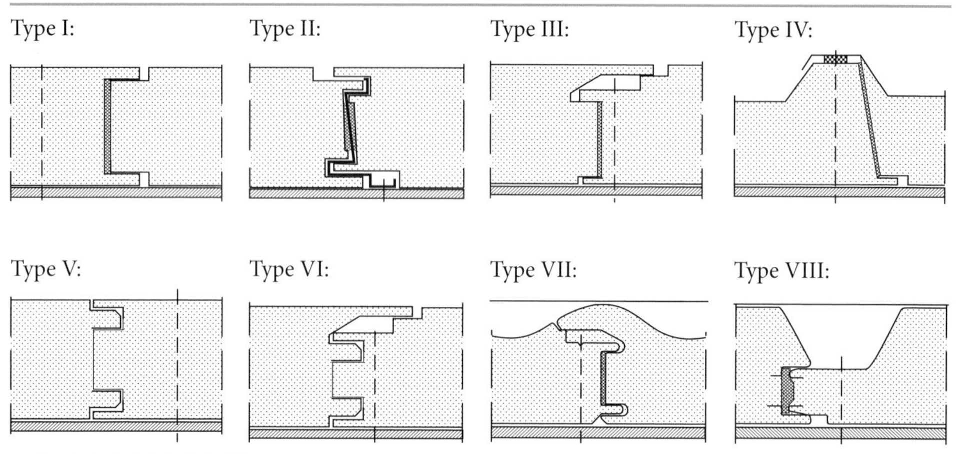

Table 4.10 Example: joint of sandwich elements, variation of displacement.

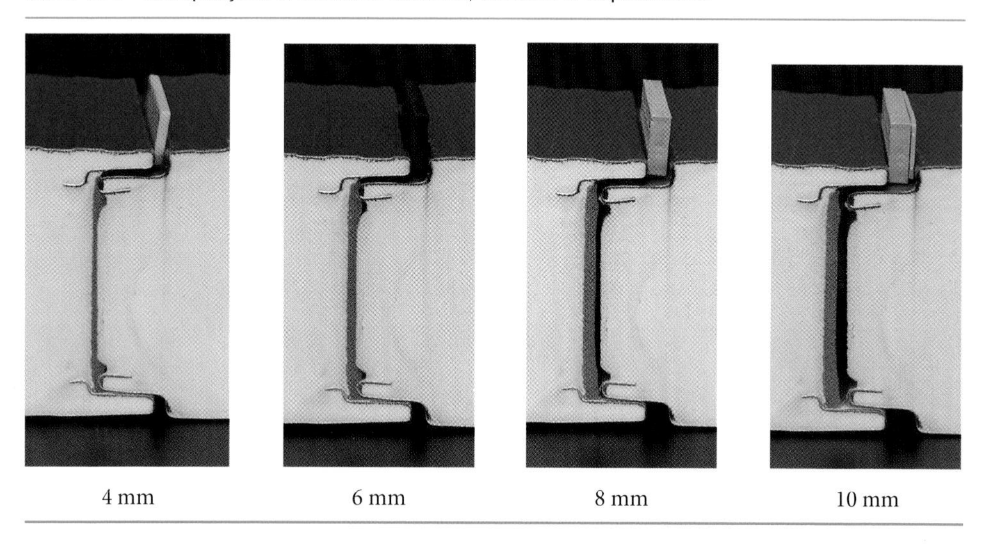

geometries for which no analytical solution is available: they are divided into a large number of simply shaped elements that are themselves calculable.

The longitudinal joints of metal sandwich panels can be designed very differently depending on the product (see Tables 4.9 and 4.10). A distinction is made between visible (Type I) and concealed fastening (Type III).

Figure 4.60 shows FE models, temperature distribution and heat flux of two different typical longitudinal joints of steel sandwich panels.

Nonrepeating thermal bridges at linear interfaces: The heat transfer within the thermal transmission area of nonrepeating linear interfaces of sandwich construction can be reduced by 75% through thermally enhanced construction details. Details and values of standard and enhanced construction can be found in the 'Thermal Bridge Atlas for Metal Sandwich Construction' [68].

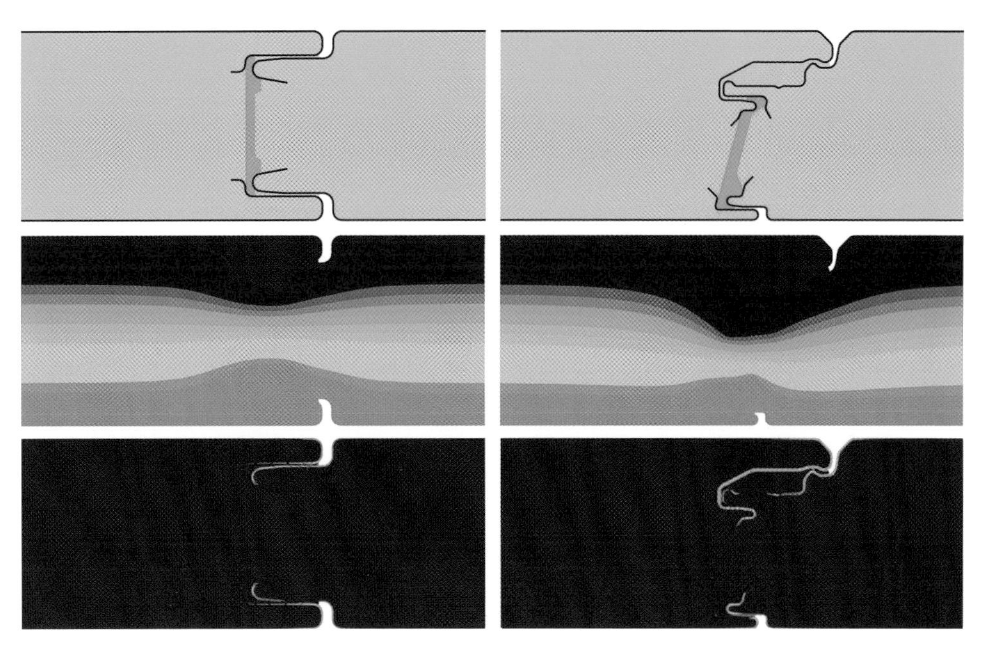

Figure 4.60 FE models, temperature distribution and heat flux of joints Type I (left) and Type III (right).

4.11.2.3 *Air-tightness of sandwich construction*

Air-tightness is an important attribute for improved energy efficiency of building envelopes. Uncontrolled ventilation losses should be minimized, and the benefits of mechanical ventilation systems are greater if buildings are air-tight. Moisture problems can also occur if warm, wet air can infiltrate into facades.

Joint air-tightness of plane elements: Various solutions for building envelopes in steel consist of plane prefabricated elements (e.g. roof and wall sandwich panels, cassette profiles, curtain walling) and their joints. They can be tested in a laboratory according to EN 12114 [69] (determination of air-leakage 'a-value' [$m^3/(h{\cdot}m{\cdot}daPa^{2/3})$]) to verify requirements on a European level.

An example for the measurement of a typical joint geometry is shown in Table 4.10 and Figure 4.61. Only if the joint is installed perfectly (4 mm) can the requirements be fulfilled.

Based on testing of products with different joint geometry by several European producers, the results show that no general statement regarding the air-tightness of sandwich element joints is possible. The joint tightness depends on the particular joint geometry and joint width that is achieved during installation. There is a critical joint width, where the joint just meets the requirements of the joint tightness. In particular, the location, size and compression of a sealing strip within the joint lead to strongly different behaviour of joints with regard to air-tightness. Relatively small variations of the joints can lead to unwanted high air-leakage rates.

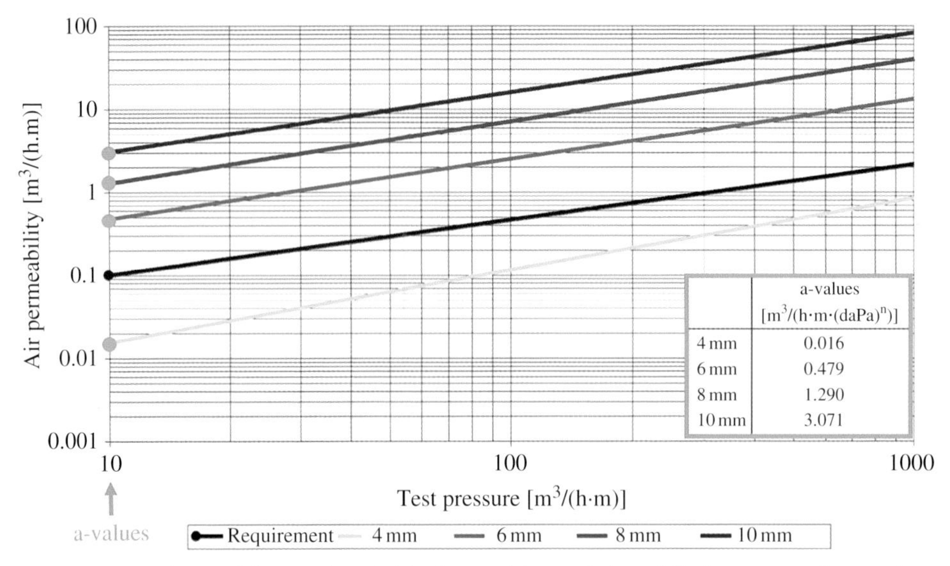

Figure 4.61 Results: air-tightness of sandwich element joint, variation of displacement and comparison with requirement (black curve).

4.11.2.4 *Conclusion*

The analysis of steel sandwich panels has shown that an accurate determination of the effects of thermal bridges caused by metallic components in the building envelope requires the use of FEM. The determination of the air-tightness of joints between steel components in a reproducible and economical way is only possible by on-site and laboratory tests. For steel sandwich panels, the key requirements are air-tightness of a joint and the minimum requirements for thermal insulation. In future, lightweight steel buildings should be all optimized in view of heat transfer and reliable air-tightness. To this end, the products and joints should be further developed to reduce thermal bridges and to tolerate imperfections in the fit of joints on site.

4.11.3 Effective thermal insulation by application of steel cassette profiles

4.11.3.1 *General*

In industrial and commercial buildings, steel cassette profiles are widely used in the construction of the building envelope. The modular design allows simple, fast and cost-effective assembly.

On the interior side, cassettes consist of steel profiles with a thickness between 0.75 mm and 1.50 mm in which the thermal insulation is placed. The thickness of the insulation corresponds to the ridge height of the cassette profile. From the outside, trapezoidal profiles are fixed with screws to the flange of these cassette profiles. By using a 3 mm thin separating strip between warm cassette profile (inside) and cold trapezoidal profile (outside), thermal bridging effects can only be reduced insufficiently (Figure 4.62, left). However, the thermal insulation property of steel

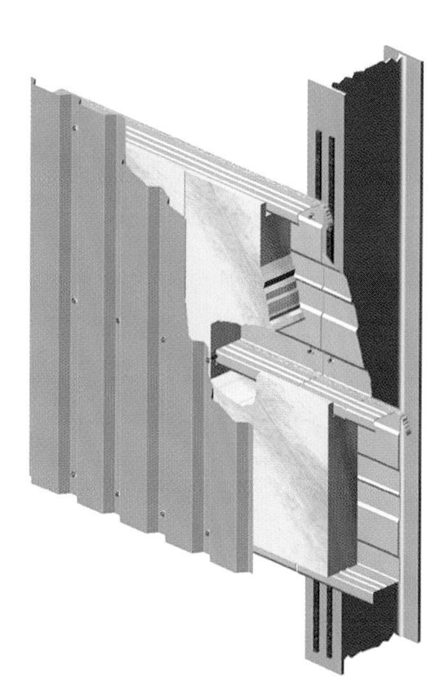

Figure 4.62 Layout for steel cassette construction.

cassette construction can be improved by the provision of an additional insulation layer between the cassette and trapezoidal profile (Figure 4.62, right).

The following thermal study is carried out according to EN ISO 10211 [67] and aims at the evaluation of heat transfer through steel cassettes with and without use of additional thermal insulation. The calculations are performed on a variety of construction variations in order to establish the thermal effectiveness of different setups. Against this background, a diagram is derived with which the thermal transmittance of different steel cassette construction can be determined directly. On the basis of these results the calculations are repeated with the Microsoft Excel tool 'IFBS Wärmedurchgang 1.3' [70], which is often used in practice. After a comparison between both methods, different uses of steel cassettes are evaluated in the context of the energy-saving ordinance. Further information on the applied principles and calculation methods can be found in [71].

In order to determine the effective thickness of the thermal insulation in steel cassettes, numerical studies of different construction variations are performed. A first thermal insulation of TCG 035 [thermal conductivity of 0.035 W/(m·K)] is applied. In further calculations, the influence and impact of the use of alternative thermal insulations of TCG 030 or 040 is determined.

4.11.3.2 *Investigation variants*

On the basis of Figure 4.63, the geometric parameters were varied. The thickness of the steel profiles are selected as $t_{Ni,min}$ = 0.75 mm and $t_{Ni,max}$ = 1.50 mm as the extreme values. Steel cassette profiles are manufactured with web heights from 90 to 240 mm, which leads also to an insulation thickness in the cassette d_c between 90 and 240 mm. Ten different cases are analysed. To reduce the thermal bridging effect of the web, two different alternatives are considered: either a conventional

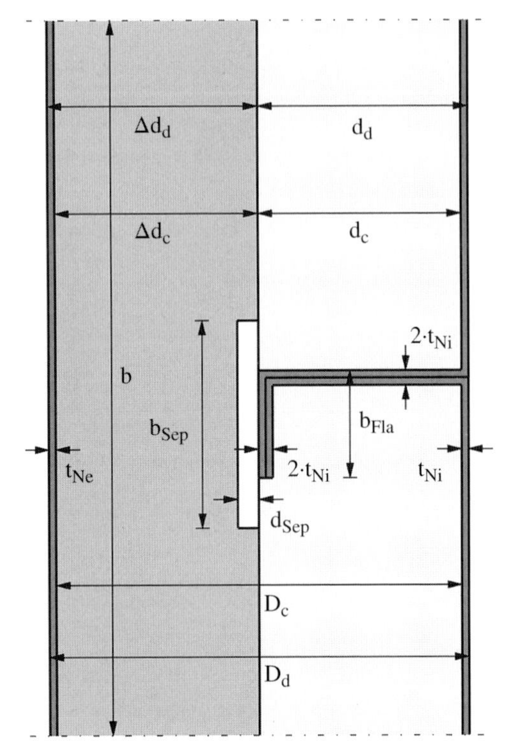

Figure 4.63 Schematic section steel cassette construction.

Table 4.11 Thermal conductivities of materials.

	Steel	Thermal Insulation	Separating Strip
Thermal conductivity λ [W/(m·K)]	50	0.035	0.055

variant with the arrangement of a separating strip ($d_{Sep} \times b_{Sep} = 3 \times 60$ mm) or an improved design with an additional insulation layer. For the additional insulation, the two thicknesses $\Delta d_c = 40$ mm and $\Delta d_c = 80$ mm were analysed.

The minimum total insulation thickness is therefore $D_{c,min} = 90$ mm for a 90 mm steel cassette profile with a separating strip, and $D_{c,max} = 320$ mm for a 240 mm steel cassette profile with 80 mm additional insulation layer.

The width of the cassette element is chosen as b = 600 mm and the flange of the steel cassette profile as $b_{Fla} = 40$ mm. The thermal conductivities of the engaged materials are defined according to EN ISO 10456 [73], DIN 4108-4 [74] and IFBS-Fachinformation 4.02 [75] (Table 4.11).

4.11.3.3 *Calculations and results*

In order to assign the design value of the thermal transmittance, the numerical calculations are carried out according to the various parameters. Based on these results, the thermal transmittances are converted into effective thermal insulation

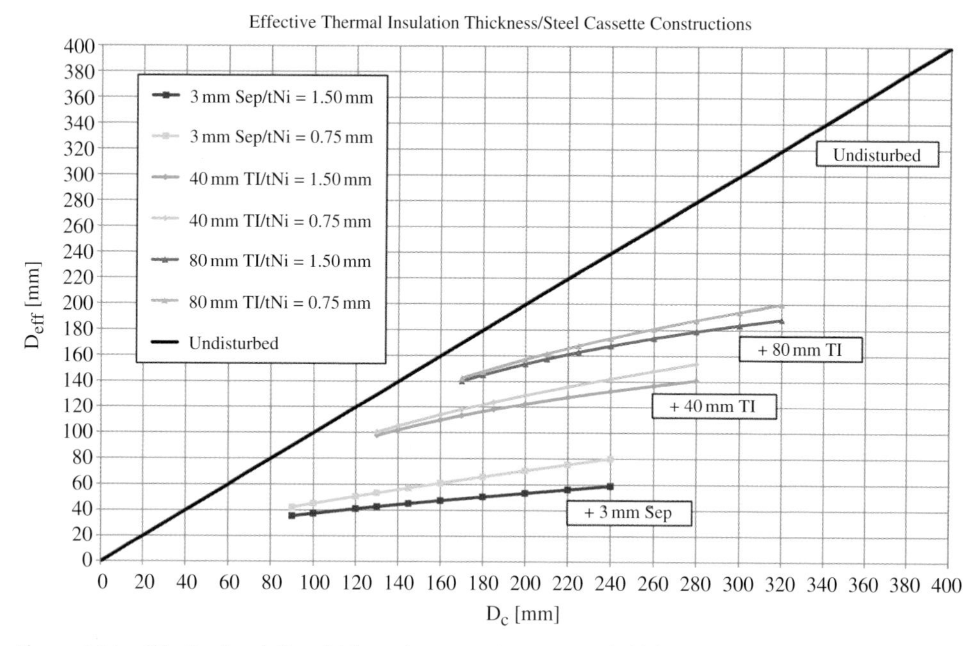

Figure 4.64 Effective insulation thickness in comparison to actual thickness.

Table 4.12 Exemplary results effective operative insulation thickness.

	Pure Insulation	3 mm Separating Strip	40 mm Additional Insulation Layer	80 mm Additional Insulation Layer
Insulation cassette d_c [mm]	—	200	160	120
Insulation additional Δd_c [mm]	—	0	40	80
Insulation summed D_c [mm]	200	200	200	200
Reading D_{eff} [mm]	200	55	120	155
Loss $D_c - D_{eff}$ [mm]	0	145	80	45
Loss of insulation effectiveness[%]	0	73	38	23
Thermal transmittance U [W/(m²·K)]	0.170	0.590	0.273	0.219

thicknesses. These are contrasted with actually used insulation thicknesses and are presented in Figure 4.64.

Using this diagram, thermal transmittances for steel cassette constructions can be determined. Moreover, the effect of thermal bridging of the cassette web for each case can be seen by this figure. For example, if 200 mm thermal insulation is used for steel cassette profiles with a steel thickness of 1.5 mm, the results listed in Table 4.12 can be obtained.

The table shows that using 200 mm insulation within the cassette profile, and with a 3 mm separation strip, leads to a loss of about 73% of the thermal

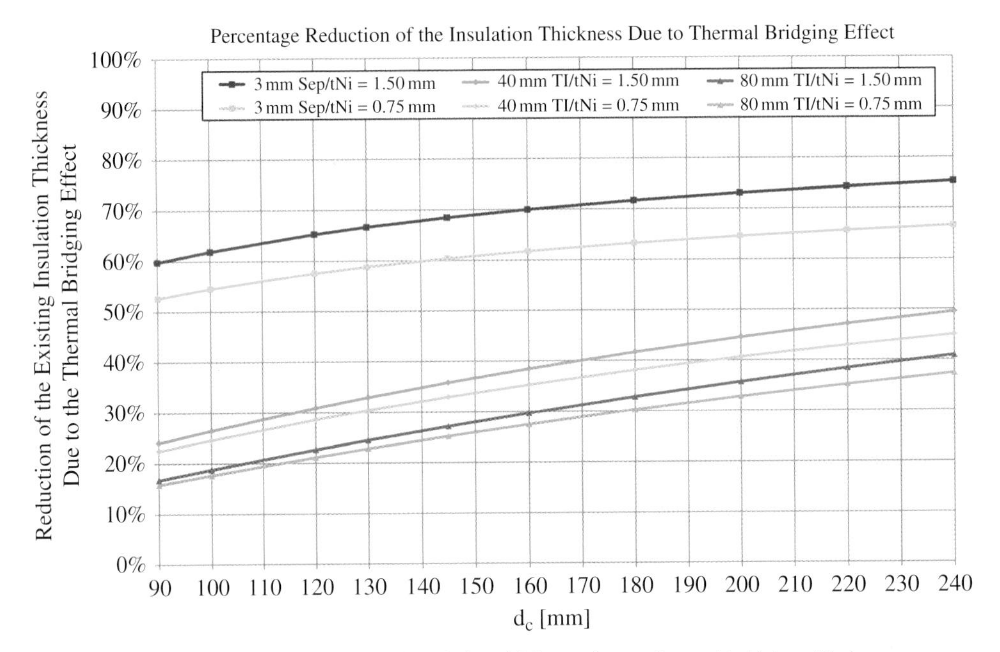

Figure 4.65 Percentage reduction of the insulation thickness due to thermal bridging effect.

performance of the insulation due to the thermal bridging effect, whereas the thermal transmittance increases by a factor of 3.5 in comparison to the undisturbed case (with no thermal bridging). On the contrary, with the use of 40 mm of the 200 mm total insulation as an additional insulation layer in front of a 160 mm steel cassette profile, the remaining thermal bridging effect reduces the actual effective insulation thickness by 38%. By using an additional insulation of 80 mm in front of a 120 mm cassette profile, only 23% of the thermal performance of the insulation is lost due to thermal bridging effects.

Figure 4.65 shows an alternative illustration of the results. The percentage loss of insulation thickness due to thermal bridging effect is plotted as a function of the insulation thickness within the cassette profile. By using 200 mm insulation with a conventional separating strip, a reduction of 65%–73% in thermal performance can be seen. The combination of a 40 mm additional insulation layer with a 160 mm steel cassette profile reduces the effective insulation thickness by 35%–38%. In contrast to this, the use of 120 mm cassette profile with 80 mm additional insulation leads to a loss of less than 25%.

4.11.3.4 *Comparison with IFBS-Fachinformation 4.05*

For the comparison between the FEM results and the calculations basing on [75] and [70], the investigations were repeated using the Excel tool. The results are also converted into effective thermal insulation thicknesses. Figure 4.66 includes a corresponding diagram containing the results of the FEM calculation as well as the results of the IFBS Excel tool calculation.

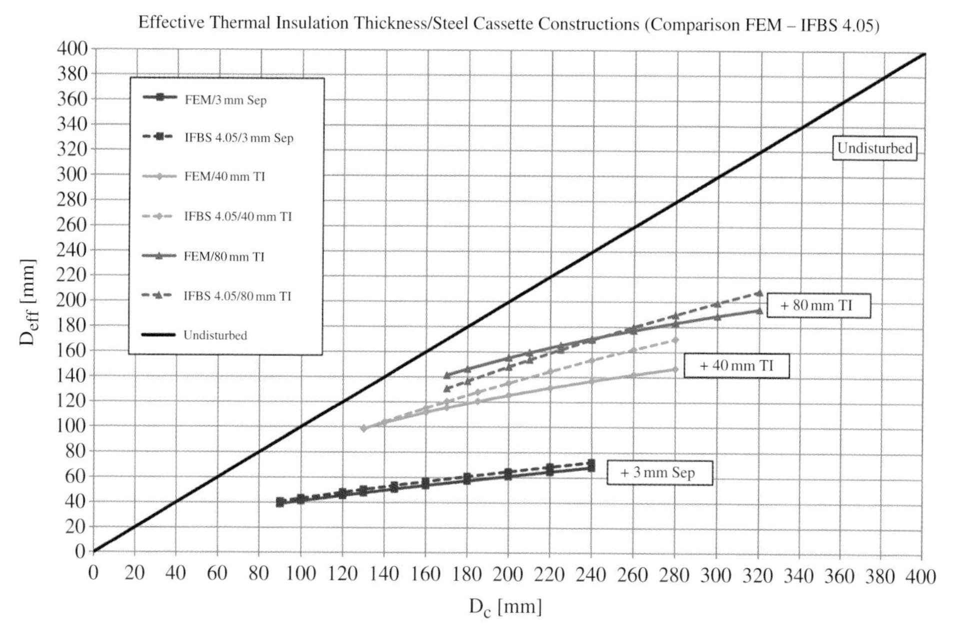

Figure 4.66 Effective operative insulation thickness – comparison of FEM calculation and IFBS 4.05 [72].

It can be seen that the effective insulation thicknesses for steel cassette with 3 mm separating strip and 40 mm additional insulation layer according to the FEM calculations are lower than the calculations using the Excel tool. For the variants with 80 mm additional insulation, no general statement can be made. Both curves intersect at a point, so that the results of IFBS-Fachinformation 4.05 [72] are partly above and below the FEM results.

In this context, it should be noted that the differences between the two methods are relatively low. For conventional construction with 3 mm separation strips, the deviations are lower than 6% or 0.04 W/(m²·K); with 40 mm additional insulation variations of about 13% or 0.03 W/(m²·K) were obtained, whereas with 80 mm additional insulation layer, a deviation less than 8% or 0.02 W/(m²·K) was obtained.

Because of the upward limitation of the occurring thermal transmittance by the chosen simplifications for the FEM calculations, the slight noticeable differences between the two methods are assessed as negligible in terms of accuracy.

4.11.3.5 *Assessment of thermal transfer*

Reference value according to the energy saving ordinance: In order to evaluate the obtained results at a component level, calculations basing on the current German energy saving ordinance EnEV 2009 and the current cabinet decision for the future EnEV from 6 February 2013 [76] are performed. Previously steel cassette construction has been used almost exclusively in nonresidential construction, so that the following statements refer to this sector. Annex 2, Table 1 of the German EnEV contains reference values for the achievement of several building components. The corresponding values for exterior walls are listed in Table 4.13.

Table 4.13 Extract from EnEV, Annex 2, Table 1 – reference values.

Building Component	Interior Target Temperatures with Heating ≥ 19°C	Interior Target Temperatures with Heating from 12°C–19°C
Exterior wall	U = 0.28 W/(m²·K)	U = 0.35 W/(m²·K)

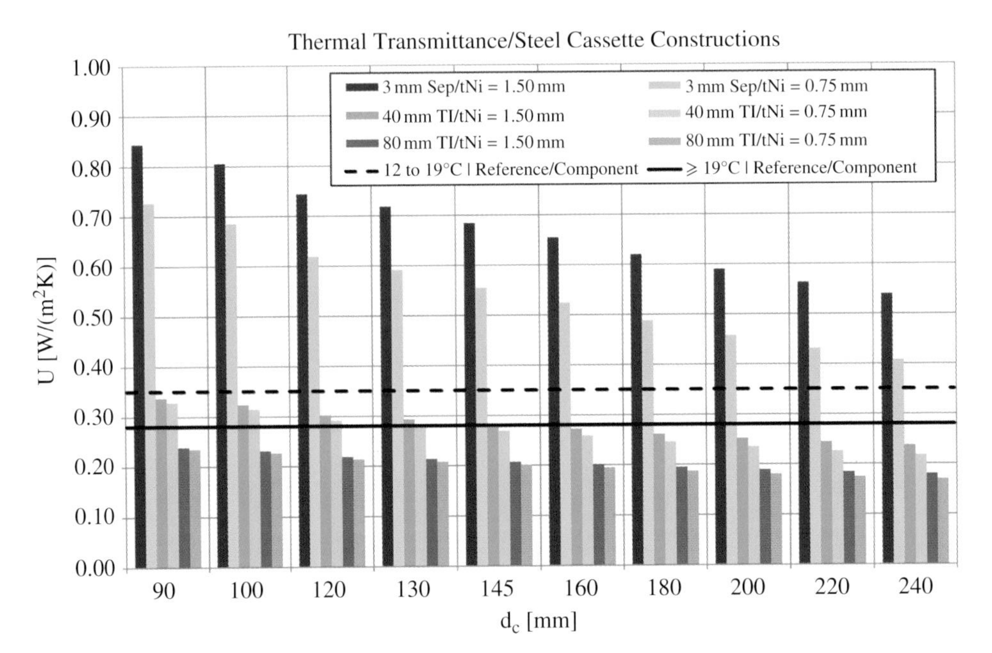

Figure 4.67 Thermal Transmittance / Steel Cassette Constructions comparison with German energy saving requirements.

Assessment of the results: The design values of the thermal transmittance (U values) of various steel cassette construction are compared in Figure 4.67 with the design requirement of 0.28 W/(m²·K) or 0.35 W/(m²·K). The thermal transmittances are presented, both with separation strips or additional insulation layers, depending on the thickness of the individual cassette profile. The reference values for exterior walls are registered as horizontal lines, according to internal target temperatures.

It can be seen that for the cases studied here, all steel cassette constructions without an additional insulation layer, i.e. all with 3 mm separation strips, exceed the reference values, and so compliance is only possible through the use of an additional layer of insulation.

For normal heated buildings, the reference level of 0.28 W/(m²·K) can be reached by using a 40 mm additional insulation on the outside of a 145 mm steel cassette profile. For a cassette profile with the lowest profile height (90 mm), a combination with 80 mm additional insulation is required. A comparison of these cassette systems shows that although both meet the reference value, the version with an 80 mm insulation layer has around 15% lower thermal transmittance, with about 10% less

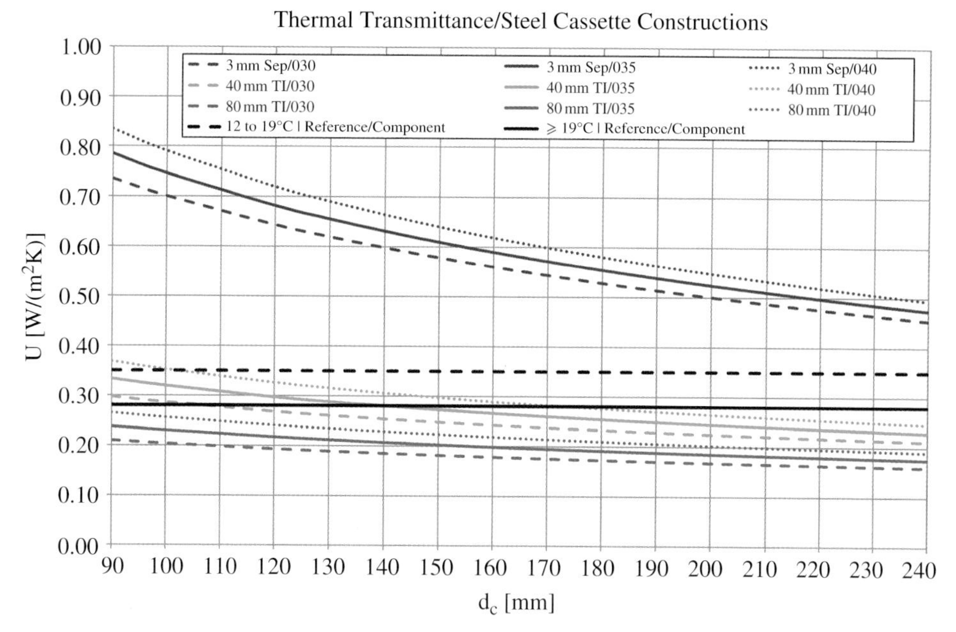

Figure 4.68 Influence of different thermal conductivities on the thermal performance of cassette cladding systems.

insulation. However, the 140 mm deep cassette can clearly span farther than the 90 mm deep cassette, which is an important factor in the optimum solution.

Influence of the thermal conductivity of the insulation: The previous discussed investigations have been performed with a thermal conductivity of the insulation of 0.035 W/(m·K). The effect of different conductivities is assessed for 0.030 W/(m·K) and 0.040 W/(m·K) (Figure 4.68). In order to keep the clarity in the diagram, the differentiation of the cassette profile thicknesses is omitted; and, average values are plotted. For a 90 mm steel cassette profile, the deviation between the different thermal insulation is 13% for 3 mm separation strips and 23% for 80 mm additional insulation, based on the mean value of 0.035 W/(m·K). For the 240 mm steel cassette profile, the variation reduces to 8% (3 mm separation strip) or 17% (80 mm extra insulation).

A comparison of the results illustrates that even a significant reduction of the thermal conductivity of the insulation is not sufficient for constructions with 3 mm separation strips to satisfy the reference value of thermal transmittance of 0.35 W/(m^2·K) or 0.28 W/(m^2·K). In contrast, it is possible for the variants with additional insulation to achieve compliance with the reference values by using a lower thermal conductivity for limit cases. Thus, steel cassettes with a profile depth of about 100–145 mm can reach the reference value of 0.28 W/(m^2·K) for normal heated buildings by the use of insulation with a thermal conductivity of 0.030 W/(m·K).

The use of an optimized thermal insulation with low thermal conductivity thus makes less sense for conventional steel construction cassettes with 3 mm separation strips. Regarding the German EnEV, it seems better to use thermal insulation with a higher thermal conductivity combined with an additional insulation layer of 40 mm.

4.12 FLOOR SYSTEMS

4.12.1 Steel as key component for multifunctional flooring systems

Markus Feldmann, Dominik Pyschny, Martin Classen and Josef Hegger

The majority of contemporary buildings have almost monofunctional characteristics. This often leads to the necessity to substantially restructure such buildings before they reach their economic life. This is because designers do not take into account the requirements of today's users and potential demands of the future. Thus, innovative solutions in multifunctional building design are required in order to promote sustainability in construction. A decisive contribution to this is represented by integrated floor-slab systems in steel-concrete composite construction.

Floor slabs not only fulfil loadbearing and stabilising functions, they also create the separation between adjoining functional units and they provide integration of building services, as well as influencing the physical properties of the building. The floor slab is designed primarily for its structural considerations, overlooking other building characteristics. The wide variety of functions a floor structure has to provide implies the need for multifunctional floor-slab structures. The most important requirements of floor-slab systems are structural capabilities, architecture, manufacturing of components, fire safety, building physics, construction, flexibility and use, dismantling and ultimate recycling. These requirements are summarised in Figure 4.69.

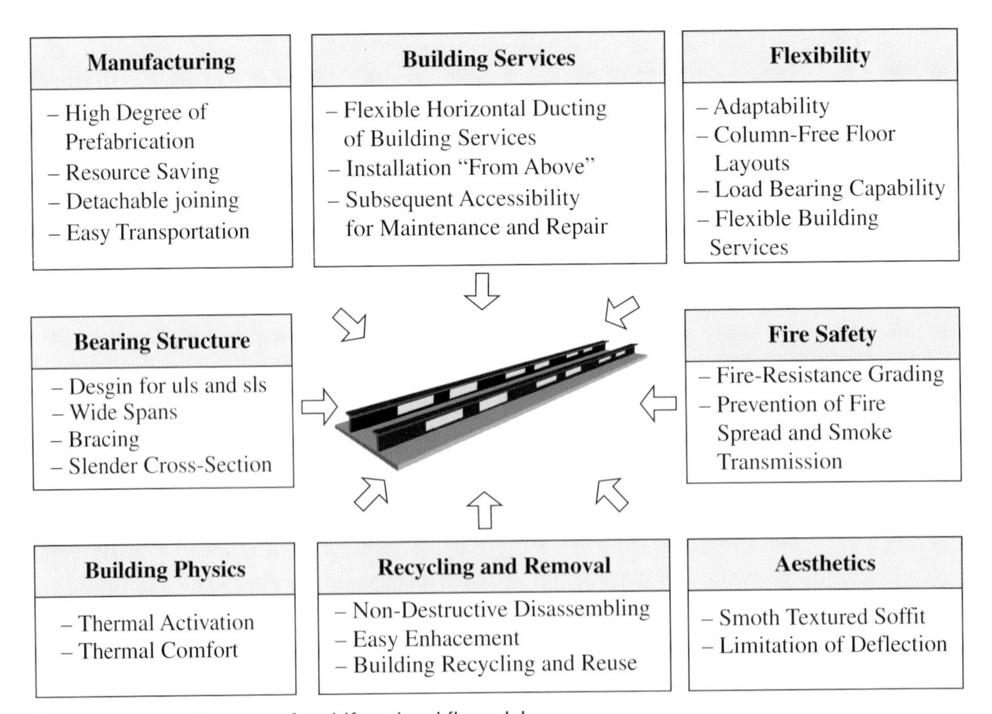

Figure 4.69 Requirements of multifunctional floor-slabs.

Based on the above-mentioned requirements, the following characteristics for definition of sustainable floor systems are

- provision of reserves in loadbearing capacity, large spans and avoidance of intermediate supports in order to achieve maximum flexibility of use;
- integration of building services in the loadbearing structure with access preferably from the top of the floor slab to its installation cavity;
- high degree of prefabrication and use of detachable steel connections for ease of extension, short building times, component reuse and building recycling;
- possible use of continuous concrete slab that is exposed to the room space for thermal capacity and to ensure fire safety of the structure.

Figure 4.70 is an example of the integrated and multifunctional floor-slab system InaDeck.

Spans of up to 16 m and a service load of 5 kN/m^2 can be achieved and the system offers a high degree of flexibility of use. The composite cross section consists of T-shaped steel profiles, which are attached to a slender prestressed concrete slab of 10 cm thickness at the bottom side of the floor system. The transfer of shear forces at the steel-concrete joint is performed via puzzle-shaped shear connectors (see [77], [79]) that are cast directly into the web of the section.

The slab system features an internal cavity, which enables the complete integration of building services in the loadbearing structure. Large web openings in the steel profiles allow for a variable and adaptable service-routing for ventilation ducts of large diameter. Ease of access of the installation floor and a convenient installation and maintenance of the components from above is achieved by the use of removable cover panels placed on top flanges of the steel profiles.

The overall height of the floor-slab structure including panels and floor covering does not generally exceed 50 cm, as all building services are integrated in the floor depth. In this way, increased space efficiency is obtained compared to conventional buildings in concrete construction that may feature downstand beams and suspended ceilings for the installation of building services.

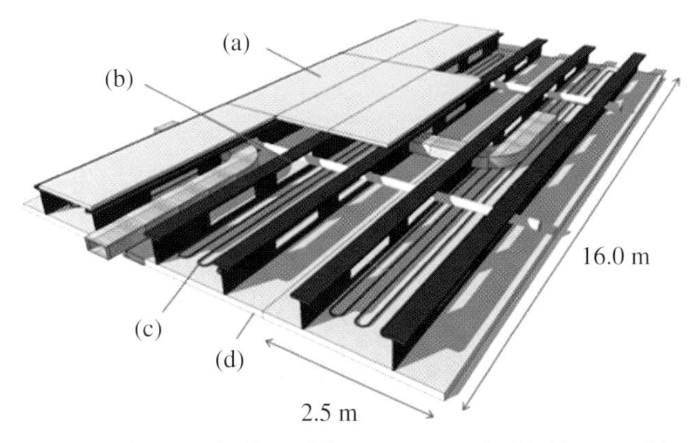

Figure 4.70 Integrated composite floor-slab system InaDeck with (a) removable cover panels, (b) large web openings, (c) integrated cooling lines, (d) concrete slab.

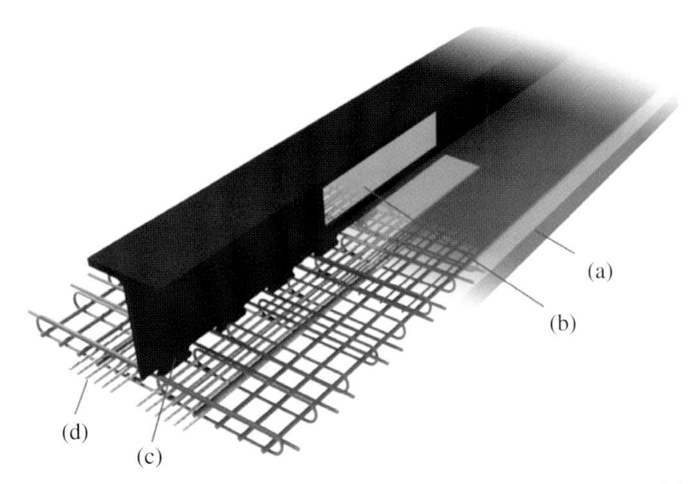

Figure 4.71 Features of the floor-slab section: (a) concrete slab in the tensile zone, (b) large web opening, (c) puzzle-shaped shear connector, (d) prestressing tendons in pretensioned concrete.

To improve the thermal comfort and energy efficiency of the enclosed living/working areas, the ceiling system is equipped with integrated heating and cooling networks for thermal activation. The concrete slab thus determines the thermal performance of the floor-slab system.

A slab thickness of 10 cm fulfils the fire-safety requirements in the event of fire exposure from the underside. Results of numerical fire simulations show that the ventilation conditions within the floor do not allow for fire spread and lead to extinguishing of the fire.

The integrated slab system allows for an easy dismantling and reuse of single components as well as for reuse by means of detachable connections and joints.

To ensure the desired multifunctionality of the integrated floor-slab system, a balanced ratio between static-structural aspects and various other requirements is obtained. This unconventional design includes the following additional design features (see Figure 4.71):

a) Concrete slab in the tension zone: In order to prevent excessive cracking, increase its stiffness and minimize its deflection, prestressed concrete is used.
b) Web openings for integration of building services: Unlike conventional composite beams, the multifunctional composite floor system has large openings in the web of the steel profiles. These openings significantly influence the shear force and deformation behaviour. The combination of composite beams with web openings and composite dowels is a further innovation.
c) Composite dowels as shear connectors: To transfer shear forces between the steel web and the concrete slab, puzzle-shaped composite dowels are used. For conventional composite structures the structural behaviour of composite dowels has been investigated under both static and cyclic loading (see [77]–[80]). However, their use in slender, prestressed or cracked concrete slabs is a new solution [81].

Experimental verification: To evaluate the global loadbearing behaviour of the integrated composite floor system, 21 beam tests were performed. The high local

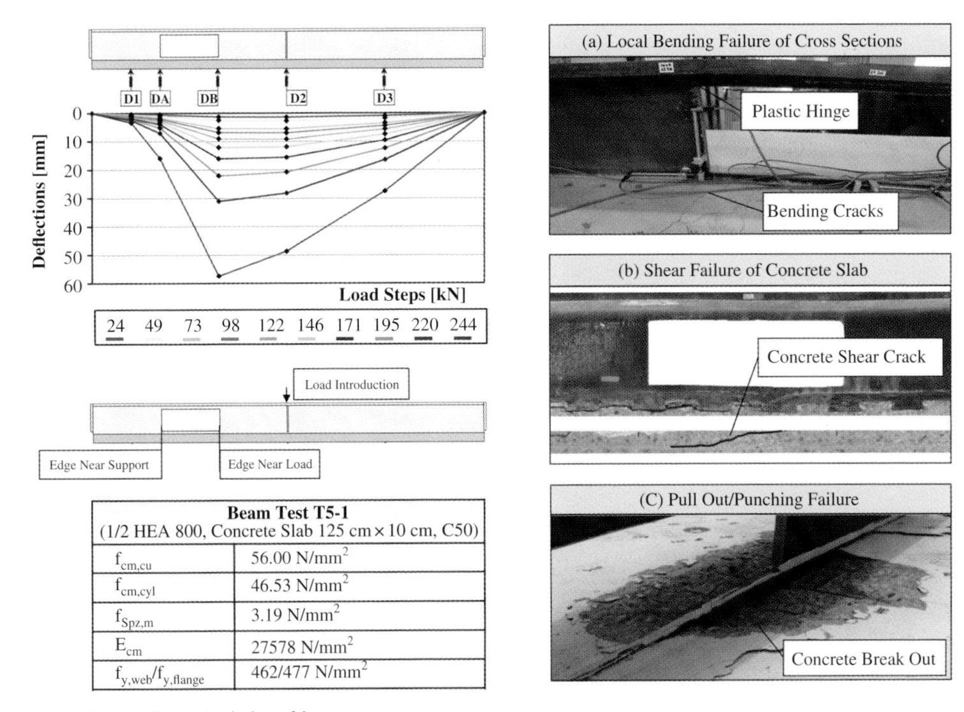

Figure 4.72 Characteristics of beam tests.

forces at the opening edges required additional tests of the composite dowels concerning their local pull-out, punching and shear behaviour (about 40 additional small-scale tests).

The composite beam tests were aimed at gaining knowledge about the load-bearing and deformation behaviour of the floor system with web openings under shear and bending. All beam tests had the cross sections of a half-slab element and were tested with loading at mid span (span length 5 m) (Figure 4.72).

At the web openings, local moments occur, which affect the stresses acting on the beam. The deflections of the composite beam will increase significantly, particularly at the edge of the opening (Figure 4.72, top left). Furthermore, shear forces from the perforated steel profile are redistributed to the concrete slab, which leads to shear cracking of the concrete slab. In addition, the redistribution causes pull-out (opening edge close to support) and punching stresses (opening edge close to load) in the composite dowels underneath the edge of the opening (Figure 4.72, bottom right).

The tests presented in [82] show that the required spans for a flexible utilisation are realistic and feasible. Based on the tests in combination with accompanying numerical parameter studies, a structural model can be derived.

Qualitative evaluation of the floor slab system with regard to sustainability criteria

Environmental quality

An investigation of the environmental benefit of flooring systems with long spans [83] showed, that the integrated composite floor system leads to a more slender load bearing structure, lighter columns and foundations. So the flooring system

Amstelcampus Boerhaave Complex

Location:	Amsterdam, Netherlands
Architects:	KPF and Studio V
Building description:	Dual use of land and optimal adaptability for multifunctional student housing.
Steel details:	The lightweight floor system allowed construction of residential units on top of the sports facilities.
Sustainability:	The 27 m^2 units can be merged into larger apartments in the future with relative ease.
Awards:	Winner of the Netherlands Sustainability Prize.

Figure 1 Amstelcampus Boerhaave complex, Amsterdam. © Slimline Buildings.

Example provided by Slimline Buildings.

yields high resource efficiency throughout the whole building structure resulting in favourable carbon footprint.

Flexibility and convertibility

The floor slab system InaDeck provides a high level of flexibility and adaptability during the whole life cycle of a building. The load bearing structure with spans of 12-16 m allows for column free and variable floor plans for different office arrangements and residential layouts. With variable arrangements of internal walls in lightweight construction (dry walls), a flexible and multifunctional space structuring is possible. By integrating the building services in the flooring structure and avoiding use of suspended ceilings, the construction height is reduced significantly. Hence, the usable space increases with increasing area efficiency of the building.

La Fenêtre

Location:	The Hague, Netherlands
Architect:	Rudy Uytenhaak
Building description:	Prestigious apartment building known as 'The Window on The Hague'.
Steel details:	The steel structure is largely visible from the outside of the building.
Sustainability:	Freedom of layout and adaptability – buyers were able to adapt their homes on the drawing board.
Awards:	Received a grant from the Dutch Ministry of Housing for being an example of an Industrial, Flexible and Deconstructable (IFD) construction project.

Figure 1 La Fenêtre, The Hague. © Slimline Buildings.

Figure 2 Construction site – the steel structure is visible. © Slimline Buildings.

Example provided by Slimline Buildings.

Steekterpoort

Location:	Alphen aan den Rijn, Netherlands
Architect:	Blok Kats, van Veen Architecten (BKVV)
Building description:	Zero energy and very slender bridge control center with thermal activation.
Steel details:	Round, nonsegmented elements for the front side of the facade.
Sustainability:	Slender exterior, lightweight construction, energy-efficient climate system.
Awards:	Nominee for the Dutch National Steel Prize 2014.

Figure 1 Steekterpoort, Alphen aan den Rijn. © Slimline Buildings.

Figure 2 The Slimline floor system enables the lean appearance of the building.© Slimline Buildings.

Example provided by Slimline Buildings.

Construction process

This composite flooring system is prefabricated without any need for an additional top concrete layer. The high grade of prefabrication allows an economically efficient production of the elements for continuously high quality. Complex formwork and reinforcement works are avoided. The prefabricated elements are safe and fast to install on site and lower the risk of delays in the construction process, reduce logistic steps, and enable "just in time" processes. Furthermore, these production methods contribute significantly to a low waste and avoid, noise, and dust on the construction site.

Ease of maintenance

All of the components relevant for maintenance of the primary construction of the integrated floor slab systems are easily accessible. Due to the removable plates on the upper side of the structure, selective and systematic maintenance can be achieved. The simple realisation of maintenance measurements decreases the costs in use, increases the lifetime of the components and secures the recycling of the load bearing structure for later purposes.

Ease of dismantling, deconstruction and end of life

By using prefabricated elements and detachable connections and bolted connections, the integrated flooring system allows for separability of each component. So at the end of a life cycle, the floor slab system allows for down cycling, recycling as well as reuse.

4.12.2 Slimline floor system

Chris Oudshoorn and Ger van der Zanden

The Slimline floor system is an innovative constructive floor system that integrates the ceiling, installation space and subfloor. The solution is based on I-profile steel beams spanning from loadbearing wall to wall, which makes the floor system self-supporting. The beams contain a standard pattern of openings, and generally two to three beams are integrated, by the lower flanges, into a thin concrete ceiling slab. The beams and slab are combined in prefabricated elements that are topped with an upper floor layer of choice. The floor elements are walkable immediately after installation. All service ducts and pipes can be easily integrated in the floor and are accessible and modifiable from above.

This system significantly reduces material use and is thinner and lighter than conventional floor systems. Various features can be included to meet client demands, such as thermal activation, acoustic strips and services for high user comfort and energy efficiency. Prefabrication facilitates a fast building process, and the voided system provides for adaptability of the building and greatly improves its economic lifespan (see Figure 4.73).

Structural properties – steel beam: The loadbearing capacity of the floor systems relies on the use of integrated steel beams. The longer the span, the higher the

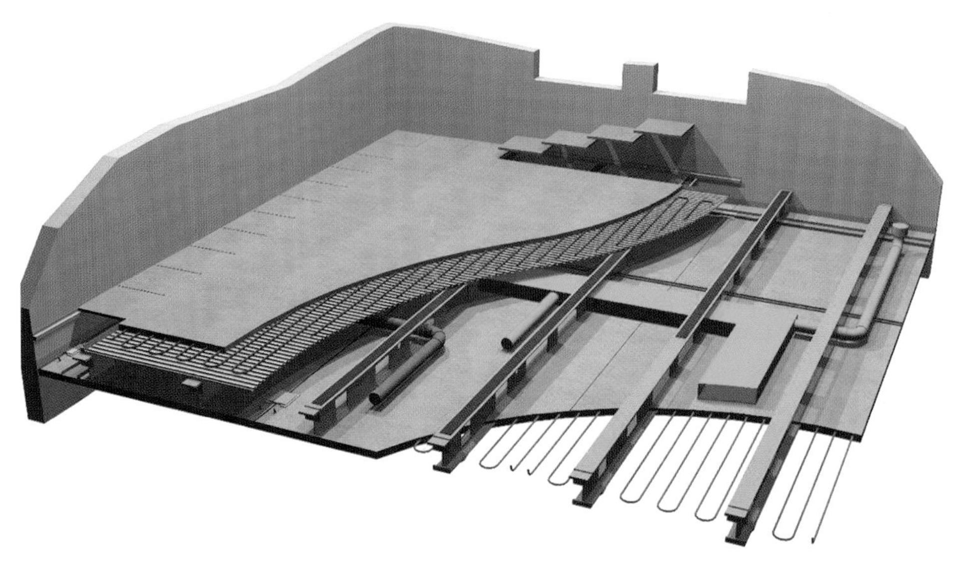

Figure 4.73 Impression of the floor system with double thermal activation (ceiling cooling and floor heating), profiled steel plate concrete screed subfloor and flex-zones along the wall and facade allowing floor access using removable tiles. © Slimline Buildings.

beam. The standard floor element uses IPE profiles positioned at a spacing of 1200 mm. It is possible to use heavier beams (HE) or reduce the beam spacing.

Opening pattern: The same opening pattern is applied to all beams and aims to offer as many openings as structurally possible. However, if a project calls for a different opening pattern, or the building's design leads to a local deviation from the standard pattern (for example near a vertical shaft), other options are available (see Figure 4.74).

Concrete slab: The precast concrete slab forms the ceiling to the room. It is factory finished and ready to be sprayed, or it can be left as concrete surface. The slab is walkable and supports the services. The concrete slab provides diaphragm stability for the construction. The slab provides the fire resistance of the Slimline floor system (more than 145 minutes) and contributes to the overall acoustic performance of the separating floor. The concrete slab can be provided with cast-in features or openings to allow passage of pipes and cables. It can also be fitted with strips of acoustic material to meet stricter acoustic requirements.

Subfloor: The subfloor covers the beams and acts as the walking surface. A number of subfloor options are available, all offering floor heating possibilities, as described below:

1 Profiled steel plate concrete screed: A profiled steel plate with concrete screed is installed on top of the steel beams in order to create a floating subfloor. This subfloor is installed on site and is placed on rubber strips to separate it from the steel beams and make it float. Separating the subfloor is necessary for the acoustic performance.
2 Profiled steel plate concrete screed with flexible access zones: A floating screed on profiled steel sheeting might be perceived as inflexible with regard to access

Venco Campus

Location:	Eersel, Netherlands
Architect:	Van Lierop Cuypers Spierings Architecten
Building description:	Egg-shaped office building with exceptional Slimline floor system application.
Steel details:	Slimline allows the hollow floor to be utilized as plenum.
Sustainability:	Low self-weight and reduced material use; all installations are permanently accessible, which results in flexible floors layouts and optimal adaptability.
Awards:	BREAAM Outstanding (five-star) certificate.

Figure 1 Venco Campus Eersel. © Slimline Buildings.

Figure 2 Slimline floor elements with steel beams allow special building shapes, like the egg-shaped Venco Campus. © Slimline Buildings.

Example provided by Slimline Buildings.

Figure 4.74 The hole pattern in the Slimline floor system's steel beams accommodates all common building installations, for example in the Bruges police station (Belgium). © Slimline Buildings.

points in the subfloor. However, only a relatively small area of the subfloor has to be accessible in order to reach the main services within the horizontal shaft. 'Flex-zones' create accessibility in locations where maintenance has to be performed, or where data and electrical connections are modified. Flex-zones along the facade, in the server room or in the hallway offer a practical solution for maximum flexibility. Provisions for additional flexible access zones can be included to allow installation of new services in the future. An accessibility framework is placed on top of profiled steel sheeting that can be conveniently opened when applicable (see Figure 4.75).

3 Raised access subfloor Support studs and a grid structure are attached to the top flange of the IPE beams. The grid spans the beams and acts as a support for (generally) 600 × 600 mm acoustic floor tiles. This is the most flexible option, as every single tile can be easily removed for maintenance or modification.

System dimensions: The span of the Slimline floor system is variable and can be adjusted to the project specifications. Column-free spans of up to 16.2 m are possible. The available precast element widths are 2.4, 2.7 and 3.0 m.

The floor system consists of a 70 mm concrete slab, in which the lower flange of the steel beams is cast. The element thickness depends on the type of steel beam, which in turn is determined by the span and floor loads. A typical floor system with IPE300 beams including subfloor is about 400 mm high. Because a suspended ceiling is not necessary to house the building services, the Slimline system can save

Figure 4.75 Construction of the profiled steel plate concrete screed subfloor with flex-zones in the Bruges police station (Belgium). © Slimline Buildings.

up to 500 mm in floor height, which allows an extra floor for every six floors in the same building height. The floor system has a self-weight of 300–350 kg/m^2, which is only half of an equivalent concrete structure.

Resource efficiency – sustainable building considerations

In modern buildings, it is necessary to rethink the way services are integrated in the loadbearing structure. Today's building principles are mostly based on traditional solutions. As a result, service ducts and pipes are often hidden from view using a suspended ceiling in office buildings or by encasing the services in the concrete floor. Adding a suspended ceiling means every storey of the building has to be constructed higher than necessary, resulting in the loss of space. Pouring concrete over the ducts and pipes in the floor results in a heavy floor and more material use. More importantly, the owner or user cannot easily change the floor layout or change the purpose of the building. Moreover, how can user requirements be met in 10 years time, taking into account new types of service installations?

A high quality building that meets the needs of current and future occupants is the best investment from an economical perspective and in sustainability as well. The longer the building lifespan, the smaller the total environmental impact of the structure. The Slimline floor system enables the realization of high quality buildings that offer significantly higher comfort levels to the occupants. An additional advantage is the ability to achieve a full functional change (for example, from office to residential units) within the existing building with relative ease. This permits design of a building to accommodate both residential and office use,

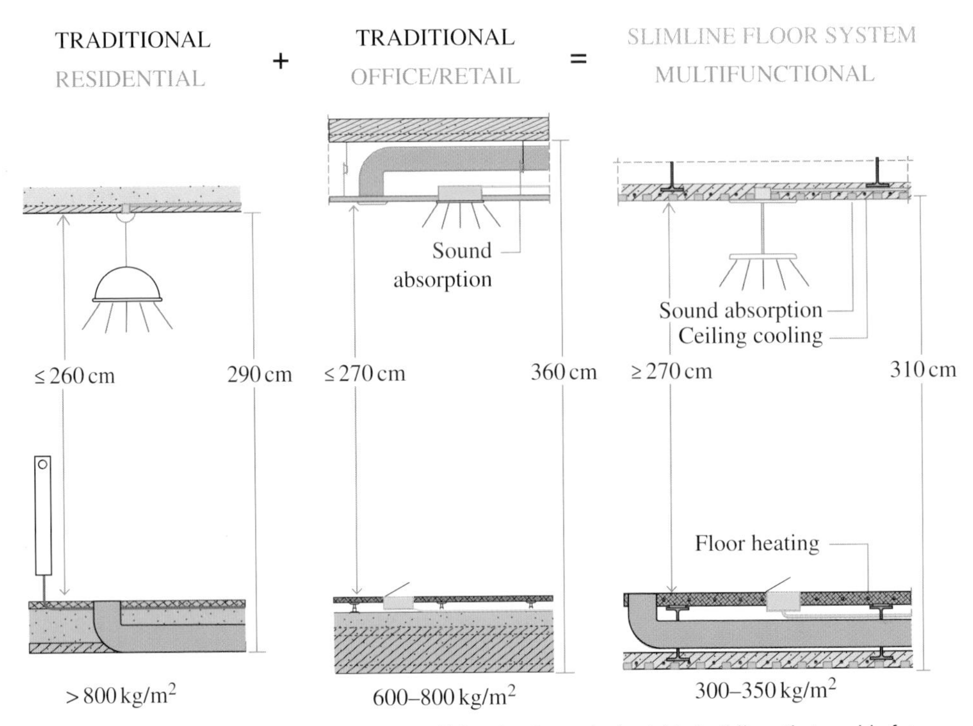

Figure 4.76 The Slimline floor system creates light, slender and adaptable buildings that enable future functional changes. © Slimline Buildings.

or a variable combination of both. The ability to change the function is primarily defined by the adaptability of the building's service installations (see Figure 4.76).

The beams for the floor system may be produced using recycled steel of S235/S275 quality. After dismantling at the end of life, they will be recycled or reused again. Lower amounts of concrete lead to weight reductions of over 50%, which means foundations can be minimized. In office buildings, substantial additional savings can be achieved on the application of expensive items like the facade, as the thin floor system with integrated installations saves building height. The floor system also consists of prefabricated elements, which minimizes the amount of waste and the risk of on-site errors.

Energy is preserved in different stages of construction and use. Lower amounts of concrete lead to a reduced CO_2 footprint and fewer transport movements, as well as potentially a smaller amount of on-site materials and equipment. The dry construction method saves the time traditionally required to dry out the poured concrete floors. For example, the Slimline floor system can be equipped with dual thermal activation, which leads to an exceptionally comfortable indoor climate in the occupied building while also offering savings in energy consumption. Running cool and warm water through the integrated pipes results in an optimal indoor climate without the air circulation of traditional air-conditioning units.

The challenges in real estate today show that 90% of the available buildings cannot accommodate the adjustments necessary for future building life cycles, primarily because of the inflexibility of the floor system used. Modern office buildings have to respond to changes in use through a high level of flexibility, which

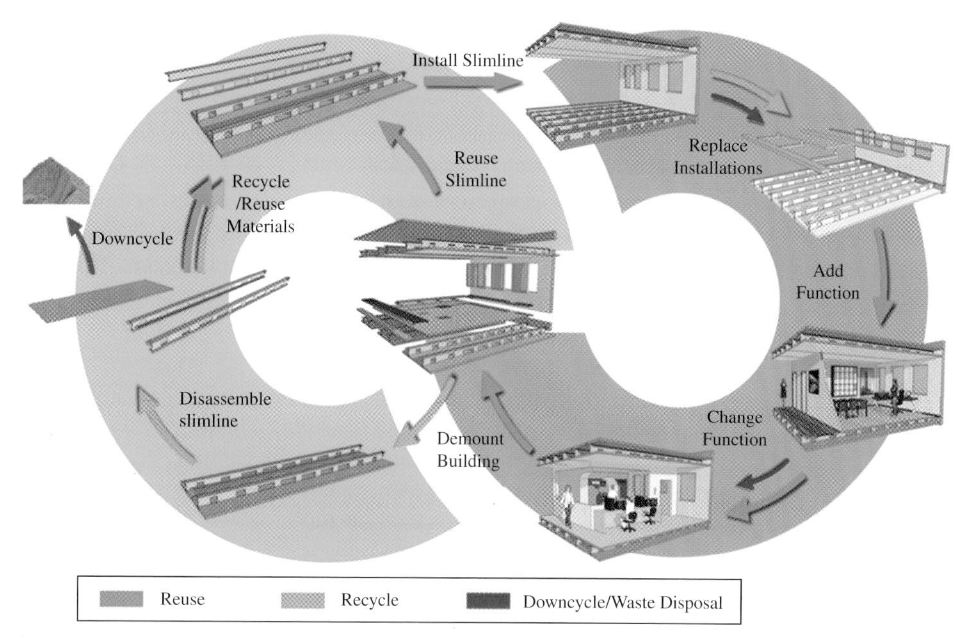

Figure 4.77 Slimline floor system aligns with the cradle-to-cradle philosophy. © Slimline Buildings.

decreases the risk of a vacancy and decreases the LCCs. When changing office buildings into residential buildings no significant modifications to the loadbearing structure are required. The floor-slab system provides a high level of adaptability during the whole life cycle of a building.

Building with such floor systems is a sustainable choice, in line with the cradle-to-cradle philosophy. Significant savings in material use, potentially reusable and recyclable floor elements, fewer transport movements and waste reductions result in an energy-efficient building with an extended lifespan. Several of these buildings have been awarded BREEAM 'Excellent' and 'Outstanding' certificates (see Chapter 5).

4.12.3 Profiled composite decks for thermal inertia

Markus Feldmann, Bernd Döring, Vitali Reger and Mark Lawson

The design of low carbon buildings requires measures to reduce the energy demand for comfort cooling and air conditioning, and therefore naturally ventilated or mixed-mode buildings. An important element to reach this goal is the thermal capacity of the building fabric. The concept of passive cooling in buildings through thermal inertia has increased in importance in recent years [87], [88].

The thermal behaviour of concrete flat slabs lends itself to a one-dimensional heat flow model and is included in standard procedures in EN ISO 13786 [89] and in various building simulation programs such as in [90], [91]. However, the treatment of slabs with ribs exposed to the room is not covered by these procedures. Therefore, more elaborate investigations are needed to determine the impact of these deck systems on the thermal behaviour.

Office H. Wetter AG

Location: Stetten, Kanton Aargau, Switzerland

Architect: Daniel Schwarzentrub, Wetter AG

Building description: The building represents typical office buildings for small and medium sized companies and fulfills their needs in regard to flexibility and cost efficiency. It is four storeys high and has a footprint of 30 × 9 m. The first floor is used for a large meeting room and the company´s restaurant. The three floors above are used as offices and meeting rooms for up to 45 employees.

Steel details: For the vertical structure a braced steel frame was used. The bracing was realized using tie rods, some of them placed in the staircase and lift structure, which is located in the front of the building. The slabs are made of prefabricated composite elements of the brand TOPfloor INTEGRAL. This construction method allows renunciation of intermediate columns, which makes the use of buildings most flexible. Due to the construction method very short mounting times of two days per storey are possible. The fire protection was realized by using intumescent coating.

Sustainability: Because of the low consumption of concrete and the high material efficiency of TOPfloor INTEGRAL, the composite structure of the building consumed less than half of the grey energy compared to a reinforced concrete structure. In addition to this, the absence of intermediate columns combined with exceptional easy accessibility to the building for equipment and installation allows a most flexible use.

Figure 1 Office building H. Wetter AG. Photographer: Felix Wey. © H. Wetter AG.

Figure 2 High flexibility thanks to renunciation of intermediate columns. Photographer: Felix Wey. © H. Wetter AG.

Figure 3 TOPfloor INTEGRAL elements with large openings for easily accessible installation. © H. Wetter AG.

Figure 4 Construction phase – rapid construction with prefabricated TOPfloor INTEGRAL elements. © H. Wetter AG.

Example provided by Martin Mensinger.

Figure 4.78 Left: cross section through a typical shallow composite slab. © Kingspan. Right: deep decking used in a long-span composite slab. © Tata Steel.

4.12.3.1 *Composite slabs in multistorey buildings*

Composite slabs and ribbed slabs are commonly used in office buildings, residential buildings, educational buildings, hospitals and car parks. These slabs consist of a profiled steel decking layer of 0.9–1.2 mm thickness supporting a concrete slab of 120–300 mm depth. There are two generic forms of composite slabs – a shallow deck slab in which the steel deck profile is 50–80 mm deep, and a deep deck slab in which the steel deck profile is 190–225 mm deep (see Figure 4.78).

In most buildings, electrical and mechanical services and a false ceiling are suspended from the composite slab, which means that in the majority of buildings, the decking is not exposed to the room air. However, there is now more interest in utilising the thermal capacity of the composite floor structure. This can be achieved either by exposing the metal surface directly (often by painting the metal surface) or by using a perforated ceiling that allows for free air movement.

4.12.3.2 *FEMs of heat flux in composite floor slabs*

FEM models were set up for composite slabs with various deck profiles by considering a sinusoidal variation of room temperature between 18°C and 26°C. The temperature distributions in the three types of composite floor slabs are shown in Figure 4.79 at four times of the day. The lowest temperatures are at around 6 h, and the highest room temperatures are experienced in the summer at about 18 h.

The total quantity of heat absorbed by the slab is the integral of the rate of heat transfer over the heating period. Figure 4.80 shows the result for the different composite slabs in comparison to conventional concrete flat slab.

The maximum heat flux was about 30 W/m^2 for all the composite slabs (except for the 90 mm slab depth), and the total heat stored over the daily cycle was between 200 and 220 Wh/(m^2·d). In comparison, the equivalent values for a solid slab were a maximum heat flux of 25 W/m^2 and a total heat storage of 175 Wh/(m^2·d). The heat storage in a composite slab is therefore 20% more than in a solid slab of 200 mm depth. Furthermore, the weight of a typical composite slab is less than half of that of a 200 mm deep flat slab, and so the heat storage per unit floor weight is 2.5 times more in a composite slab

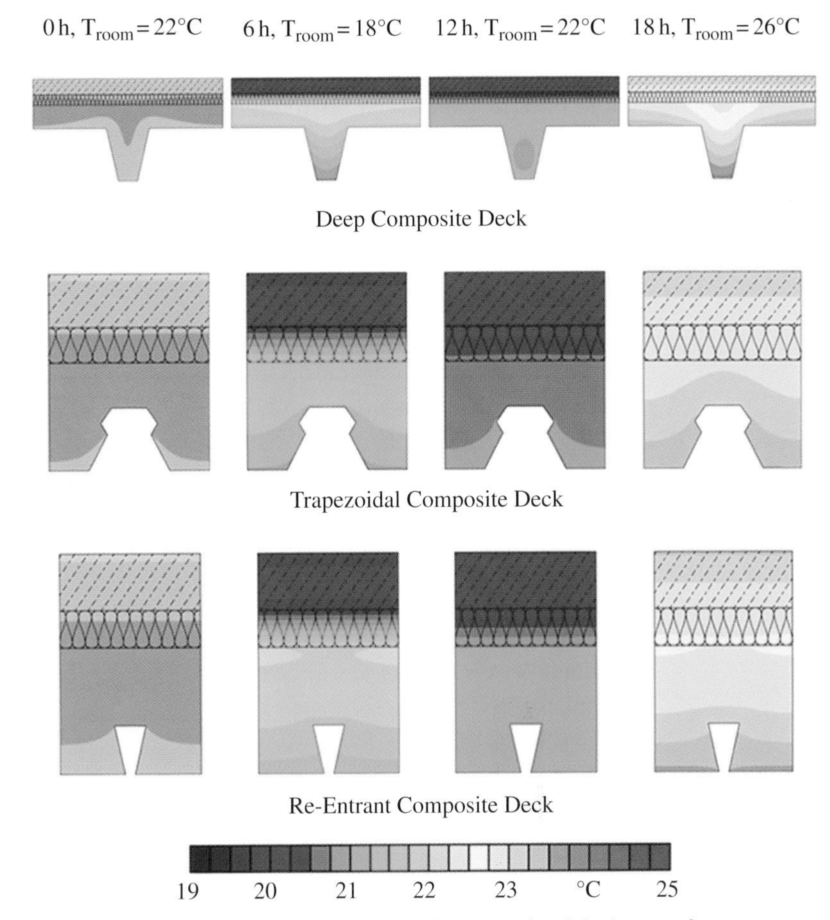

Figure 4.79 Temperature distribution within the composite slab elements from FEM calculations at four times of the day.

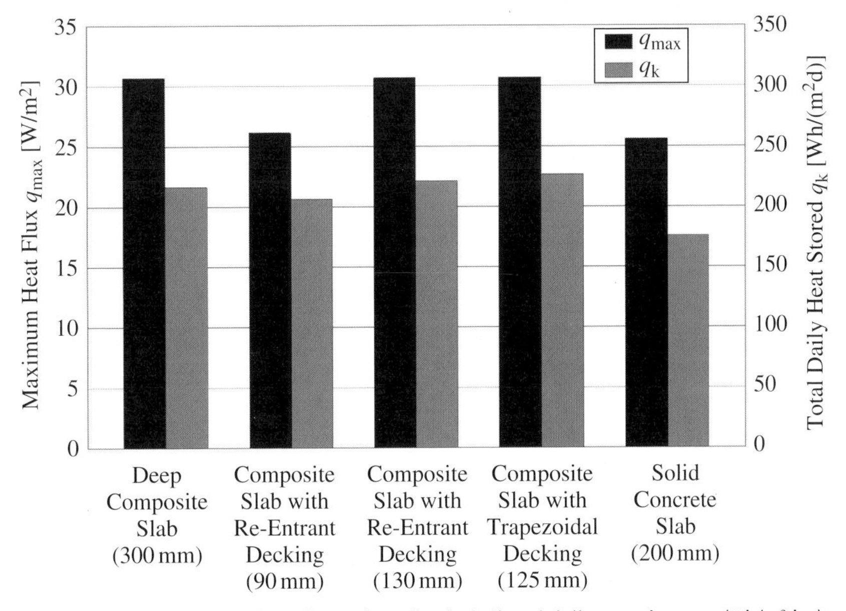

Figure 4.80 Comparison of maximum heat flux (W/m²) and daily stored energy (Wh/m²day).

4.12.3.3 *Application in building simulation methods*

Using the previously calculated thermal properties, it is possible to run a thermal building simulation over a whole year taking account of two climatic conditions in London and Berlin. The following example was made for a typical room (4 m wide and 5 m deep, on the south-facing side) in a representative office building.

The maximum room temperatures considering a flat concrete slab and a composite slab are similar, but in the temperature range between 23°C and 25°C, a positive effect of the improved heat transfer can be identified. For example, in a London climate, the number of hours above 25°C is reduced from 360 to 300 (or a reduction of 20%), as shown in Figure 4.81. The differences at 26°C and 27°C are very small. In the Berlin climate, the effects at higher room temperatures are more noticeable. For example, the number of hours above 26°C is reduced from 330 to 295 (or a reduction of 10%).

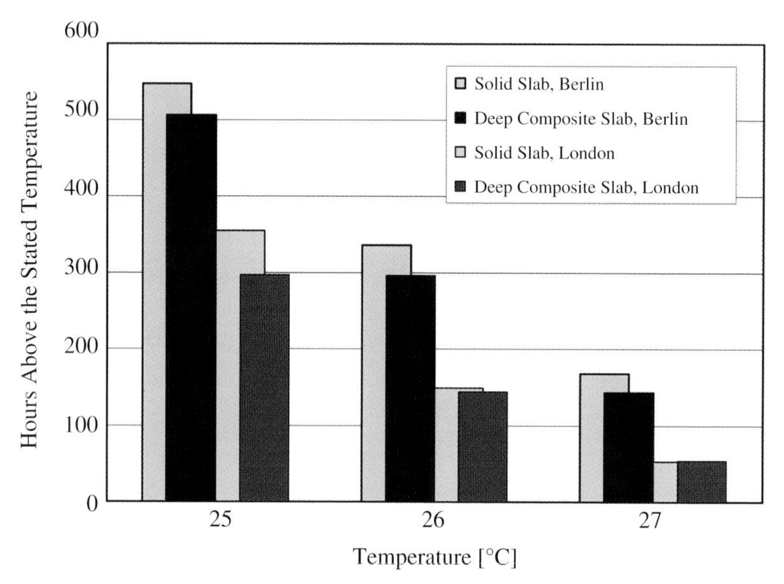

Figure 4.81 Number of hours with room temperatures above 25°C–27 °C for a deep composite slab and for a flat concrete slab (location: Berlin and London).

4.12.4 Thermal activation of steel floor systems

Markus Feldmann, Bernd Döring, Vitali Reger, Mark Lawson and Jyrki Kesti

4.12.4.1 *Introduction*

The integration of a piping system into the floor slab has become very popular in recent years. The option to improve the thermal comfort significantly with moderate additional investment costs and low energy costs are the main reasons for this development. For deck systems in steel, various solutions have been developed. Two examples are shown below.

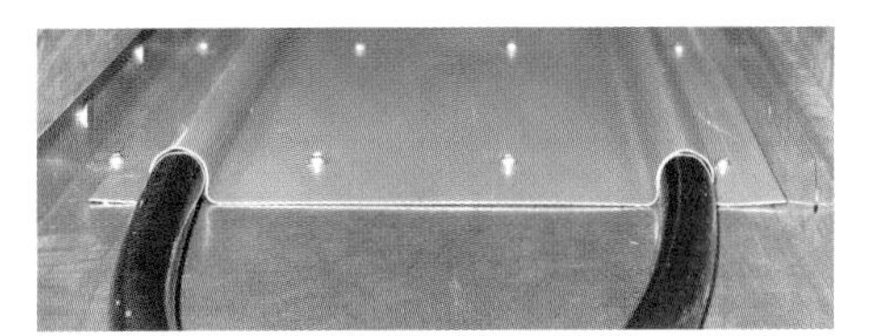

Figure 4.82 Left: test specimen, composite double layer floor. Right: Ω-profile used to contain the pipes.

Figure 4.83 Composite floor with embedded pipes.

4.12.4.2 Technical solutions

Double layer flooring element with integrated heating/cooling: Figure 4.82 shows a prototype of double layer flooring element with integrated radiant heating/cooling. The 200 mm deep double layer flooring system spans up to 7 m between secondary beams and has a 100 mm concrete topping. Certain ribs in the decking system are concrete filled and reinforced as a series of T-sections (Figure 4.82, left). The integration of the piping system is shown in Figure 4.82, right. The decking system may be prefabricated with its services installed in the factory, but the concrete will generally be placed on site [93].

Composite floor (deep deck) with integrated heating/cooling system: Figure 4.83 shows a composite deck system with trapezoidal steel sheets and an embedded water piping system, which is placed on site.

4.12.4.3 *Performance data*

The performance of the double layer flooring element with integrated radiant heating/cooling was tested. A FEM analysis showed (Figure 4.84, left) that the thermal contact from the pipes to the lower steel sheet is essential, and therefore the pipes are fixed by an omega-profile (Ω, see Figure 4.84 left, bottom). Using this detail, the temperature distribution as shown in Figure 4.84, right, was obtained by thermal modelling. The cooling capability of this solution amounts to 35 W/m^2 at a temperature difference of 8 K between the room and cooling water.

Figure 4.85 shows the thermal performance of a composite floor using a deep trapezoidal steel sheet, in which the pipes are parallel to the deck ribs. This solution has a cooling capability of 50 W/m^2 at a temperature difference of 8 K between the room and the water in the pipes [94].

4.12.5 Steel decks supporting zero energy concepts

Jyrki Kesti, Bernd Döring, Vitali Reger, Markus Kuhnhenne and Mark Lawson

The zero energy approach requires a combination of energy conservation and energy generation techniques. The focus will be on systems where the building fabric and structure participates actively in the energy balance of the building and therefore reduces the building's energy demand. In different European countries, demonstration projects have been realised that reach the nZEB-approach (nearly zero energy building) to a certain level [94], [95]. To be successful, a combination of three main components is essential:

- optimising the building envelope for reducing heat transmission and providing the optimum of solar gains and daylight;

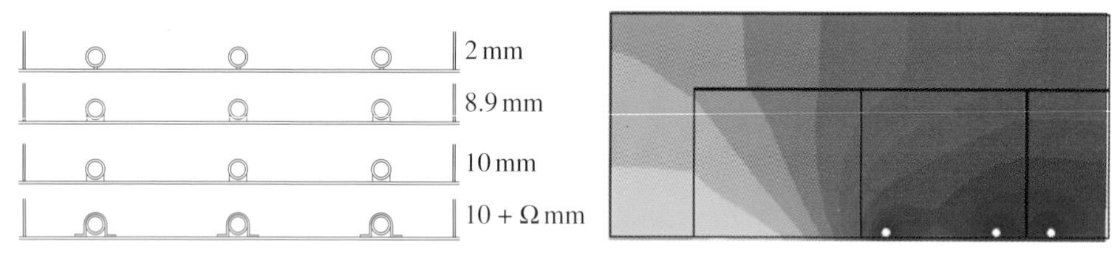

Figure 4.84 Numerical testing of double layer flooring element with radiant heating/cooling.

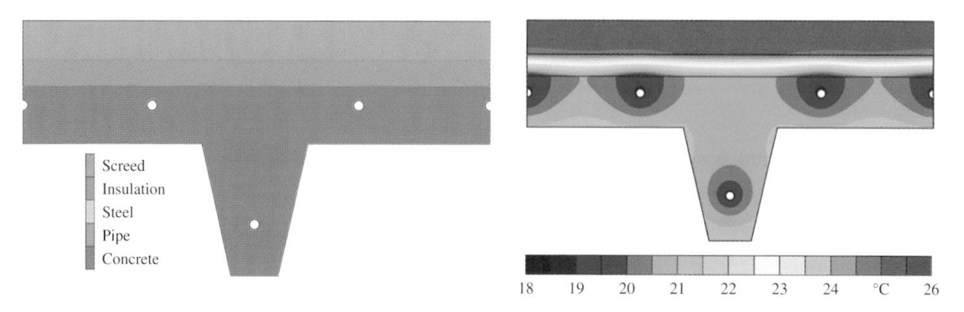

Figure 4.85 Numerical testing of composite deck with trapezoidal sheet with radiant heating/cooling.

- efficient heating and cooling technologies, including thermal storage (focusing on building fabric and ground heat exchanger);
- integration of active solar components (focusing on PV elements for generating electricity).

In particular for mechanical system and the heat storage, new solutions were developed and investigated. An alternative structural frame solution using the innovative double deck system (see Section 4.12.3) was developed and was combined with steel energy piles that act as the foundation of the building but also provide for ground heat exchange or storage.

4.12.5.1 *Reference building*

The investigations of the energy performance of a multistorey office building were made using a real building as a reference plan form. This building is placed in Jyväskylä, Finland. Figure 4.86 shows the ground plan of the building, which has six storeys and a net floor area of 9400 m^2.

This building was 'virtually' placed in three different European climatic locations: north (Helsinki), west/maritime (London) and south (Bucharest). For the thermal simulations, weather datasets on an hourly base were taken for these three sites.

The relevance of the facade design (and generally the building envelope in total) on the energy performance of a building is evident. This includes the heat transfer coefficient (U-value) but also aspects of solar gains (positive in heating period, negative in cooling period), daylighting and ventilation. For reaching the zero energy target, ambitious values for the relevant parameters were defined, depending on the local climate (Table 4.14).

In the energy pile, a vertical heat exchanger system is installed inside the pile. The energy pile system works in a similar way as a traditional ground heat source and can be utilized for heating by use of a heat pump, and for cooling either with free ground cooling or with use of a chiller.

For the study, the six-storey office building (see Figure 4.86) has 50 energy piles all with a diameter of 220 mm and length of 30 m.

For the determination of energy consumption, simulations with the thermal building simulation tool TRNSYS were performed. The challenge was the optimisation of the energy supply system for the nearly zero energy solution, at which a preselection of the components was made:

- ground source heat exchanger in the form of energy piles, as a heat or, alternatively, cold source for the heat pump and the chiller;
- solar collector, mainly domestic hot water (DHW), if useful – this is also for heating or reinjection of heat into ground source heat exchanger;
- heat pump for panel heating, AHU and DHW;
- chiller for air handling unit;
- free ground cooling through energy piles to radiant panels when possible;
- auxiliary heater (electrical), if heat pump and solar collector cannot fulfil the demand;
- three thermal storages (heating, DHW, cooling).

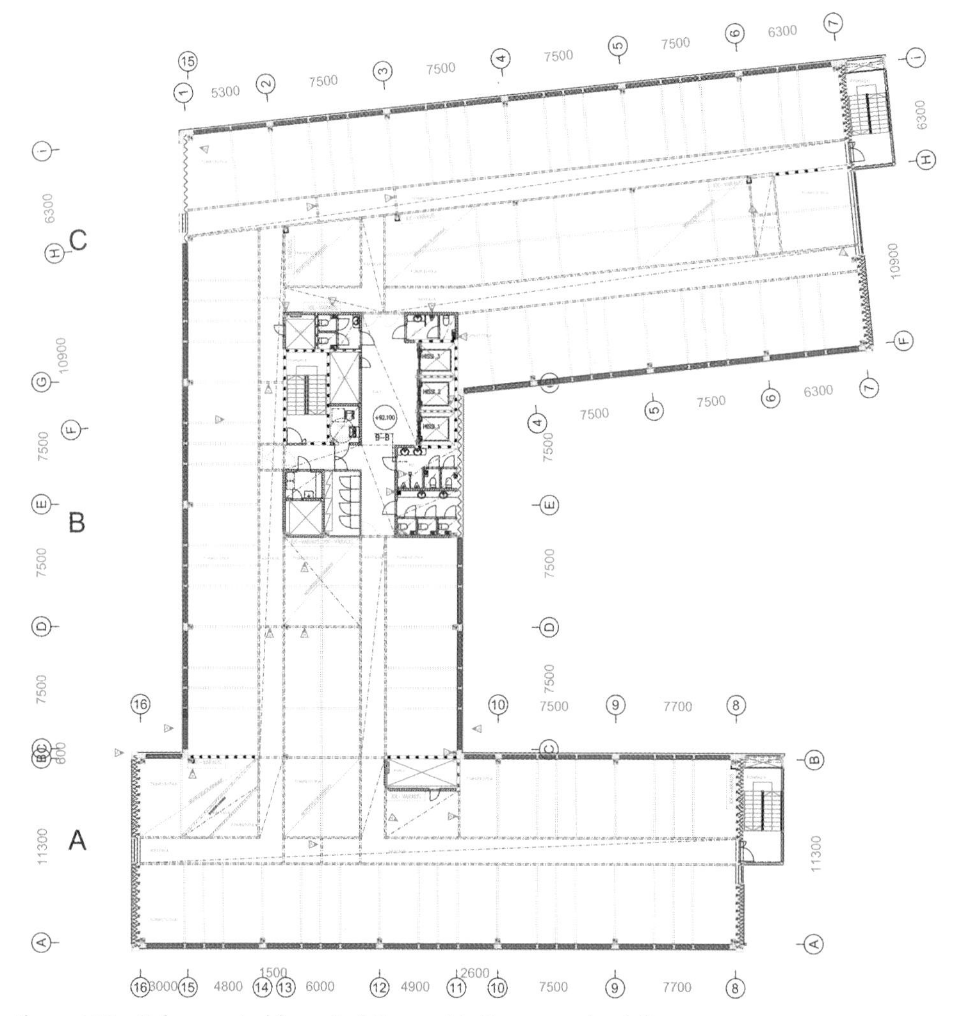

Figure 4.86 Reference steel frame building used in the energy simulations.

Table 4.14 Main characteristics of buildings, depending on location.

Component	Helsinki	London	Bucharest
Wall U value, W/m^{2}K	0.12	0.1	0.15
Roof U value, W/m^{2}K	0.09	0.1	0.1
Floor U value, W/m^{2}K	0.1	0.15	0.25
Air-tightness, n_{50} [1/h] or q_{50} [m^3/hm^2]	n_{50}: 0.5	q_{50}: 2.0	n_{50}: 0.6
Window glazing U value, W/m^{2}K	0.45	0.5	0.45
Window g value	0.24	0.2	0.24
Ventilation air flow control type	VAV	VAV	VAV
Heat recovery	0.85	0.85	0.80

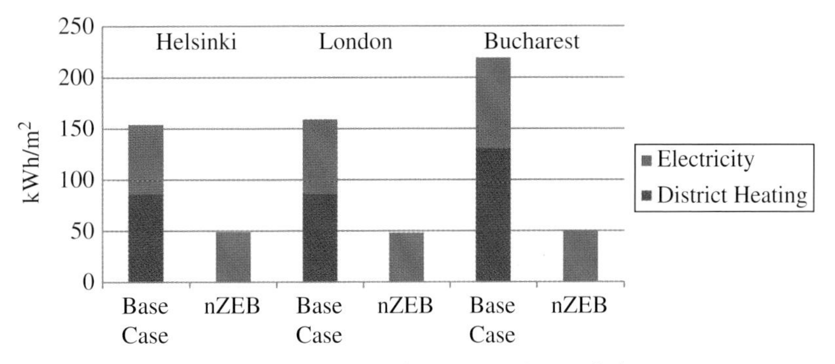

Figure 4.87 Final purchased energies of the base case and nZEB designs.

The energy production, especially using renewable sources for the zero energy solution and their influence on the energy consumption, was investigated in the three locations. The size of the solar collector and the capacity of the heat pump and the chiller were optimised for the different climates.

The structure, floor slab and MEP (mechanical, electrical and plumbing) routings of the nearly zero energy solution enhance the efficiency of the building. The small energy demand of the optimised solution maximises the share of renewable energy from the energy piles. According to the energy simulations, it is possible to reduce the heating energy consumption with an optimal nearly zero energy solution by more than 90%, the cooling electricity by 60%–70%, and electricity use by 40% in each of the locations compared to a 'base case' of an office building with a more conventional design. This leads to a reduction of the purchased energy of 68% in Helsinki, 70% in London and 77% in Bucharest (see Figure 4.87). For the three sites, a use of 50 kWh/m^2 (electricity) per year remains for the nZEB case, in which the demand for heating, cooling and ventilation amounts to only about 10 kWh/m^2 per year (for all investigated climates).

The yearly generation of electrical energy through the PV panels may be expressed per unit floor area and is 18.5 kWh/m^2 (for Helsinki, as an example). Almost all of the generated electricity is consumed by the building rather than being fed into the public grid. This generation represents a notable part of the energy consumption of the building. The energy demand for HVAC can be covered, hence the nZEB-approach is fulfilled.

4.12.6 Optimisation of multistorey buildings with beam-slab systems

Richard Stroetmann

In order to optimise the design of supporting structures, the important components of the supporting structure have to be considered together because the optimum designs of the individual elements may not necessarily correspond to the overall optimum (see [29]). For example, by reducing the number of columns in multistorey buildings, the column loads are higher and the costs with respect to the load-carrying capacity are lower. However, this is also associated with longer beam spans and heavier cross-sections. By using centre columns, the span of the floor beams is reduced, but it may be necessary to provide additional downstand beams in the longitudinal direction of the building, which increases the number of beam connections and thus the workshop and installation time.

Figure 4.88 shows four sections of floor plans of buildings with different arrangements of beams and columns together with the direction of span of the floor slab. For the variants without internal columns (top half of figure), the beams span between the outer columns as a single-span system. If the beam spacing is smaller than the column spacing, edge beams are needed to transfer the load to the columns. If the spacing is the same, the floor beams can be connected directly to the columns so that edge beams are not required (Figure 4.88, right). For deeper building plans, rows of internal columns are used to reduce the beam spans (Figure 4.88, lower half of figure). To create the necessary free space for a central corridor and to permit various forms of office arrangements, an asymmetrical arrangement of the columns is often appropriate (see [29], [30], [96]).

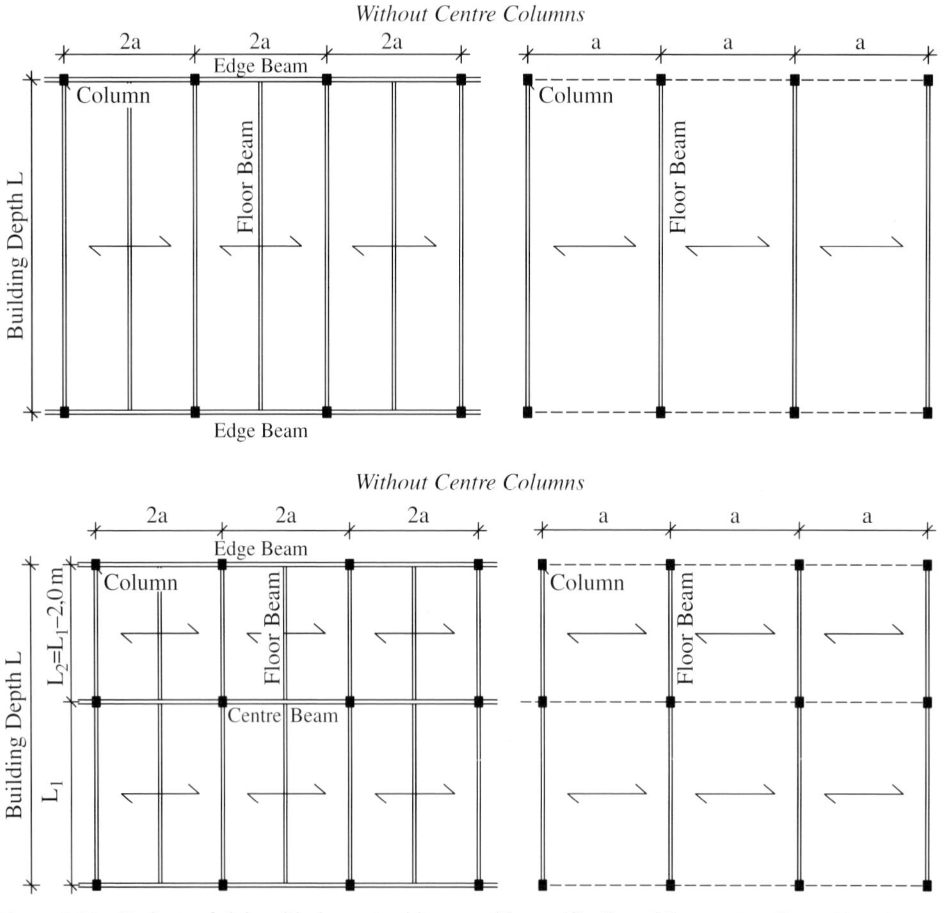

Figure 4.88 Variants of slabs with downstand beams with specification of the construction grid and components [36].

Parametric studies were carried out [29] in order to establish the use of materials associated with the different structural variants and which construction grid and system designs have the most positive effect on sustainability. Superimposed dead loads and live loads of 1.5 and 3.0 kN/m² (characteristic values) were used in the calculations. Deflections were limited to L/250 under total loading, taking account of precambering of the steel beams (see [30]).

IPE sections in steel grade S460 were considered for the beams (see [30], Figure 4.89). Concrete grade C30/37 was used to design the reinforced concrete slab. Construction grids with a = 2.4 m, 3.6 m and 4.8 m were included in the study. The column spacing was initially chosen to be twice the grid dimension 'a' (Figure 4.88, left). This also corresponds to the spacing of the edge and centre beams in the longitudinal direction of the building. In addition, an investigation was made for a = 4.8 m for the same beam and column spacing in order to be able to eliminate the centre and edge beams (Figure 4.88, right). The normal forces determined for a five-storey office building with a height between floors of 3.5 m were used to design the steel columns, and HE-type steel sections of the strength class S460 were chosen.

Figure 4.89 shows the weight of steel sections (in kg) needed for a floor area L × 2a of the variants shown in Figure 4.10 for building depths (spacings of outer columns) between 10 m and 16 m. By considering 4 × 6 different grids, the graph presents a total of 24 variants. A differentiation is made between the steel needed for the floor, edge and centre beams and for the columns. It is apparent that the weight of the beams increases with increasing building depth and beam spacing. This is particularly clear in the long-span systems without centre columns (blue and black bars). However, for the variants with internal columns, the costs for the floor beams are lower because of the shorter spans, but the costs for the edge and centre beams are higher. This also applies to the columns because the loads from the area considered are supported by three columns instead of two. For the variants without edge or centre beams (see columns for 4.8* in spacing in Figure 4.89), whereas the steel required for the beams remains the same (compare the results for 4.8* with those for 4.8 in Figure 4.89), for the weight the columns increases because twice the number of columns is used to transfer almost the same load. However, the total steel requirement is lower because the edge and internal beams are eliminated.

The requirement for constructional materials for an area of L × 2a is given in Figure 4.90 (divided into structural steel, reinforcing steel and concrete). It should be noted that the ordinate values for concrete have to be multiplied by a factor of 10 to obtain the requirement in kilograms. As the thicknesses of the slabs are only dependent on the beam spacing a, the same masses are obtained for the concrete when the values of a and L are the same. The amount of reinforcing steel is given by the slab design, which includes the steel requirement for transfer of shear forces from the composite beams into the slab, the reinforcement needed to resist the hogging moments of the composite continuous beam (including the necessary ductility reinforcement) and the reinforcement needed to limit crack widths (w_k = 0.4 mm). Most of the reinforcing steel weight results from the design of the slab in bending, as it can be seen by comparing systems with and without central

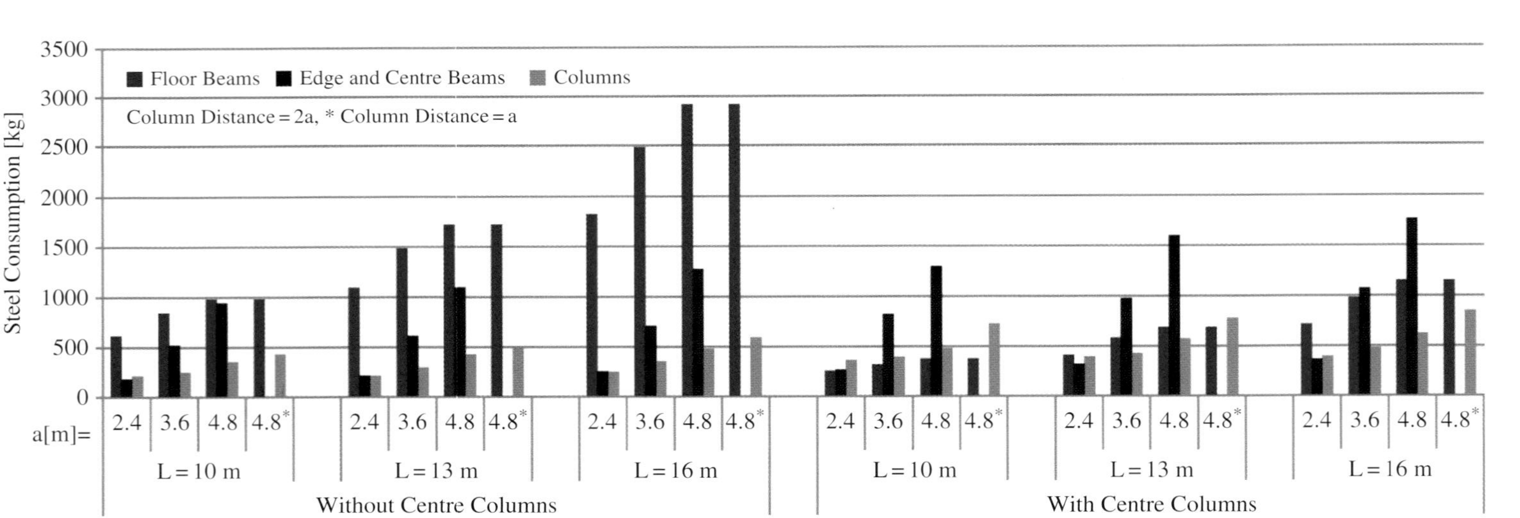

Figure 4.89 Steel consumption per slab (L × 2a) for selected variants of slabs with downstand beams [36].

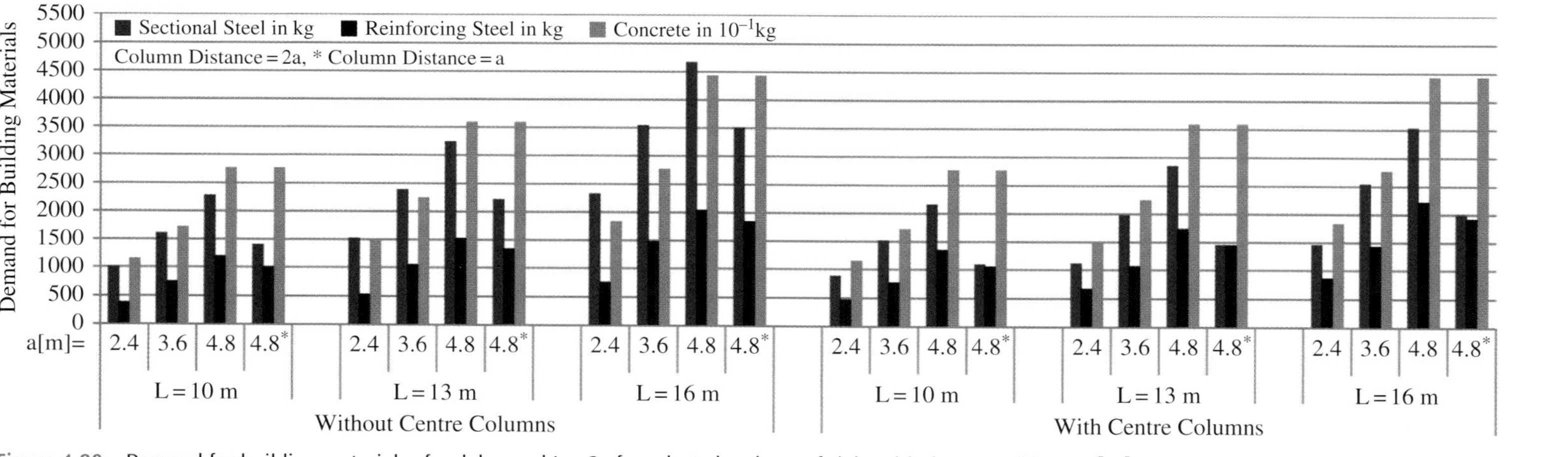

Figure 4.90 Demand for building materials of a slab panel L × 2a for selected variants of slabs with downstand beams [36].

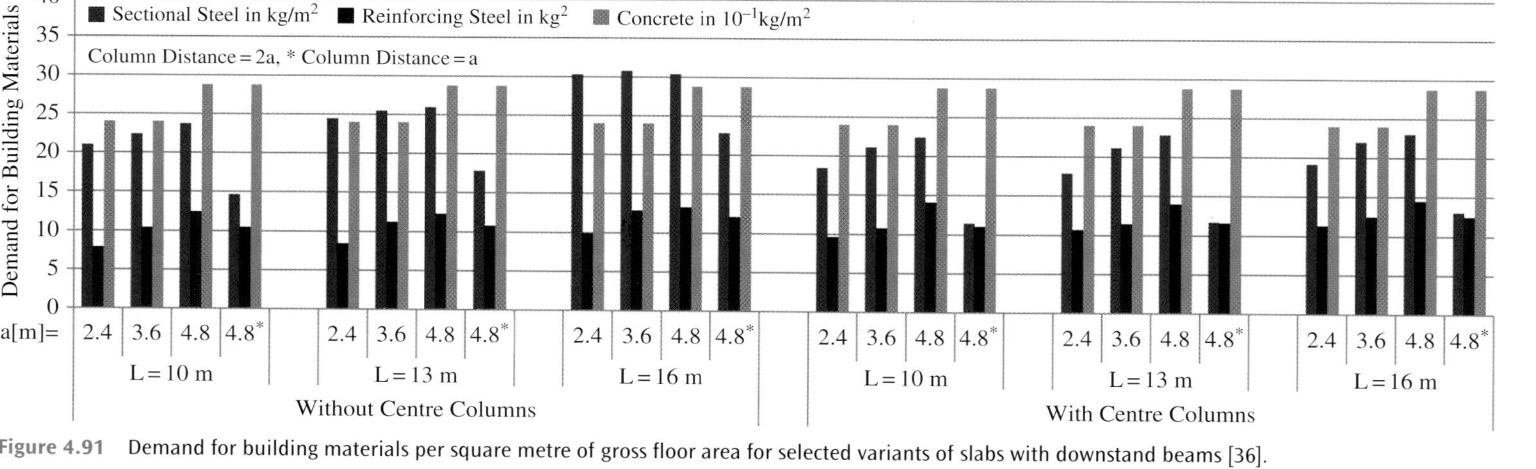

Figure 4.91 Demand for building materials per square metre of gross floor area for selected variants of slabs with downstand beams [36].

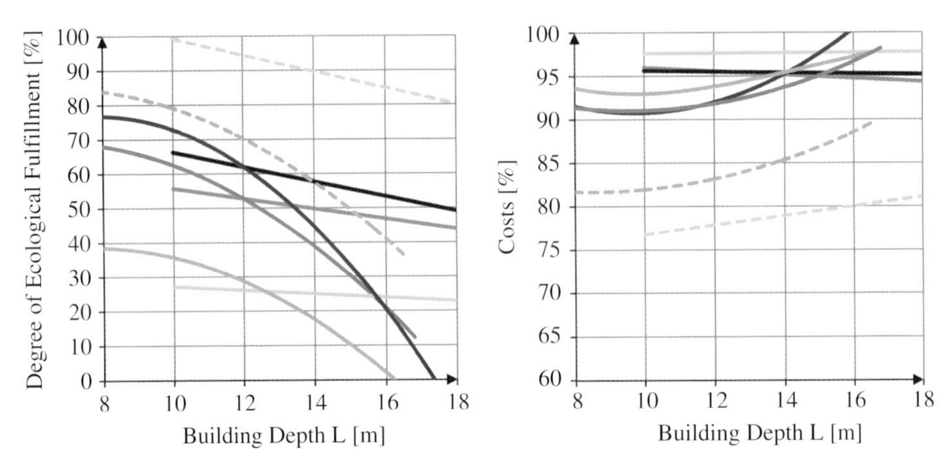

Figure 4.92 Degree of ecological fulfilment and costs of structural systems with downstand beams [36].

columns for the same values of a and L. The quantities of structural steel given in Figure 4.90 correspond to the sum of the amounts used for the individual components of the variants in Figure 4.89. By comparing systems with and without centre columns, it can be seen that with small building depths, the costs are similar, but for the variants without internal columns, disproportionately more steel is required with increasing span, L. The variants with a = 4.8 m in which there is a row of columns in every supporting axis (Figure 4.88, right) require significantly less steel than the variants with the column spacing 2a = 9.60 m, where deeper edge and internal beams are needed because of the longer spans.

Figure 4.91 shows the quantities of constructional materials required per square metre of gross floor area. By relating the quantities to the area, a large part of the differences due to the column spacings in Figures 4.89 and 4.90 is evened out. The diagram only contains two different quantities for concrete per square metre in accordance with the two slab thicknesses for a = 2.4 m and 3.6 m and for a = 4.8 m (see [8]). For reinforcement, the quantities are between about 8 and 16 kg/m^2 floor area. The values for the systems with centre columns are higher than those for the systems without them. The structural steel requirement for the variants investigated is between 14 and 38 kg/m^2 floor area. The variants with centre columns in which the column spacing is the same as the beam spacing (Figure 4.88, to right in lower half) show the lowest figures and the variants without centre columns but with edge beams (Figure 4.88 to right in upper half) show the largest figures. For L = 10 m, the quantities are similar.

An evaluation of the impact of the variants on the environmental outcome and the percentage costs of the variants is given in Figure 4.92, not including the effect of the connections. The method of evaluation is described in Chapter 3.3. The blue lines represent the variants without centre columns. It can be seen that with increasing building depth, the ecological benefit decreases and the costs increase. The same applies for the variants with centre columns, but the values are more favourable above a certain building depth. The best results are obtained for the variants with centre columns when the column spacing is the same as the beam spacing (dashed lines in Figure 4.92).

4.13 SUSTAINABILITY ANALYSES AND ASSESSMENTS OF STEEL BRIDGES

Tim Zinke, Thomas Ummenhofer and Helena Gervasio

4.13.1 State of the art

4.13.1.1 Buildings and infrastructures

Sustainability assessment of buildings has been extensively discussed in the last 10 years, and a large number of assessment schemes have been developed. An overview of existing approaches is presented in Chapter 5. The outcome of a sustainability assessment for (office) buildings is often a label (for example, platinum, gold, silver and bronze) or a 'seal of approval'. These results can be used to communicate a suitable building standard and environmental and operational advantages to tenants or other stakeholders. Building users can benefit from lower energy consumption, thus reducing the operation costs. Also, the efforts and costs for the implementation of a sustainability assessment are used to verify the building value and to obtain a higher rent or selling price.

In contrast to buildings, bridges are a key part of the transportation network and have to ensure functionality and to provide for an uninterrupted traffic flow. As a consequence, they fulfil a supportive role to society and cannot be analysed in a detached manner from the network. If the network is affected by the complete or partial closure of a single bridge, high economic impacts can occur. Also, a long construction period or an intensive maintenance affects traffic. As a consequence, time delays of the infrastructure users, increased accident rates and air pollution occur. In comparison to office buildings, these dependencies require the expansion of the system boundaries, the adaption of the assessment indicators and the adjustment of the assessment scheme. The external effects most important for sustainability assessment of bridges and usually not considered within the assessment systems for buildings are visualised in Figure 4.93.

A further characteristic is given by the legal framework, as bridges are regularly planned, built and operated by a public authority whereby (office) buildings often belong to private clients. This is the same if an infrastructure component is operated within a public private partnership. Regulations, as specifications and targets for national traffic routes, environmental assessments and cost-benefit analyses have to be followed. As a result, a market for sustainable bridges does not exist, and awarding of an assessment label will not increase its value. Therefore,

New Bridge over Po River, Highway #9

Location:	Lodi and Piacenza, Italy
Client:	Anas spa
Structural and architectural design:	MCA Engineering – Prof. Eng. Michele Mele
Building description:	The 8000 t steel structure extends itself over 800 m and is supported by seven existing piles in masonry and three new ones, made of reinforced concrete.
Steel details:	An orthotropic deck collaborating with the spatial steel structure beneath has allowed a great reduction of the loads acting on the historical piers. To maximize the simplicity and easiness of fabrication of the inferior nodes, nine alternatives were compared. The durability and the reduced maintenance costs were guaranteed through a specific scenario of efficient coating sequences, the choice of the members' shape itself and the implementation of forms preventing water stagnation.
Sustainability:	The best solution was chosen on the basis of an analytical evaluation of various alternatives, to the aim of estimating the more sustainable one. The chosen solution emits 10% less CO_2 per year of life than the other alternatives.
Awards:	The bridge is one of the first cases in Europe of an LCA analysis applied to a big infrastructure. Those factors permitted the start-up of the procedure to internationally certify the bridge as an 'environmental product'. The procedure is actually in progress at Anas district of Milan. The bridge, its deconstruction and reconstruction were the subject of a National Geographic special and the infrastructure has been awarded as the 'Job of the year' in Amsterdam's World Demolition Awards.

Figure 1 Bridge over Po River, Lodi and Piacenza, Italy. © MCA Engineering.

Figure 2 To maximise the simplicity and easiness of fabrication of the inferior nodes, nine alternatives were compared. © MCA Engineering.

Figure 3 Using an orthotropic deck collaborating with the spatial steel structure beneath allowed a great reduction of the loads acting on the historical piers. © MCA Engineering.

Example provided by Fondazione Promozione Acciaio.

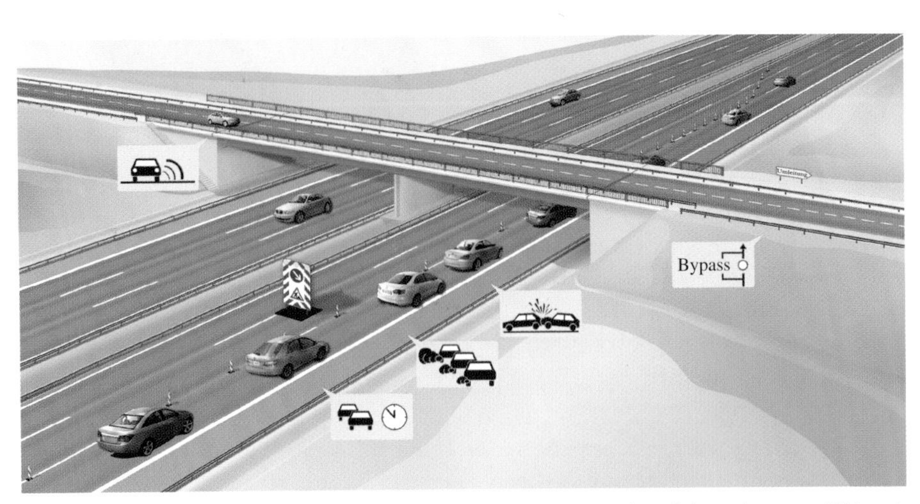

Figure 4.93 External effects important for the assessment of bridges: time delays of users, additional exhaust emissions, changed accident rates and accident likelihood, necessary detours and noise emissions, for example, generated by transition joints.

the goal of a sustainability assessment of bridges should not be the creation of some kind of certificate or, rather, label to find a sustainable design by comparing different construction variants in an early planning stage.

4.13.1.2 *Existing assessment approaches*

Sustainability assessment systems for infrastructure projects, such as bridges should be designed differently compared to buildings. The systematic aspects have been described in the previous sections. Additionally, the results of calculation methods applied can strongly depend on the position of the bridge within the network and the level of the average daily traffic as well as topography and geology of its location. Further, normally a service life of 100 years is required, and it is difficult to generate an assessment system because of all these particularities. As a result, only a few assessment approaches exist for bridges and infrastructures compared to buildings.

Internationally there are two types of approaches: general schemes applicable for different types of civil engineering works, and specialised analyses for one type of infrastructure. Furthermore, the application area can be divided into three classes: assessment systems active on the market, indicator frameworks with a focus on practical applicability and research projects. The worldwide existing approaches for civil engineering works and bridges in particular are compiled in Figure 4.94. The assessment systems and the indicator frameworks shown are applicable for different kinds of civil engineering works, and the research projects named are all focusing on bridges.

The oldest infrastructure assessment system is CEEQUAL [97], launched in 2003. Envision [98] and the Infrastructure Rating Scheme [99] are more recent and entered the market in 2012. All of them have been developed independently of the assessment systems for buildings by independent organisations. An example for a proprietary system is SPeAR [100], operated by the consulting company Arup. The

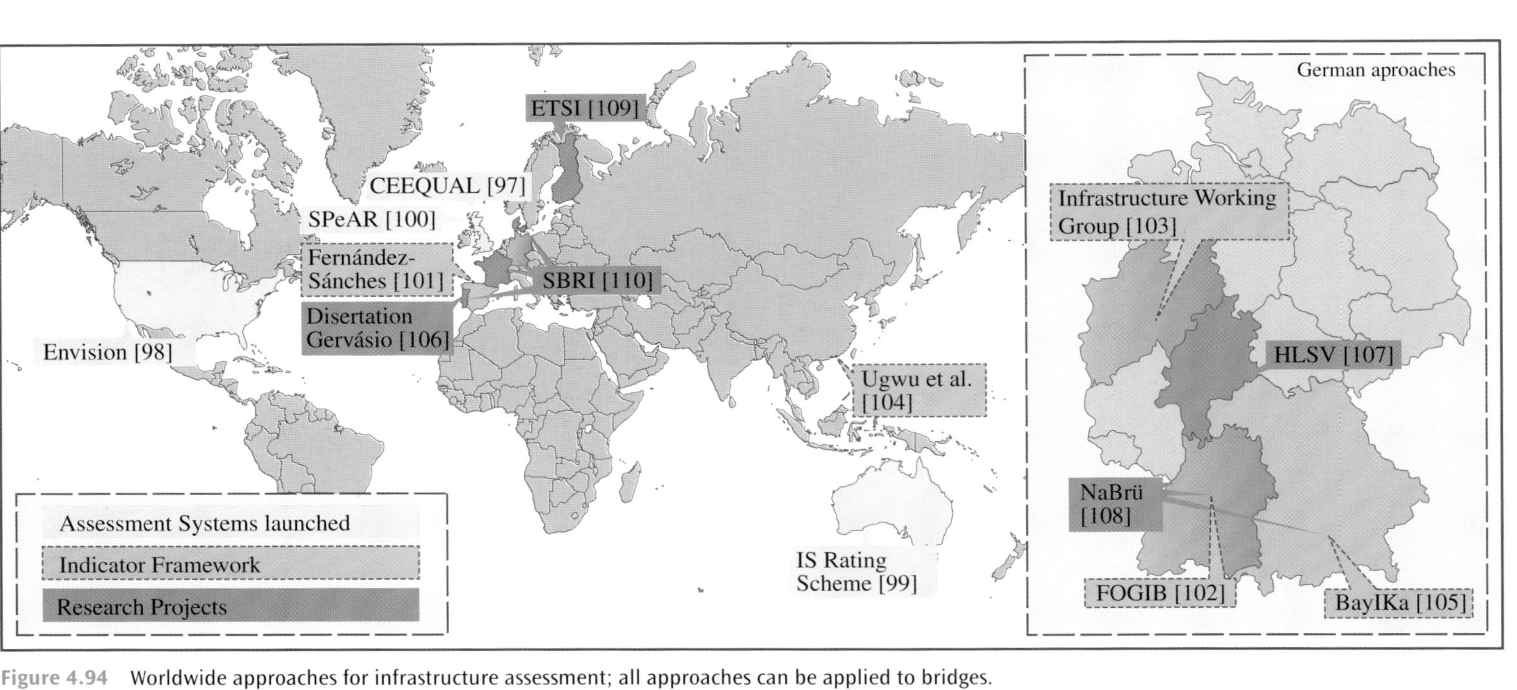

Figure 4.94 Worldwide approaches for infrastructure assessment; all approaches can be applied to bridges.

different indicator frameworks are normally the result of research projects and aim at creating a decision supporting system ([101]–[104]). The indicators proposed by BayIKa [105] are the basis for a competition, are aimed at the tendering phase, and focus on practical applications. In recent years, several research projects have been developing approaches and methods especially for the sustainability assessment of bridges ([106]–[110]). In the future, the findings should be transferred into standards and practical applications.

For sustainable buildings, many standards are under development by ISO and CEN, and an overview of them can be found in Chapter 2. For infrastructure projects, a few approaches exist. All of them aim at civil engineering works and are not specialised for use on bridges. On the ISO level, the working group ISO TC59 SC17 WG5 has launched the standard ISO/TS 21929-2: 2015, which provides a framework for the development of indicators for civil engineering works [111]. On the CEN level, the working group CEN TC350 WG6 is currently developing a document dealing with specific principles and requirements for sustainability assessment of civil engineering works, which is intended to be part of the standard series EN 15643.

4.13.2 Methods for bridge analyses

4.13.2.1 Life-cycle costs and whole-life costs

The method of the life-cycle cost (LCC) calculation with its application for bridges is well known. Many mostly methodologically oriented approaches (e.g. [112], [113]) and several practically oriented calculations for reference bridges (e.g. [114], [115]) have been published. However, the application of LCC for bridges often takes place detached from other indicators used for sustainability assessment of bridges. For an integrated assessment, the use of a consistent framework and the same input data for different methods is necessary. Such a consistent approach is often chosen in research projects where all methods use coordinated input data.

Another inconsistency that exists in the LCC calculation method is the consideration of different components. It can be concluded that all directly occurring bridge costs during the complete life cycle are regularly included. In some studies, cost components such as administrative costs or income are additionally considered. In some cases, user costs are also calculated as a part of the LCCs. Overall, no internationally accepted standardised framework exists. In this chapter, the classification proposed in ISO 15686-5 [116] and presented in Figure 4.95 is used for all calculations presented in Chapter 6.6.

As shown in Figure 4.95, all costs that can be directly assigned to the bridge are taken into account in the LCCs. For maintenance costs, this includes rehabilitation measures like the renewal of expansion joints or the concrete edge beams. Other cost components necessary for the management (i.e. authorities for office facilities, operation of bridge management systems, etc.) and potential costs for financing structures are classified as nonconstruction costs. Additionally, income and revenues are normally treated separately from the costs. As for bridges, usually no direct incomes result, so this component can be neglected. Finally, external

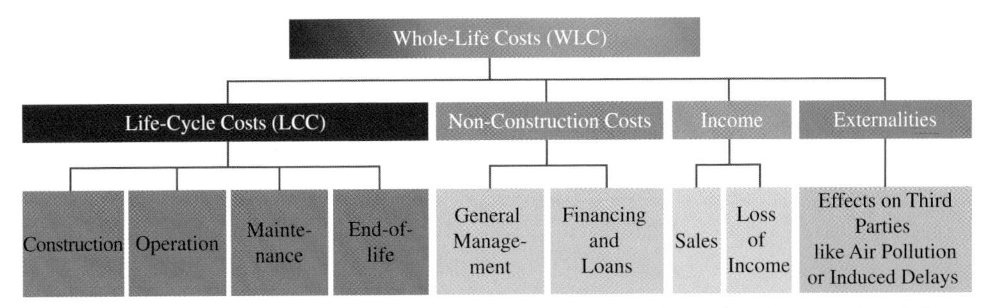

Figure 4.95 Differentiation of the terms 'whole-life costs' and 'life-cycle costs' and presentation of the different cost components included according to ISO 15686-5.

effects (or externalities) describe impacts on third parties. This cost component is discussed in detail in Section 4.13.3.

All cost components can be summarised under the term whole-life costs (WLCs). It is evident that WLCs represent a superordinated aggregation class and permit classification of the subordinated groups more precisely. For instance, LCCs can be defined with narrow system boundaries, and all costs accruing outside of the defined boundaries can be captured in other cost groups. This naturally results in a more comprehensive assessment system.

4.13.3 External effects and external costs

External costs are rooted in the field of transportation and infrastructure financing. In economics, it is a method to describe the mutual influence of different economic operators. External effects occur if the actions of one economic operator lead to an impact on other economic operators [117]. For instance, the construction of a bridge over an existing highway will lead to an obstruction of the highway users and to longer travel times, as well as increased exhaust emissions. These emissions will affect third parties in terms of poor air quality along with effect on health and accelerated deterioration of buildings. When transferring these effects into monetary units (monetisation), the results are external costs. Calculations of external costs allow direct comparison of different effects and with the LCCs of a bridge [118].

The basic interrelations explained above show that a disregard of external effects can lead to an overall welfare loss [119]. Figure 4.96 (left) visualizes the basic economic background. Marginal costs describe the change of the costs related to the increase of one unit output. A private marginal cost function contains costs like fuel and maintenance of vehicles, while a social marginal cost function additionally includes, for example, health costs resulting from air pollution. The different output quantities m_1 and m_2 when using different marginal cost functions are the reasons for the occurrence of external costs; see Figure 4.96. Since the macroeconomic optimum is smaller than the optimum resulting from a private optimization, a welfare loss occurs if external costs are neglected.

Calculating the overall impacts for a national economy, the total external costs for road traffic in Germany can be determined; see Figure 4.96 on the right side.

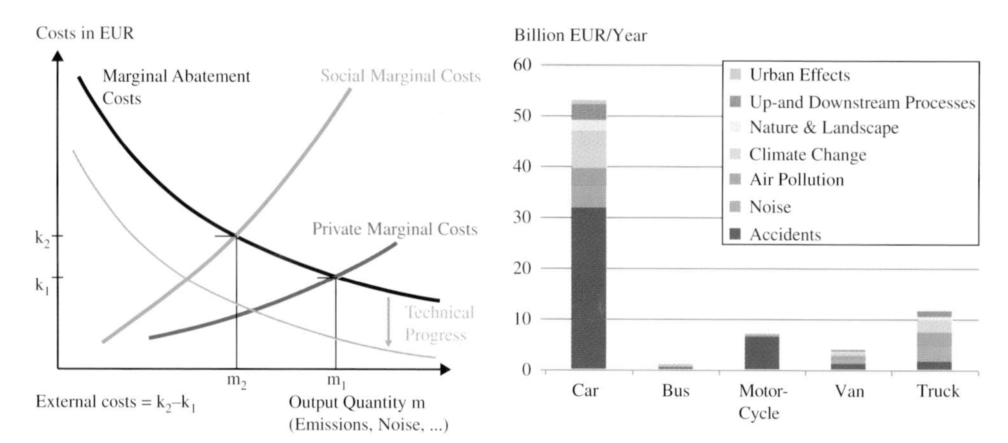

Figure 4.96 Development of external costs depending on private and social marginal cost functions (left, based on [120]); overall external costs of the road traffic in Germany, year 2005 (right, from [121]).

In 2005, about €80 billion of total external costs resulted from road traffic. The majority of external costs are caused by accidents, but also noise, air pollution and climate change are relevant cost categories.

In recent years, several studies have been carried out to determine marginal and total external costs for road traffic. According to [118] they can be classified into three groups, as follows. *Detailed analysis* deals with one specific aspect, for example, application of different monetisation methods, description and improvement of assessment aspects, analysis of a specific region and its special characteristic. Many of these analyses exist, and so no specific sources are mentioned here. Since 2000, a variety of *studies on external costs in the traffic sector* have been published. They focus on the national level [121] or the European level [122]–[125] and normally distinguish between the different transport modes of road, rail, aviation and shipping. *Meta-analysis* studies are based on the results of existing studies and compare them. A European reference project is the *Handbook on Estimation of External Costs in the Transport Sector* [127]. National approaches also exist, for example, in Germany [126]. In meta-analyses, one important issue is apparent: the uncertainty of input parameters and results. One main result of all studies dealing with external costs are marginal cost rates for single external effects. All these rates are subjected to variations. For example, the marginal cost for climate change is specified with a range of €17–€70/(t CO_2) in [127] and with a range of €20–€280/(t CO_2) in [126], both depending on different boundary conditions. Since uncertainty is an important topic for the correct interpretation of calculations, this aspect is briefly described in Section 4.13.5.

4.13.4 Life-cycle assessment

One of the first complete LCA calculations for bridges was performed by Lünser [128] in 1998. In recent years, several projects have been conducted that analyse environmental impacts for different types of bridges and bridge components.

LCA is normally not used by the infrastructure assessment systems active on the market or the indicator frameworks created for decision support (both displayed in Figure 4.94).

Just as with the results of LCC calculations, the results of an LCA depend on the system boundaries defined and the type of analysed structure. Table 4.15 summarises the results for the impact category global warming potential (GWP) of some European studies and gives the most important bridge characteristics. With the exception of [128], all bridges analysed are classified as small to medium span bridges. Influencing parameters, such as foundation conditions, country-dependent design specifications and different databases used as input parameters, are not indicated. All these parameters can influence the results. Therefore, a direct comparison should be made carefully. Nevertheless, Table 4.15 allows estimation of the basic range of GWP generated by a bridge within its life cycle of 100 years.

The description of the methods for the calculation of LCC and LCA shows that the methodological principles applied for buildings can also be used for the bridge structure. System boundaries have to be adapted and input parameters have to be specified, but the basic concept is transferable. To meet the assessment requirements for bridges, external effects have to be incorporated additionally. As explained, many effects can be converted into external costs. On the other hand, external environmental impacts (i.e. generated by exhaust emissions) can also be integrated in the impact categories of the LCA. This procedure has not often been performed in the past, but it can help to ensure an assessment that is aligned with the society's needs.

4.13.5 Uncertainty

In the field of bridge sustainability analysis, the challenge to simulate the complete life-cycle, normally defined by 100 years, must be taken into account. At present, not all future developments can be determined accurately. Therefore, the integration of uncertainties is a promising path. Several approaches to consider uncertainties already exist in civil engineering and especially in bridge engineering. Uncertainties often support analyses in the field of reliability theory [132]–[134] and risk assessment [135], [136]. For sustainability analysis [137], it becomes progressively important.

The scenario analysis represents one essential tool to handle uncertainties. Different aspects and parameters can be modelled with scenario analyses. Based on [138], the aspects are

- *varying the total service life and residual life:* working with different assumptions that change the contribution of impacts from erection and impacts within the life-cycle;
- *considering changes of central assessment parameters:* for example, the discount rate used in LCC as well as energy prices and traffic intensity used for external cost calculations can be mentioned;

Table 4.15 Comparison of the impact category GWP for different bridges [130].

Source	Country and Year	Construction	Bridge Length [m]	Effective Bridge Width [m]	Bridge Area [m^2]	Construction Processes	GWP [kg CO_2 Eq.]/m^2
ETSI [109]	Finland 2009	1-span steel hollow section	43	8	340	Included (estimated)	765
ETSI [109]	Finland 2009	1-span reinforced concrete hollow section	39	11	420	Included (estimated)	595
Gervásio [106]	Portugal 2010	3-span steel composite bridge	78	11	936	Included	1118
Gervásio [106]	Portugal 2010	3-span reinforced concrete bridge	82	10	777	Included	1080
Gervásio [106]	Portugal 2010	2-span reinforced concrete hollow section	60	7	425	Included	856
Graubner et al. [130]	Germany 2011	2-span steel composite bridge	67	11	905	Included (estimated on the basis of [129])	1580
Graubner et al. [130]	Germany 2011	2-span prestressed concrete bridge	66	12	927	Included (estimated on the basis of [129])	1210
Lünser [128]	Germany 1999	Reinforced concrete bridge with continuous beam	618	13	7849	Included	1116
Zinke et al. [131]	Germany 2010	2-span steel composite bridge	58	13	741	Included (estimated on the basis of [129])	1180
Zinke et al. [131]	Germany 2010	Integral steel composite bridge	33	12	403	Included (estimated on the basis of [129])	1110

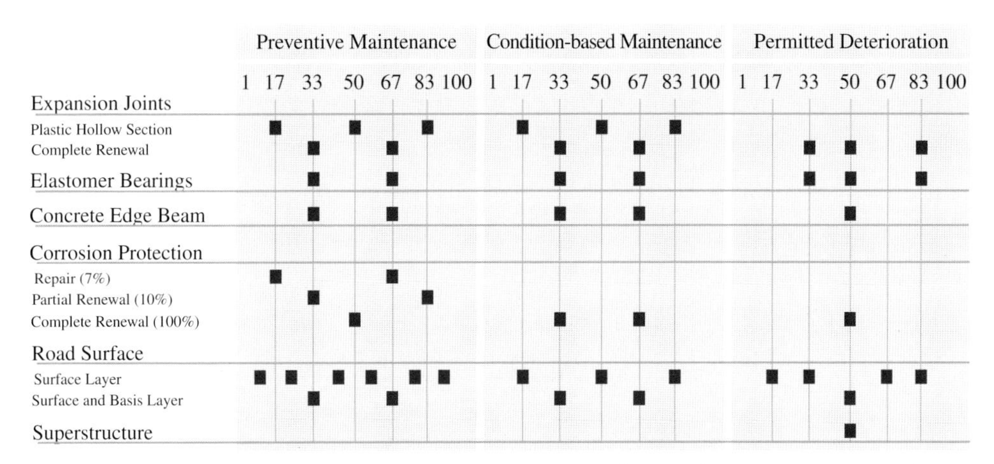

Figure 4.97 Scenarios for maintenance strategies specified for different bridge components; data for expansion joints valid for joints with more than one plastic hollow section [138].

- *modeling different erection processes:* dependent on usage of prefabricated elements or the in situ erection of a bridge, the building process can have very different influences on the external effects;
- *defining strategy-based maintenance and rehabilitation scenarios:* by a detailed modelling of the service life with component-based rehabilitation cycles, different possible developments within the life cycle can be considered.

An example for the implementation of different maintenance and rehabilitation strategies is shown in Figure 4.97. In this figure, the strategy-dependent renewal cycles of the single bridge components are highlighted. The condition-based maintenance strategy can be classified as a reference scenario that aims at a minimum of rehabilitation measures by planning coordinated actions. Preventive maintenance aims at obtaining a high condition index, and within the strategy of permitted deterioration, only absolutely necessary and safety-relevant measures are carried out.

4.14 STEEL CONSTRUCTION FOR RENEWABLE ENERGY

Anne Bechtel, Peter Schaumann, Natalie Stranghöner and Jörn Berg

In the past, cost effectiveness and constructional aspects of structures were the main design drivers for new constructions. The increasing importance of environment-friendly products and the reduction of CO_2 emissions have an impact on the application of holistic design concepts in all industries. In particular, renewables – producing 'green' energy – follow sustainable concepts. Hence, current holistic designs include sustainability aspects, which dominate the decision process and the cost effectiveness of future renewable construction. A rating system for the steel construction used in renewable energy systems is

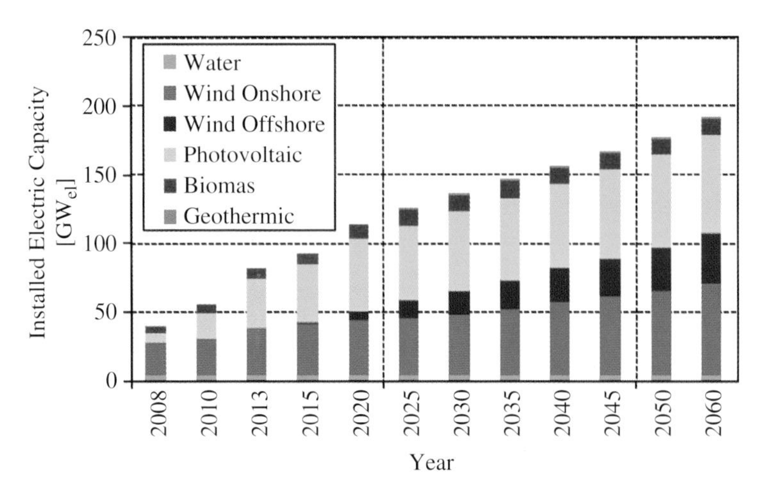

Figure 4.98 Forecast of installed electric capacity for Germany according to [147].

introduced here. The project included over 30 industrial partners to quantify and develop sustainability issues for renewable energies that are of significance (see [140]).

Motivated by current targets concerning the renewable expansion, and the steel quantities used for the support structure of renewable energy systems, carefully selected renewable energy constructions were analysed. Figure 4.98 presents the expected growth of the national renewable energy market in Germany and the resulting expansion. The wind energy market will mainly contribute to the regenerative electric power in the future [141]. In addition, the number of biogas plants will increase. Although photovoltaic generators are supposed to have a larger expansion in use than biogas plants, the construction for biogas plants contains larger amounts of steel. Therefore, the steel support structures for offshore wind energy turbines and steel digesters for biogas power plants are in the focus of the following case study.

The annual installation of onshore and offshore wind energy in Germany from 1990 to 2030 shows a growing market (Figure 4.99). The peak for annual installation regarding onshore wind energy turbines was in the year 2002, whereas the peak for offshore wind turbines is expected to be in 2022. The onshore wind power in future will be mainly affected by repowering. Until 2030, the cumulated capacity regarding offshore wind is expected to reach 15,000 MW, provoking a huge expansion of the offshore wind sector.

For an average steel demand of 150 t/MW for onshore and 250 t/MW for offshore wind turbines, and a supposed 80%–100% market share for steel, the annual steel demand of about 400,000 tonnes of steel in 2022 indicates the huge potential of wind energy construction with steel. The expected offshore capacity of 15,000 MW in 2030 means, as from 2014, an average annual installation of 180 wind turbines with 5 MW capacity per turbine, leading to a demand of 320,000 tonnes of steel every year. To reach the targeted expansion, the offshore industry

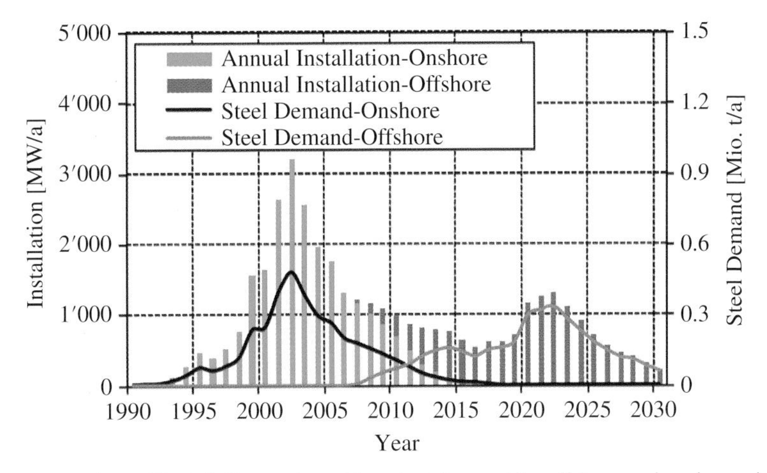

Figure 4.99 Annual installation and resulting steel demand for offshore and onshore wind energy in Germany.

needs to adopt strategies for production and installation. For the support structures for wind turbines, an optimised design leads to an increase of the overall efficiency. A modest optimisation of structural details can increase the total efficiency significantly. Mass production ensures the success for the offshore expansion with a significant optimisation potential.

The growing potential for renewable energy arising from biogas is presented in Figure 4.100. Since 2010, the number of biogas power plants and the installed electricity capacity in Germany have increased extensively. Furthermore, across Europe, the number of biogas power plants has increased by over 30% between 2009 and 2010. In 2010, German biogas production delivered 61% of biogas production across Europe [142], [143]. Consequently, a huge potential for biogas power plants can be recognised in Germany and across Europe.

Currently, the main part of biogas digesters of existing biogas power plants in Germany is manufactured from concrete with a market share of approximately 90%–95% [144]. The extensive manufacturing and installation process for concrete digesters could be replaced by the application of more economic steel shells in the future. For this reason, a huge potential for the steel part in manufacturing biogas digesters can be recognised. Approximately 120 t of structural steel are necessary for manufacturing of a 500 kW_{el} biogas power plant with two main digesters, one slurry tank and one tank for digestate, corresponding to a structural steel demand of 240 t/MW_{el}. Due to the rising potential of bio energy in Germany, an average annual increase of at least 230 MW_{el} and up to nearly 500 MW_{el} can be estimated up to the year 2020, if the positive trend of the biogas sector up to the year 2012 is considered as shown in Figure 4.100. Assuming a scenario with 100% market share of structural steel digesters of all future biogas digesters, the average annual structural steel demand can be estimated to 55,000 t up to 120,000 t. These numbers illustrate the high potential for manufacturing steel digesters of biogas power plants.

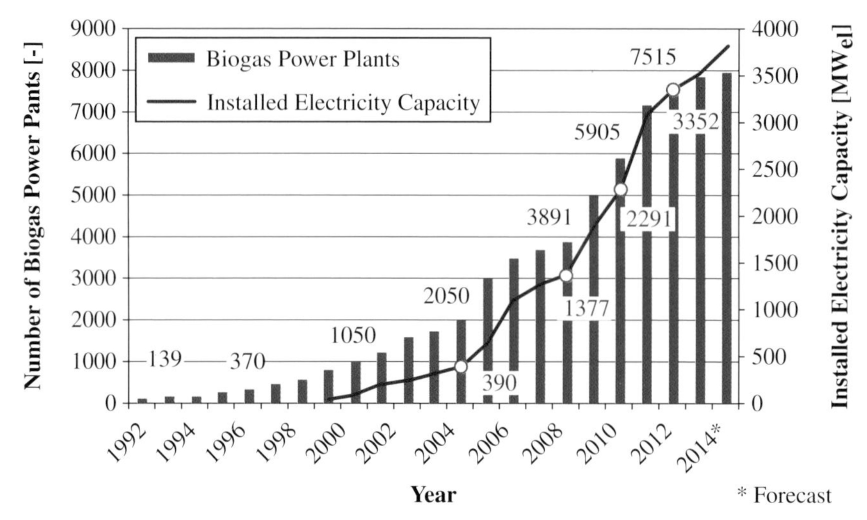

Figure 4.100 Number of biogas power plants and installed electricity capacity in Germany [143].

4.14.1 Sustainability assessment concept

In light of existing rating systems for buildings, such as the German assessment system for sustainable federal buildings (BNB) [26] and the rating system of the German Sustainable Building Council (DGNB) [86], a sustainability rating system for steel construction of regenerative energy sources was developed.

The rating systems of the BNB and DGNB (see Chapter 5) consist of the six sustainability categories: ecology, economy, sociocultural and functional, technical aspects, process, and local effects. Each of these categories is defined by a certain number of subcriteria and indicators reflecting the impact of manufacturing processes as well as building characteristics on the sustainability. Due to the basic understanding of sustainability reflecting the elements of ecology, economy, sociology, process and technology, the sustainability assessment system for steel construction of renewables is also based on these elements. Each category consists of numerous criteria and indicators describing certain effects of the steel structure. Some indicators of the DGNB or BNB rating system have been transferred to evaluate the impact of steel structures for renewable energy systems, but indeed most of the indicators are too close to the building concept. Therefore, investigations concentrated on establishing criteria reflect the needs for steel structures of renewables. Due to this, additional criteria describing the sustainability for steel construction of renewables were established. In a first step, proven indicators originating from the building industry and characteristics reported in literature were used to determine new criteria. Subsequently, 200 possible criteria were analysed regarding their applicability to steel construction for renewables. Special focus was set on wind energy converters and biogas plants. Finally, 35 criteria were identified to be best fit for the sustainability approach regarding renewables, shown in Table 4.16. For the criteria in italics, no assessment method has been defined so far. For each criterion, the steel structure itself was defined as the functional unit.

Table 4.16 Overview of sustainability criteria for renewable energy systems.

Ecology	U1	primary energy, nonrenewable
	U2	total primary energy and renewable
	U3	abiotic depletion potential
	U4	water demand
	U5	global warming potential
	U6	ozone depletion potential
	U7	photochemical ozone creation potential
	U8	acidification potential
	U9	eutrophication potential
	U10	*fine dusts**
	U11	risk for human and environment
	U12	*noise pullution**
	U13	recycling potential
	U14	waste
Economy	Ö1	life-cycle costs
	Ö2	expenditure for research & development
	Ö3	*employment effects**
Sociology	S1	occupational safety
	S2	apprenticeship quota
	S3	*qualification index**
	S4	advanced training
	S5	*staff fluctuation**
	S6	family friendliness
	S7	social engagement
Technology	T1	corrosion protection
	T2	operating life time
	T3	monitoring friendliness
	T4	*varied usability**
	T5	removal, recycling friendliness and reusability
Process	P1	quality management
	P2	material and resource efficiency
	P3	implementation of management systems
	P4	construction
	P5	realisation time
	P6	transport

italic written indicators are not elaborated

Table 4.17 Defined life-cycle stages for the sustainability assessment of steel support structures for renewables.

Planning	Product	Construction	Operation	Removal/ Recycling	Benefits
A0	A1-3	A4-7	B	C	D
A01 Planning	A1 Raw material procurement	A4 Transport to manufacturing site	B1 Operation	C1 Removal	Possibilities to...
A02 Project management	A2 Transport	A5 Manufacturing	B2 Maintenance	C2 Transport	...Reuse
	A3 Production	A6 Transport to construction site	B3 Remedial works	C3 Waste management	...Recycling potential
		A7 Erection	B4 Exchange	C4 Disposal	...Recovery

Focusing on environmental aspects, 10 criteria were taken from the standardized LCA according to DIN EN ISO 14040 [139] and combined with four new criteria, resulting in 14 criteria defining the environmental characteristics within the new assessment. Three criteria cover economic effects, such as LCCs, according to EN 15643-4 [147] and expenditure for research and development. The social performance is mainly reflected by company-related criteria, for example, family friendliness, social engagement, work safety and advanced training. In total, seven criteria represent the social part within the sustainability assessment. Technical and process elements are described by product-related criteria reflecting technical and logistical solutions. Five technical and six process criteria complete the sustainability assessment method. For each of all criteria, a profile was written to provide necessary information about the criteria and relevance to the user of the method. In combination with a detailed method description for each criterion, the sustainability benefits for steel construction of renewable energy systems can be evaluated.

Additionally, a tool based on Excel was established, enabling a practical application for the user. For each category, a polar diagram can be presented with the calculated single criteria values plotted on the axis of the diagram. The centre of the diagram displays the value zero as basis. Hence, applying the values for different constructional solutions leads to different spanned areas showing the sustainability impact. In the end, the sustainability for different steel structure solutions can be presented and compared by five diagrams for the sustainability characteristics ecology, economy, sociology, technology and process.

For all categories and criteria, the assessment has to encompass the decisive life-cycle stages (see Table 4.17): A0, planning; A1–3, product; A4–7, construction; B, operation; and C, end of lifetime in terms of removal, which is followed by stage D, the lifetime-exceeding stage including benefits resulting, for example, from recycling of the material. These life-cycle stages were defined according to EN 15978 [145] (see also Chapter 2).

4.14.2 Sustainability characteristics

4.14.2.1 Ecological characteristics

The sustainability assessment in terms of ecological aspects is commonly based on criteria that refer to LCAs. Within an LCA, several environmental impacts, tracing back to emissions of pollutants or the consumption of resources, caused by a product system considering its entire life cycle are gathered. LCA results are currently used within existing rating systems such as the BNB or the DGNB. The results of specific products are also published in EPDs (see Chapter 3.6). The values used for the Chapter 6 case studies originate from the ÖKOBAUDAT [146] (see Chapter 3.10). The methodology of LCA is standardized in EN ISO 14040 [139] and is internationally accepted.

The criteria presented in Table 4.16 are generally accepted. Hence, they were selected for the described sustainability assessment system for steel construction of renewables. In addition to the typical LCA-based criteria, further relevant criteria regarding steel construction of renewable energy systems were considered with detailed research. Because steel is a nearly complete recyclable material, it is very important to assess the inserted share of recycling material during the steel production. The use of recycling material allows for the reduction of the values for all LCA-based criteria, because of the fact that the energetic effort and the resulting emissions of pollutants are significantly lower when using scrap material instead of primary resources (see Chapter 3.9). The recycling potential after the use of the construction also plays an important role. If a part of the steel scrap can be reused or recycled, it is possible to afford a credit item for the benefit of the investigated product system.

4.14.2.2 Economic characteristics

The economic effect of steel structures for renewable energy systems is based on the financial costs originating from material and production costs including construction and maintenance. The economic aspect is conventionally a key parameter and leads to relevant decisions for the choice of a product or constructional solution. Due to the comparatively high material costs for steel, the design has to be optimised. However, costs increase for transportation and installation. Therefore, a detailed LCC analysis has to be performed within the sustainable evaluation of steel structures for renewables.

Beside the LCCs, additional indicators representing economic effects were generated to reflect other economic effects that increase costs but are still production worthy. Costs arising during research and development of the structures are one of these indicators affecting the LCC assessment negatively. With regard to the development of new products, high costs arise from research and development during an early stage of the product development. In order to integrate these costs that are to be assessed positively, the category 'expenditure in research and development' was chosen.

The effect of increasing product demand and industry sector growth should be integrated in the sustainable design in order to evaluate growth positively. The category 'employment effects' reveals the effect of growing markets and product demand and reflects the position of the company in the market. The category includes two indicators: effect of the product on the employment growth, and the company growth compared to the growth of the industry sector.

4.14.2.3 *Social characteristics*

Complementary to the economic and ecological aspects, the social effect reflects the third well-known parameter regarding sustainability. The social impact encompasses the effect of the product over the life cycle to workers, users and persons concerned. With regard to the steel structure of renewable energy systems, the effects during the production, manufacturing and construction processes to the employees and persons concerned can be differentiated. Regarding the installation of steel structures, safety plays an important factor, and so the criterion 'occupational safety' was generated. This criterion implicitly covers safety precautions, including safety training of the staff. To quantify the effects of safety precautions, the numbers of accidents or the accident rate can be calculated.

Furthermore, the criterion 'qualification index' takes into account the number of trained and untrained employees needed to fulfil selected life-cycle stages. By the application of the criterion 'family friendliness', the work–life balance can be evaluated. In addition to the social indicators reflecting the effect on the worker, the general social orientation of the company can be taken into account considering social commitment. The unit to measure the social impact is presented by qualitative statements.

4.14.2.4 *Technical characteristics*

Technical aspects reflect the technical quality of the product regarding monitoring, repair and inspections for the construction over its lifetime. The lifetime of the structure is one of the categories reflecting the operating life of the structure and the investment needed for a longer lifetime. The lifetime of steel structures can be affected by the protection system, and so the criterion 'corrosion protection' considers the planned and applied corrosion system.

In addition to the maintenance and inspection intervals needed, the planned and applied monitoring concept is evaluated, including the maintenance friendliness of the construction. This indicator affects the planning and operation stage concerning constructional details.

With regard to the end of life and the removal of the construction, additional criteria are developed to involve these impacts. The ease of removing the structure is taken into account and evaluated qualitatively.

Special constructional solutions regarding the variety of uses and the multiple uses of construction that are conceivable should be scored positively. Renewable energy systems are primarily used to generate energy. A secondary use of waste products that arise during the production of energy or the use of the construction for a second utilization could be elements of the indicator 'varied usability'.

4.14.2.5 *Process effects*

The last category illustrates the procedural aspects. This criterion includes constructional and company-related effects of the construction and the process. Process impacts are, for example, the time to plan, manufacture and construct the steel structure of renewables. This criterion is called the 'realisation time'. In many cases, this criterion is accompanied by costs and therefore linked to the economy of the construction.

The material and resource efficiency reflects the effective use of material. The aim of this indicator is to minimize the waste of material that could have been avoided. The mass of the material displays the unit of this indicator. The efficiency can be presented by comparing total mass and used/not-used material mass.

Furthermore, the effect of construction site and transportation is included in the sustainability rating system for steel structures of renewable energy systems. The noise and waste production on site is included in the indicator 'construction'. This category was developed based on the identical parameter within the BNB system for buildings.

The transportation of products and construction to the location of the energy plant can be decisive for the sustainability assessment. Therefore, the criterion 'transport' gives the opportunity to analyse the number of transportations, the type of transport carrier and the transport distance. Long transportation distances affect the environment and costs but are part of the constructional process.

Regarding the company effects, the criterion 'implementation of management systems' depicts the integration of management systems, for example, regarding the environment.

REFERENCES

[1] EPD-BFS-20130094. (2013) *Environmental Product Declaration: Structural Steel: Open Rolled Profiles and Heavy Plates*. Berlin: Institute Construction and Environment (IBU).

[2] Larsson, N. (2004) *Integrated Design Process*. Ottawa: iiSBE.

[3] Kibert, C.J. (2005) *Sustainable Construction: Green Building Design and Delivery*. New York: Wiley.

[4] Sev, A. (2011) A comparative analysis of building environmental assessment tools and suggestions for regional adaptations. Civil Engineering and Environmental Systems 28(3), pp. 231–245.

[5] American Institute of Steel Construction. (2014) Structural steel solutions: Why do designers & owners choose structural steel? Available at https://www.aisc.org/content.aspx?id=3792. Accessed 29 October 2014.

[6] Støa, E. (2012) Adaptable housing. In: Smith, S.J., ed., *International Encyclopedia of Housing and Home*. San Diego: Elsevier, pp. 51–57.

[7] Schmidt III, R., et al. (2010) What is the meaning of adaptability in the building industry? O&SB, p. 233.

[8] Gosling, J., et al. (2013) Adaptable buildings: A systems approach. Sustainable Cities and Society 7, p. 44–51.

[9] Tweed, C., Sutherland, M. (2007) Built cultural heritage and sustainable urban development. Landscape and Urban Planning 83(1), pp. 62–69.

[10] Conejos, S., Langston, C., Smith, J. (2014) Designing for better building adaptability: A comparison of adaptSTAR and ARP models. Habitat International 41, pp. 85–91.

[11] Gosling, J., et al. (2008) Flexible buildings for an adaptable and sustainable future. Proceedings 24th Annual ARCOM Conference. Cardiff, UK: Association of Researchers in Construction Management.

[12] Slaughter, E.S. (2001) Design strategies to increase building flexibility. Building Research & Information 29(3), pp. 208–217.

[13] Durmisevic, E. (2006) *Transformable Building Structures*. Delft: Delft University of Technology.

[14] Burgan, B.A., Sansom, M.R. (2006) Sustainable steel construction. Journal of Constructional Steel Research 62(11), pp. 1178–1183.

[15] Blok, R., Koopman, E. (2012) Comparing and quantifying the structural flexibility of three buildings in the Netherlands. In: *Concepts and Methods for Steel Intensive Building Projects*. Munich: Multicomp.

[16] EN 15643-1. (2010) *Sustainability of Construction Works - Sustainability Assessment of Buildings - Part 1: General Framework*. Brussels: European Committee for Standardization.

[17] Kreislaufwirtschaft Bau c/o Bundesverband Baustoffe – Steine und Erden e.V. (2013) Mineralische Bauabfälle Monitoring 2010. Berlin

[18] German Federal Statistical Office. (2013) Environment – disposal of wastes in 2013, by waste category. Available at http://www.statistik-portal.de/Statistik-Portal/de_jb10_jahrtabu12.asp. Accessed 25 August 2014.

[19] Directive 2008/98/EC of the European Parliament and of the Council on waste and repealing certain directives. *Official Journal of the European Union*.

[20] Thienel, K.-Ch. (2011) Mauersteine und Mörtel. In: *Bauchemie und Werkstoffe des Bauwesens*. München: Univ. München Institut für Werkstoffe des Bauwesens.

[21] Stoll, M. (2010) Recycling von mineralischen Abfällen aus Sicht der Wirtschaft, S. 22, 'Grafik zur RC-Baustoff-Verwendung des BRB'. Duisburg; Statistisches Bundesamt, Bonn, Fachserie 19, Reihe 1, 2010.

[22] EPD-BFS-20130173. (2013) *Environmental Product Declaration: Hot-Dip Galvanized Structural Steel: Open Rolled Profiles and Heavy Plates*. Berlin: Institute Construction and Environment (IBU).

[23] Industrieverband Feuerverzinken, Feuerverzinken. (2008) Internationale Fachzeitschrift 37.Jahrgang Special Nachhaltigkeit. Düsseldorf.

[24] RIWIS – Regional Real Estate Economic Information System; Institute of the German Economy Cologne. Available at www.riwis.de. Accessed 29 October 2015.

[25] Clamor, T., Haas, H., Voigtländer, M. (2011) Büroleerstand – ein zunehmendes Problem des deutschen Immobilienmarktes, IW-Trends – Vierteljahresschrift zur empirischen Wirtschaftsforschung aus dem Institut der deutschen. Wirtschaft Köln 38(4).

[26] Bundesministerium für Verkehr, Bau und Stadtentwicklung (BMVBS): Bewertungssystem Nachhaltiges Bauen (BNB), Version 2011. Available at www.bnb-nachhaltigesbauen.de. Accessed 29 October 2015.

[27] Institut für internationale Architektur-Dokumentation GmbH &Co. KG. (2013) *Büro/Office – Best of DETAIL*. München: Edition DETAIL.

[28] Eisele, J., Staniek, B. (2005) *BürobauAtlas – Grundlagen, Planung, Technologie, Arbeitsplatzqualitäten*. München: Callwey

[29] Mensinger, M., Stroetmann, R., Eisele, J., Feldmann, M,. Lingnau, V., Zink, J., et al. (2014) *Nachhaltigkeit von Stahl- und Verbundkonstruktionen bei Büro- und Verwaltungsgebäuden*, Final Report, AiF project No. 373 ZBG, Düsseldorf.

[30] Stroetmann, R., Podgorski, C. (2014) 'Zur Nachhaltigkeit von Stahl- und Verbundkonstruktionen bei Büro- und Verwaltungsgebäuden – Tragkonstruktion en Teil 1'. Stahlbau 83(4), pp. 245–256.
[31] Eisele, J., Staniek, B. (2005) *BürobauAtlas*. Darmstadt: Georg D.W. Callwey GmbH & Co. KG.
[32] Mensinger, M., Stroetmann, R., Eisele, J., Feldmann, M., Lingnau, V., Zink, J., et al. (2011) Nachhaltige Büro- und Verwaltungsgebäude in Stahl- und Stahlverbundbauweise. Stahlbau 80(10).
[33] EN 1991-1-1: Eurocode 1. (2002) *Actions on Structures – Part 1-1: General Actions – Densities, Self-Weight, Imposed Loads for Buildings*. Brussels: European Committee for Standardization.
[34] Feldmann, M., Heinemann, Ch. (2010) Neues Verfahren zur Bestimmung und Bewertung von personeninduzierten Deckenschwingungen. Bauingenieur 85, pp. 36–44.
[35] Maier, C. (2005) Großversuch zum Einfluss nichttragender Ausbauelemente auf das Schwingungsverhalten weitgespannter Verbundträgerdecken'. Darmstadt: Eigenverlag Fachgebiet Statik und Dynamik der Technischen Universität Darmstadt.
[36] Stroetmann, R., Podgorski, C., Faßl, T. (2014) Zur Nachhaltigkeit von Stahl- und Verbundkonstruktionen bei Büro- und Verwaltungsgebäuden - Tragkonstruktionen Teil 2. Stahlbau 83(9), pp. 599–607.
[37] Masconi, H.-W., Muess, J., Schmitt, H., Seidel, U. (1990) Schraubenloser Verbundbau beim Neubau des Postamtes 1 in Saarbrücken. Stahlbau 59(3), pp. 65–73.
[38] Ameling, D. (2008) Ressourceneffizienz – Stahl ist die Lösung. Vortrag mit Schriftbeitrag beim Dresdener Leichtbausymposium im Juni 2008.
[39] Ameling, D., Endemann, G. (2009) Ressourceneffizienz: Gute Argumente für den Stahl. Stahl und Eisen 127(8).
[40] Stranghöner, N. (2006) Werkstoffauswahl im Stahlbrückenbau. Habilitation, Technische Universität Dresden.
[41] Stroetmann, R., Deepe, P. (2009) Anwendung höherfester Stähle zur Steigerung der Ressourceneffizienz im Bauwesen, VDI-Berichte 2084, November 2009 – Bauen mit innovativen Werkstoffen, pp. 267–305, Köln.
[42] Stroetmann, R. (2011) High strength steel for improvement of sustainability. In: EUROSTEEL 2011, 6th European Conference on Steel and Composite Structures, Proceedings Volume C, pp. 1959–1964. Budapest.
[43] Stroetmann, R. (2011) Sustainable design of steel structures with high-strength steels. Dillinger Colloquium on Constructional Steelwork on 1 December 2011, Dillingen.
[44] Schröter, F. (2007) Stähle für den Stahlbau – Anwendung moderner Baustähle und Neuerungen im Regelwerk. In: *Stahlbaukalender 2007*. Berlin: Ernst.
[45] prEN 10025-6. (2014) *Hot Rolled Products of Structural Steels - Part 6: Technical Delivery Conditions for Flat Products of High Yield Strength Structural Steels in the Quenched and Tempered Condition*. Brussels: European Committee for Standardization.
[46] voestalpine Grobblech GmbH. (2013) Hochfeste und ultrahochfeste Grobbleche: Gewicht sparen bei optimaler Schweißbarkeit. Firmenschrift, Ausgabe 06.03.2013. Linz.
[47] prEN 10025-4, (2014) *Hot Rolled Products of Structural Steels - Part 4: Technical Delivery Conditions for Thermomechanical Rolled Weldable Fine Grain Structural Steel*. Brussels: European Committee for Standardization.
[48] EN 10149-2. (2013) *Hot Rolled Flat Products Made of High Yield Strength Steels for Cold Forming - Part 2: Technical Delivery Conditions for Thermomechanically Rolled Steels*. Brussels: European Committee for Standardization.

[49] EN 1993-1-12: Eurocode 3. (2007) *Design of Steel Structures Part 1–12: Additional Rules for the Extension of EN 1993 up to Steel Grades S 700*. Brussels: European Committee for Standardization.

[50] Stroetmann, R., Deepe, P., Rasche, Ch., Kuhlmann, U. (2012) Bemessung von Tragwerken aus höherfesten Stählen bis S700 nach EN 1993-1-12. Stahlbau 81(4), pp. 332–342.

[51] Stroetmann, R., Deepe, P. (in preparation) Tragwerke aus höherfesten Stählen – Planung, Berechnung und Ausführung. Düsseldorf: bauforumstahl.

[52] EU Council Directive 96/61/EC of 24 September 1996 concerning integrated pollution prevention and control.

[53] Tom Woolley, B. (2018) *Galvanizing and Sustainable Construction: A Specifiers' Guide*. West Midlands: Sutton Coldfield.

[54] Parametrix Inc. (1995) *Persistence, Bioaccumulation and Toxicity of Metals and Metal Compounds*. Ottawa: ICME, International Council on Metals in the Environment.

[55] van Assche, F., Green, A. (2006) Review of bioavailability studies in the European Union risk assessment for zinc. Edited Proceedings of 21st International Galvanizing Conference, Naples, Italy.

[56] Preussag Stahl AG. (1996) UPE – Eine intelligente Alternative Firmenschrift Ausgabe November 1996

[57] Bauforumstahl. (2014) Bemessungshilfe Stützen im Geschossbau Bemessungstabellen nach DIN EN 1993-1-1 und Nomogramme für den Brandfall nach DIN EN 1993-1-2, Nr. B 503, Düsseldorf.

[58] Hauke, B., Gündel, M. (2007) Aperture plates as ductile shear connectors for high performance composite members. Conference Proceedings, Connections between Steel and Concrete, 2007.

[59] Hegger, J., Döinghaus, P., Goralski, C. (2003) Zum Einsatz hochfester Materialien bei Verbundträgern unter positiver Momentenbeanspruchung. Stahlbau 72.

[60] Pilarski, J. (2006) Technisch-wirtschaftlicher Vergleich von Hochleistungs-Verbundträgern mit traditionellen Bauweisen im Hoch- und Industriebau. Diploma thesis, University of Applied Sciences Frankfurt, 2006.

[61] Hauke, B. (2008) Economic application of composite beams with moderate high strength materials. Conference Proceedings, Eurosteel, 2008.

[62] Interessengemeinschaft Stahl-Brandschutzbeschichtung (IGSB). (2014) *IGSB – Info 1 Fire Protection Coatings for Structural Steelwork*. Düsseldorf: IGSB.

[63] Interessengemeinschaft Stahl-Brandschutzbeschichtung (IGSB). (2015) *IGSB – Info 2 Fire Protection Coatings: Design to Perfection*. Düsseldorf: IGSB.

[64] EN ISO 13789: 2008-04. (2008) *Thermal Performance of Buildings – Transmission and Ventilation, Heat Transfer Coefficients – Calculation Method*. Brussels: European Committee for Standardization.

[65] Möller, R., Pöter, H., Schwarze, K. (2011) *Planning and Building with Trapezoidal Profiles and Sandwich Elements, Vol. 1: Basics, Construction Methods, Dimensioning with Examples*. Berlin: Ernst.

[66] EN 14509. (2013) *Self-Supporting Double Skin Metal Faced Insulating Panels – Factory Made Products – Specifications*. Brussels: European Committee for Standardization.

[67] EN ISO 10211. (2008) *Thermal Bridges in Building Construction – Heat Flows and Surface Temperatures - Detailed Calculations*. Brussels: European Committee for Standardization.

[68] IFBS 4.03. *Building Physics – Thermal-Bridge Atlas*. Available at www.ifbs.de. Accessed 29 October 2015.

[69] EN 12114. (2000) *Thermal Performance of Buildings – Air Permeability of Building Components and Building Elements - Laboratory Test Method.* Brussels: European Committee for Standardization.
[70] IFBS (2004). IFBS-Wärmedurchgang – Software Version 1.3 – Zur Ermittlung der Wärmeverluste an zweischaligen Wand- und Dachaufbauten in Metallleichtbauweise. Available at http://www.ifbs.de/shop.php?action=viewProduct&pid=94. Accessed 29 October 2015.
[71] Kuhnhenne, M., Feldmann, M., Doering, B. (2010) Grundsätze und Lösungen zur Wärmebrückenreduktion im Metallleichtbau. Stahlbau 79(5), pp. 345–355.
[72] IFBS. (2006) IFBS-Fachinformation 4.05: Bauphysik – Ermittlung der Wärmeverluste an zweischaligen Dach- und Wandaufbauten. Düsseldorf.
[73] EN ISO 10456. (2010) *Baustoffe und Bauprodukte – Wärme- und feuchtetechnische Eigenschaften.* Berlin: Deutsches Institut für Normung e.V.
[74] DIN 4108-4. (2013) *Wärme- und Energie-Einsparung in Gebäuden – Teil 4: Wärme- und feuchteschutztechnische Bemessungswerte.* Brussels: European Committee for Standardization.
[75] IFBS. (2004) *IFBS-Fachinformation 4.02: Bauphysik – Fugendichtheit im Stahlleichtbau.* Düsseldorf.
[76] German Energy Saving Ordinance 2014. (2013) Unofficial reading version according to the cabinet decision of the German federal government, from 6 February 2013.
[77] Heinemeyer, S., Gallwoszus, J., Hegger, J. (2012) Verbundträger mit Puzzleleisten und hochfesten Werkstoffen. Stahlbau 81, pp. 595–603.
[78] Gallwoszus, J., Classen, M., Hartje, J. (2015): Ermüdung von Verbundkonstruktionen mit Verbunddübelleisten (lokales Tragverhalten). Beton- und Stahlbetonbau 110.
[79] Feldmann, M., Hechler, O., Hegger, J., Rauscher, S. (2007) Neue Untersuchungen zum Ermüdungsverhalten von Verbundträgern aus hochfesten Werkstoffen mit Kopfbolzendübeln und Puzzleleiste'. Stahlbau 76, pp. 826–844.
[80] Classen, M., Hegger, J. (2014) Verankerungsverhalten von Verbunddübelleisten in schlanken Betongurten. Bautechnik 91(12).
[81] Classen, M., Gallwoszus, J., Hegger, J. (2014) Zum Tragverhalten eines integrierten Verbunddeckensystems. Bauingenieur 89(3), p. 91.
[82] Classen, M., et al. (2014) Nachhaltigkeitsbewertung von Deckensystemen mit großen Spannweiten. Bauingenieur 89(3), p. 125.
[83] Feldmann, M., Pyschny, D., Döring, B. (2014) Ermittlung der thermischen Leistungsfähigkeit eines neuartigen multifunktionalen Verbunddeckensystems. Bauingenieur, März.
[84] Hegger, J., et al (2013) Entwicklung einer integrierten Verbunddecke für nachhaltige Stahlbauten. Stahlbau 82(1).
[85] Schaumann, P., Sothmann, J., Weisheim, W. (2014) Untersuchungen zum Tragverhalten eines integrierten und nachhaltigkeitsorientierten Verbunddeckensystems im Brandfall. Bauingenieur, März.
[86] German Sustainable Building Council - DGNB e.V. (2012) *Neubau Büro- und Verwaltungsgebäude – DGNB Handbuch für nachhaltiges Bauen Version 2012.* Stuttgart.
[87] Zimmermann, M. (2003) *Handbuch der passiven Kühlung.* Stuttgart: Fraunhofer IRB Verlag.
[88] Concrete Centre. (2006) *Utilisation of Thermal Mass in Non-Residential Buildings.* London.
[89] EN ISO 13786. (2007) *Thermal Performance of Building Components – Dynamic Thermal Characteristics – Calculation Methods.* Brussels: European Committee for Standardisation.

[90] 'TRNSYS' Transient System Simulation Tool. Available at http://www.trnsys.com. Accessed 23 October 2015.
[91] 'Tas for Engineering' software. Available at http://www.EDSL.net. Accessed 23 October 2015.
[92] European Commission. (2015) *Zero Energy Solutions for Multifunctional Steel Intensive Commercial Buildings (ZEMUSIC)*, RFSR-CT-2011-00032. Final report to be published, Brussels.
[93] Döring, B. (2008) Einfluss von Deckensystemen auf Raumtemperatur und Energieeffizienz im Stahlgeschossbau. Diss. RWTH Aachen.
[94] Steinbeis-Transferzentrum Energietechnik. (2006) Passivhaus ENERGON Ulm Büro-gebäude im Passivhaus-Standard, Schlussbericht Monitoring. Ulm.
[95] Peters, T. (2011) Ecology beyond buildings – performance-based consumption and zero-energy research. Architectural Design 81(6), pp. 124–129.
[96] Stroetmann, R., et al. (2014) Ganzheitliche Planung nachhaltiger Bürogebäude in Stahl- und Stahlverbundbauweise. Stahlbau 83(7).
[97] CEEQUAL. (2010) *CEEQUAL: Scheme Description and Assessment Process Handbook, Version 4.1 for Projects*. Surrey.
[98] Institute for Sustainable Infrastructure. (2011) Draft of the envision assessment system, Version 1.0., Washington, D.C.
[99] Sprigg, A. (2012) Infrastructure Sustainability (IS) rating scheme. Australian Environmental Review, pp. 30–34.
[100] Arup. (2012) *SPeAR Handbook 2012*. External Version. Sustainable Project Appraisal Routine, Version 1.1, London.
[101] Fernández-Sánchez, G., Rodríguez-López, F. (2010) A methodology to identify sustainability indicators in construction project management – application to infrastructure projects in Spain. Ecological Indicators 10, pp. 1193–1201.
[102] Fogib. (1997) *Ingenieurbauten - Wege zu einer ganzheitlichen Bewertung, Band 1–3*. Universität Stuttgart. Final report of the DFG research group civil engineering structures. Stuttgart.
[103] Graubner, C.-A., Baumgärtner, U., Fischer, O., Haardt, P., Knauff, A., Putz, A. (2010) Nachhaltigkeitsbewertung für die Verkehrsinfrastruktur. Bauingenieur 85, pp. 331–340.
[104] Ugwu, O., Haupt, T. (2007) Key performance indicators and assessment methods for infrastructure sustainability – a South African construction industry perspective. Building and Environment 42, pp. 665–680.
[105] BayIKa. (2011) Dokumentation des Ideenwettbewerbs. Entwurf einer Straßenbrücke nach ganzheitlichen Wertungskriterien. Bayerische Ingenieurekammer-Bau. Wiesbaden: Wiederspahn.
[106] Gervásio, H. (2010) Sustainable design and integral life-cycle analysis of bridges. PhD diss., Institute for Sustainability and Innovation in Structural Engineering, Departamento de Engenharia Civil Faculdade de Ciências e Tecnologia da Universidade de Coimbra.
[107] HLSV. (2010) *Nachhaltiges Bauen – Ökobilanzierungen und Lebenszykluskosten von vier Straßenbrücken*. Research report TU Darmstadt. Wiesbaden: Hessisches Landesamt für Straßen- und Verkehrswesen.
[108] Kuhlmann, U., Beck, T., Fischer, M., Friedrich, H., Kaschner, R., Maier, P., Mensinger, M., Pfaffinger, M., Sedlbauer, K., Ummenhofer, T., Zinke, T. (2011) Ganzheitliche Bewertung von Stahl- und Verbundbrücken nach Kriterien der Nachhaltigkeit. Stahlbau 80, pp. 703–710.

[109] Salokangas, L. (2009) ETSI Project (Stage 2). *Bridge Life Cycle Optimization*. Technical report nr. TKK-R-BE3.

[110] SBRI. (2013) *Sustainable Steel-Composite Bridges in Built Environment*. Research project of the Research Fund for Coal and Steel RFSR-CT-2009-00020. Edited by Kuhlmann et al. Final Report 2013.

[111] ISO/TS 21929-2. (2015) *Sustainability in Building Construction – Sustainability Indicators – Part 2: Framework for the Development of Indicators for Civil Engineering Works*.

[112] Chandler, R.F. (2004) *Life-Cycle Cost Model for Evaluating the Sustainability of Bridge Decks: A Comparison of Conventional Concrete Joints and Engineered Cementitious Composite Link Slabs*. University of Michigan, Center for Sustainable Systems. Report No. CSS04-06.

[113] Frangopol, D.M., Liu, M. (2007) Maintenance and management of civil infrastructure based on condition, safety, optimization, and life-cycle cost. Structure and Infrastructure Engineering 3, pp. 29–41.

[114] Heitel, S., Koriath, H., Herzog, C. S., Specht, G. (2008) Vergleichende Lebenszykluskostenanalyse für Fußgängerbrücken aus unterschiedlichen Werkstoffen. Bautechnik 85, pp. 687–695.

[115] Jodl, H.G. (2010) Lebenszykluskosten von Brücken - Teil 1 und 2 - Berechnungsmodell LZKB. Bauingenieur, pp. 221–240.

[116] ISO 15686-5. (2008) *Buildings and Constructed Assets – Service-Life Planning – Part 5: Life-Cycle Costing*.

[117] Baum, H., Esser, K., Höhnscheid, K.-J. (1998) *Volkswirtschaftliche Kosten und Nutzen des Verkehrs*. Bonn: Kirschbaum.

[118] Zinke, T., Schmidt-Thrö, G., Ummenhofer, T. (2012) Entwicklung und Verwendung von externen Kosten für die Nachhaltigkeitsbewertung von Verkehrsinfrastruktur. Beton- und Stahlbetonbau 107, pp. 524–532.

[119] Bartling, H., Luzius, F. (2008) *Grundzüge der Volkswirtschaftslehre: Einführung in die Wirtschaftstheorie und Wirtschaftspolitik*. München: Vahlen.

[120] INFRAS. (2007) Externe Kosten des Verkehrs in Deutschland. Aufdatierung 2005. B1669A1. Zürich

[121] ARE/BAFU. (2008) Externe Kosten des Verkehrs in der Schweiz. Aktualisierung für das Jahr 2005 mit Bandbreiten. Bundesamt für Raumentwicklung und Bundesamt für Umwelt. Zürich.

[122] CAFÉ. (2005) *Methodology for the Cost-Benefit Analysis for CAFE. Volume 1: Overview of Methodology*. AEA Technology Environment. AEAT/ED51014. Didcot, United Kingdom.

[123] COMPETE. (2006) *Analysis of the Contribution of Transport Policies to the Competitiveness of the EU Economy and Comparison with the United States*. TREN/05/MD/S07 .5358 5. Karlsruhe.

[124] GRACE. (2005) *Information Requirements for Monitoring Implementation of Social Marginal Cost Pricing*. Deliverable 1. ITS, University of Leeds. FP6-006222.

[125] HEATCO. (2006) *Developing Harmonised European Approaches for Transport Costing and Project Assessment*. Deliverable D5: Proposal for Harmonised Guidelines. IER, Stuttgart. FP6-2002-SSP-1/502481.

[126] UBA. (2007) *Ökonomische Bewertung von Umweltschäden: Methodenkonvention zur Schätzung externer Umweltkosten*. Berlin: Umweltbundesamt.

[127] Maibach, M., et al. (2008) *Handbook on Estimation of External Costs in the Transport Sector. Internalisation Measures and Policies for All External Cost of Transport*

(IMPACT). CE Delft - Solutions for environment, economy and technology. Publication number: 07.4288.52. Delft.

[128] Lünser, H. (1998) *Ökobilanzen im Brückenbau: Eine umweltbezogene, ganzheitliche Bewertung*. Basel: Birkhäuser.

[129] Friedrich, H., Kuhlmann, U., Krill, A., Lenz, K., Maier, P., Mensinger, M., Pfaffinger, M., Sedlbauer, K., Ummenhofer, T., Zinke, T. (2013) Nachhaltigkeitsanalyse von Verbundstraßenbrücken am Beispiel einer Autobahnüberführung. VDI-Bautechnik Jahresausgabe 2013/2014, pp. 111–119. Düsseldorf-

[130] Graubner, C.-A., Knauff, A., Pelke, E. (2011) Lebenszyklusbetrachtungen als Grundlage für die Nachhaltigkeitsbewertung von Straßenbrücken. Stahlbau 80, pp. 163–171.

[131] Zinke, T., Diel, R., Mensinger, M., Ummenhofer, T. (2010) Nachhaltigkeitsbewertung von Brückenbauwerken. Stahlbau 79, pp. 448–455.

[132] Frangopol, D.M (2011) Life-cycle performance, management, and optimisation of structural systems under uncertainty: Accomplishments and challenges. Structure and Infrastructure Engineering 7, pp. 389–413.

[133] Schnetgöke, R. (2008) Zuverlässigkeitsorientierte Systembewertung von Massivbauwerken als Grundlage für die Bauwerksüberwachung. Diss., Technische Universität Braunschweig.

[134] Xia, H., Ni, Y., Wong, K., Ko, J. (2012) Reliability-based condition assessment of in-service bridges using mixture distribution models. Computers & Structures 106–107, pp. 204–213.

[135] Decò, A., Frangopol, D.M. (2011) Risk assessment of highway bridges under multiple hazards. Journal of Risk Research 14, pp. 1057–1089.

[136] Hu, X., Chen, H.-Y. (1992) Probabilistic analysis of uncertainties in seismic hazard assessment. Structural Safety 11, pp. 245–253.

[137] Gervásio, H.M., Silva, S.D. (2012) A probabilistic approach for life-cycle environmental analysis of motorway bridges. In: Strauss, A., Frangopol, D.M., Bergemeister, K., eds., *Life-Cycle and Sustainability of Civil Infrastructure Systems*. Proceedings of the Third International Symposium on Life-Cycle Civil Engineering, Vienna, Austria, 3–6 October 2012. Leiden: CRC Press/Balkema, pp. 657–664.

[138] Zinke, T., Ummenhofer, T. (2013) Uncertainty of Maintenance Strategies for Bridge Sustainability Assessment. In: IABSE, ed., *Assessment, Upgrading and Refurbishment of Infrastructures*. Proceedings of the IABSE Conference, Rotterdam, Netherlands, 6–8 May 2013, pp. 66–67.

[139] EN ISO 14040. (2009) *Environmental Management – Life Cycle Assessment – Principles and Framework*. Brussels: European Committee for Standardization.

[140] Nitsch, J. (2014) *Scenarios for the German Energy Supply in the Light of the Agreement by the Grand Coalition*. Short expertise for the Federal Association Renewable Energies e.V. (Bundesverband Erneuerbare Energien e.V.), Stuttgart.

[141] Schaumann, P., Bechtel, A., Wagner, H.-J., Baack, C., Lohmann, J., Stranghöner, N., Berg, J. (2011) Sustainability of Steel Constructions for Renewables, Stahlbau 80 (2011), Issue 10, pp. 711–719. Berlin: Ernst & Sohn.

[142] Fachverband Biogas. (2014) *Branchenzahlen 2013 und Prognose 2014*, Fachverband Biogas e. V., June 2014. AEBIOM.

[143] European Biomass Association (AEBIOM). (2012) *European Bioenergy Outlook 2012 – Statistical Report*. Brussels.

[144] Stranghöner, N., Berg, J., Gorbachov, A., Schaumann, P., Bechtel, A., Eichstädt, R., Wagner, H.-J., Baack, C., Lohmann, J. (2013) Nachhaltigkeitsbewertung stählerner Tragkonstruktionen Erneuerbarer Energien. Stahlbau 82 (2013), Issue 1, pp. 42–48.

[145] EN 15978. (2011) *Sustainability of Construction Works – Assessment of Environmental Performance of Buildings – Calculation Method*. Brussels: European Committee for Standardization.
[146] ÖKOBAUDAT. (2011) Data source version 2011, Federal Ministry of Transport Building and Urban Development. Available at www.oekobaudat.de. Accessed 27 October 2015.
[147] EN 15643-4. (2012). *Sustainability of Construction Works – Assessment of Building – Part 4: Framework for the Assessment of Economic Performance*. Brussels: European Committee for Standardization.

Chapter 5

Sustainability certification labels for buildings

Raban Siebers, Thomas Kleist, Simone Lakenbrink, Henning Bloech, Jan-Pieter den Hollander and Johannes Kreißig

A building consists of a vast number of products, and each product fulfils a purpose not just on its own but in combination with the other materials. Various factors influence the use of construction products: different designs, statutory and normative requirements, the interdependence of materials and the preferences of building owners and architects. It is apparent that it is impossible to judge the sustainability of a single construction product on this basis. This chapter deals with the common certification schemes for sustainable construction and uses a typical steel construction to demonstrate the factors that contribute to a good assessment. The assessment criteria used for certification schemes, such as LEED, BREEAM or DGNB/BNB, offer support for sustainable planning and construction and also serve as a tool for quality assurance. These assessment schemes are presented briefly. Approaches that lead to a good assessment will be highlighted and the weighting within the systems is presented.

Certification schemes were developed with the aim of testing and evaluating buildings in accordance with defined criteria and to make this visible by a 'label'. Compliance with the basic principles of sustainability when planning a building always means the quality of a project is improved. In the private sector, it is solely for the building owner to decide whether a building should additionally be awarded a label for its sustainability. In Germany, it is compulsory to use the appropriate certification scheme for federal buildings if the building costs exceed €2 million, and the result of assessment must be at least 'silver'. In addition to this obligation, marketing and prestige aspects in particular can contribute to a decision to go through the certification process. Certification of buildings to satisfy sustainable criteria provides evidence of a property's high quality, so that it can gain higher rents in the marketplace. In addition, the systematic approach of certification schemes provides support for processes such as strategic planning, defining the objectives, the bidding process, awarding the contract

Sustainable Steel Buildings: A Practical Guide for Structures and Envelopes, First Edition.
Edited by Bernhard Hauke, Markus Kuhnhenne, Mark Lawson and Milan Veljkovic.

and quality assurance. Using certification schemes to prepare documentation for the necessary evaluation during the construction phase is also helpful if certification is to be carried out at a later date.

5.1 MAJOR CERTIFICATION SCHEMES

Sustainability or 'green-building' certificates provide assurance that a building is constructed according to the corresponding certification scheme. Buildings are assessed in accordance with defined criteria. In recent years many certification schemes have been developed, and of these, BREEAM (Building Research Establishment Environmental Assessment Method), LEED (Leadership in Energy and Environmental Design) and DGNB (German quality label for sustainable construction, awarded by the German Sustainable Building Council) are the best known in Europe. The criteria, verification methods and requirements concerning documentation differ among the various certification schemes, and sometimes considerably. In the United States, there were no statutory requirements regarding thermal insulation, whereas, for example, in Germany the Heat Insulation Ordinance had already been introduced in the 1970s.

When statutory requirements are taken into consideration, it can be understood why resources, energy, emissions and waste and so forth are dealt with effectively. This also applies to the associated processes, products and technologies. By international comparison, Europe is therefore one of the leaders in sustainable construction, which was recognised by the developers of the DGNB label. It has thus been able to concentrate on the other aspects of sustainability in addition to the environmental criteria. In other parts of the world, certificates as voluntary incentive schemes replace the legislation and standards that already exist in many European countries. These consist primarily of proposals for individual measures and modular component systems (see Sections 5.1.2 and 5.1.3). Due to these measures, green building has now become widely accepted in the United States. In future, the existing measures will be extended step by step toward an integrated system.

The criteria, methods of verification and regulations relating to documentation differ between the various certification schemes, considerably in some cases. These differences are mainly the result of the diverse backgrounds that have led to the development of the individual schemes. Despite the schemes differing on many points, similarities may be identified:

- The use of all schemes is voluntary for private building owners and investors.
- If building owners decide in favour of certification, the compilation of the documents required for certification is undertaken by specialists, who act as advisors during the planning and construction phases [DGNB Auditor and Consultant, BREEAM Assessor and Accredited Professional, LEED Accredited Professional (voluntary)].
- Ecological and social performance is evaluated in all schemes.

The system developed by DGNB is the one that currently takes economic criteria into consideration most comprehensively and offers an equal balance between the different performances in the final assessment. It is difficult to achieve the global standardisation of certification demanded by the market in order to achieve better comparability. Also defining what is sustainable depends strongly on local boundary conditions (climatic, cultural, political, etc.). The following sections highlight the similarities and differences between the assessment schemes mentioned.

5.1.1 DGNB and BNB

The basic system for the German Sustainable Building Council's quality label (DGNB) was developed by the German Sustainable Building Council (also DGNB) in cooperation with the Federal Construction Ministry. After successfully completing several pilot certifications, the BMVBS and the DGNB have gone their separate ways. The DGNB is voluntary and is used for construction projects carried out by private investors. For public buildings, the Ministry has developed its own derivate for evaluation – Assessment System for Sustainable Federal Buildings (BNB). As both of these systems are very similar as a result of their common history and their development, only the DGNB label will be considered here. Most of the aspects covered, however, also apply to the BNB. In order to quantify the sustainability of a building, DGNB initially defined over 60 individual criteria that can affect it. There are now only 36 criteria for the DGNB 'New Office and Administration Building, Version 2015' label. The reason for this is the continual development of the system that results in the aggregation, addition and discontinuation of criteria. The criteria are divided into environment, economy and social aspects. It was not possible, though, to assign many of the criteria to only one of these three aspects of sustainability. Two additional groups of general performance, 'Technical Quality' and 'Process Quality', were therefore defined in addition to the three pillars 'Ecological Quality', 'Economic Quality' and 'Socio-Cultural and Functional Quality'. The aim of these two broadly based measures of quality is to ensure basic levels of technical and organisational performance are achieved in a sustainable manner across all three pillars. The five qualities mentioned thus far are applied to the building and end at the boundary of the building plot. In addition, the choice of location is also assessed (see Figure 5.1).

A precertificate is first prepared for buildings that are in the planning stage. The final certificate can be issued no later than 1 year after completion, if it can be shown that the planning has been implemented. Precertification also enables improvements to be made to the building design, which means it is then possible for a precertificate in 'gold' to achieve 'platinum' status. It should be noted that it is mainly only the values used for calculations at the planning stage that are used for certification. Measurements are only carried out for some criteria after the building has been constructed.

In the assessment, 10 points can be attained in each of the criteria. Different impact factors from 1 to 7 determine the respective weighting of the criteria. Furthermore, points scored in all areas, except the quality of the location, are added together. The overall assessment of the building is based on this total and

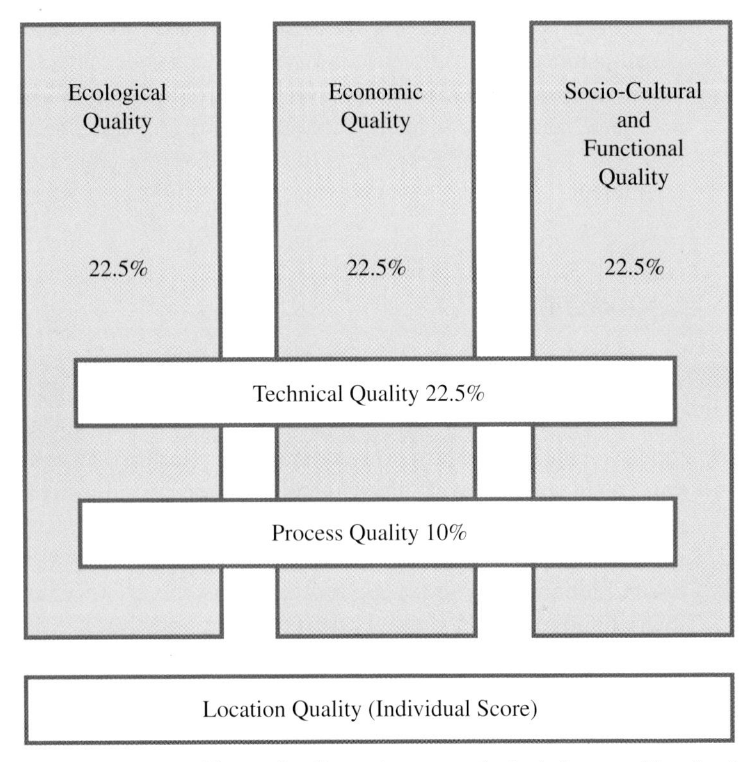

Figure 5.1 The DGNB certificate takes five main groups of criteria into consideration for assessing the quality of a building. The choice of location is also evaluated.

determines whether a silver, gold or platinum certificate is awarded. In 2015 a change in the reachable DGNB certifications was made. New buildings can no longer achieve bronze, silver and gold but now silver, gold and platinum. The DGNB certificate in silver is awarded for an overall performance of 50% provided at least 35% of the maximum points are reached in all areas. Gold is awarded for 65% and the DGNB certificate in platinum is awarded for an overall performance of 80% or more. Only in DGNB label for building stock is it still possible to achieve bronze from an overall performance of 35%.

The DGNB certification system is internationally applicable. Its flexible structure allows precise tailoring for various building types. The international implementation of the DGNB system is based on its adaptation to country-specific conditions. The international DGNB system provides the basis for two international certification routes that vary in scope. Generally, the DGNB certification is possible in every country. English-language criteria based on current European norms and standards can be applied quickly and easily anywhere in the world. Auditors who are familiar with both the DGNB system and local conditions recommend adjustments for the local application that are verified by DGNB. The conformity assessment and certification is carried out by DGNB in Germany. This international certification route is currently being implemented, for example, in Greece, Croatia, Slovenia, Spain, Turkey and the Ukraine.

DSV Logistics Building

Location:	Krefeld, Germany
Architect:	YUHA GmbH
Building description:	170 employees work in the logistics building when it is running at full capacity. The ground floor and mezzanine level are primarily used. The offices are integrated in the mezzanine level.
Steel details:	The building was erected using the modular construction system by GOLDBECK GmbH. The supporting structure is constructed from concrete supports and steel truss girders. The floor slabs are made up of a steel fibre bed with a mineral wear layer on a loosely laid PE foil. The roof sealing consists of trapezoidal steel sheeting, a PE vapour barrier, Class A1 insulation and a mechanically attached PVC sealing sheet.
Sustainability:	The logistics building is heated by gas tube heaters. The offices, conference and break rooms and the sanitary facilities are heated by a gas condensing boiler via radiators. These logistics buildings have been awarded a Multiple Certificate in Gold by the German Sustainable Building Council (DGNB). Each building that is built using this system therefore automatically meets the requirements for a DGNB certificate in Gold.
Awards:	The property was awarded the Certificate in Gold by the DGNB in 2015.

Figure 1 Sandwich facade on the DSV logistics building in Krefeld, Germany. © Goldbeck.

Figure 2 The supporting structure is constructed from concrete supports and steel truss girders. © Goldbeck.

Example provided by Goldbeck.

DSV Office Building

Location: Krefeld, Germany

Architect: AK83 Arkitektkontoret A/S

Building description: The building consists of two wings (approx. 70 × 11 m) that are connected by an atrium and built partly with a basement. Storage and building services rooms are located in the basement. The canteen with the kitchen area and office space as well as the atrium are all accommodated on the ground floor. Office space is provided on the other floors. Each floor has two recessed balconies. The entrances are barrier free, the lifts are adapted to the needs of the disabled and accessible to wheelchairs.

Steel details: The building was constructed using the modular system components by GOLDBECK GmbH. The main loadbearing elements are provided in the form of exterior wall elements and ceiling panels in a composite steel construction. The interior columns consist of steel. System partition walls are installed as dividing walls in the office areas. These can be supplemented, converted or removed in any grid or preferably a grid of 1.25 m with unrestricted operations.

Sustainability: The owner attached importance to sustainable planning and construction of the building from the outset. A thermal building simulation and a daylight calculation were carried out in order to guarantee the economic and ecological quality. The results led to an optimized design of the building services engineering.

The building is cooled and heated via a VRF heat pump with radiators and ceiling-mounted cooling cassettes. The entire building is aerated and ventilated in order to meet the hygienic outdoor air requirements for personnel and minimize energy losses through window ventilation.

The final energy demand was reduced by approx. 18% compared with ENEV 2009 (German Energy Saving Ordinance 2009).

Awards: The property was awarded the Certificate in Gold by the DGNB in 2015.

Figure 1 The office is located near the logistics building – 6570 m² office space, 300 m² basement. © Goldbeck.

Figure 2 Aluminium cassette façade. © Goldbeck.

Figure 3 A workplace of 275 people on four floors with a basement (basement, ground floor, first–third floors). © Goldbeck.

Example provided by Goldbeck.

House of Logistics and Mobility (HOLM)

Location:	Frankfurt/Main, Germany
Architect:	AS&P, Albert Speer & Partner GmbH
Building description:	The building consists of a rectangle (approx. 104 × 39 m) with an atrium and an inner courtyard, which were designed for the use of the employees and are traversed by an airy window facade. The canteen and the kitchen area, lecture rooms, the conference area and office space as well as the foyer and the gallery, which can be used for receptions and exhibitions, are accommodated on the ground floor and first floor. Office space is provided on the other floors. The building services centre is located on the roof.
Steel details:	The property was constructed using the modular system components by GOLDBECK GmbH. The main loadbearing elements are provided in the form of exterior wall elements and ceiling panels in a composite steel construction. The interior columns consist of steel. System partition walls are installed as dividing walls in the office areas. These can be supplemented, converted or removed in any grid or preferably a grid of 1.25 m with unrestricted operations. The steel structure is largely visible from the outside of the building.
Sustainability:	The building was constructed as a certified passive house. The final energy demand was reduced by approx. 33.2% compared with ENEV 2009 (German Energy Saving Ordinance 2009). The building is supplied by a district heating system. Cooling is provided by a water chiller unit. In this respect, the offices and conference rooms are heated and cooled via heating and cooling ceilings. The entire building is aerated and ventilated in order to meet the hygienic outdoor air requirements for personnel and minimize energy losses through window ventilation.
Awards:	The property was awarded the Certificate in Silver by the DGNB in 2014. This certificate corresponds to the DGNB Gold Standard 2015.

Figure 1 This building is located in the 'Gateway Gardens' district in the direct vicinity of Frankfurt. © Goldbeck.

Figure 2 Aluminium cassette facade – the facade changes colour from grey to red depending on the perspective. © Goldbeck.

Figure 3 750–850 people disperse over nine floors here every day. © Goldbeck.

Example provided by Goldbeck.

5.1.2 LEED

The LEED (Leadership in Energy and Environmental Design) rating system, which originates from the United States, was developed prior to the German system. LEED places high importance on the health and wellbeing of building occupants, but, as the name implies, the system assesses primarily the environmental quality of a building. To do this, it defines the individual environmental protection objectives. To promote the material loop points are awarded for a particularly high (10%–20%) recycled content in the building products used measured against the overall cost of the construction materials. The assessment criteria are divided into the following categories: Integrative Process, Location and Transportation, Sustainable Sites, Water Efficiency, Energy and Atmosphere, Materials and Resources, Indoor Environmental Quality, Innovation, and Regional Priority. As with the DGNB system, different schemes are used for specific building types, for example, commercial property or schools. There are four levels of certification: certified, silver, gold and platinum. It should be noted that the LEED system is based on the standards and units of measure valid in the United States, which in some cases differ markedly from European standards and regulations, such as the Energy Saving Ordinance. It is possible to adapt the system to make it easier to use in other countries. Most of the US specific requirements in LEED have alternative compliance paths based on European standards, which can be used in LEED project outside of the United States.

LEED offers a framework for determining the potential, implementation and assessment for sustainable construction, planning and operation. It is a tool that can be used voluntarily and serves as a guide as well as a means of evaluation. As with DBNB and BREEAM, the system is continually being developed and was mainly based on the needs of the U.S. building sector. Only recently has the U.S. Green Building Council recognized the need to adjust LEED for regional markets and allow local compliance paths and Pilot Credits for certain credits. The various LEED schemes address commercial, institutional and residential buildings, as well as neighbourhood developments. In this section, consideration will be given to the system for Building Design and Construction (BD + C) v4 in the New Construction category [7]. There are additional guidelines for interior design, operations and maintenance and neighbourhood development. The aim of all of these systems is to optimise the use of natural resources by promoting regenerative strategies. Any negative environmental and health impacts of construction should be minimised and high quality interior spaces created for the inhabitants of buildings. For the USGBC, the main emphasis is on meeting seven objectives:

1. Reversing the contribution to climate change.
2. Improving the health and well-being of the individual.
3. Protecting and restoring water resources.
4. Protecting and restoring the diversity of species and the ecosystem.
5. Promoting sustainable and regenerative material loops.
6. Establishing a more ecological economy.
7. Improving social justice, environmental awareness and quality of life.

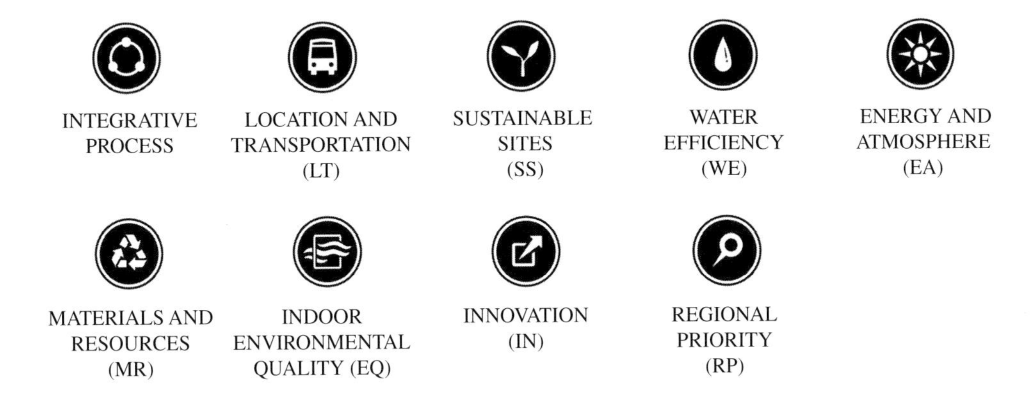

Figure 5.2 Categories in which LEED points can be scored.

These objectives form the basis of the LEED requirements and the points that can be scored. In the current Building Design and Construction (BD + C) certification scheme, it is possible to score points in the following categories (shown in Figure 5.2): Location and Transportation (LT), Sustainable Sites (SS), Water Efficiency (WE), Energy and Atmosphere (EA), Materials and Resources (MR), Indoor Environmental Quality (EQ), Innovation (IN) and Regional Priority (RP). The seven objectives mentioned above determine the weighting of the points within the categories. Depending on the points total achieved, there are four different levels of certification:

- Certified: 40–49 points
- Silver: 50–59 points
- Gold: 60–79 points
- Platinum: 80+ points

LEED aims to create a balance between the current state of the art and the development of innovative concepts. The aim is to consolidate the technical basis for sustainable construction and at the same time promote new strategies. To this end, minimum requirements that require less effort to achieve are defined for each category and points are awarded for additional measures.

Using the LEED scheme's Building Design and Construction – BD + C v4 category it will be shown how and where using steel as the building material can influence the assessment. The particular benefits and capabilities of steel construction will be highlighted.

5.1.3 BREEAM

The first certificate for evaluating the sustainability of buildings was the Building Research Establishment Environmental Assessment Method (BREEAM) from the UK. The system was developed and managed by the British Building Research Establishment (BRE). BREEAM can be regarded as

Torre Diamante

Location:	Milan, Italy
Architect:	Kohn Pedersen Fox Associates Pc
Building description:	With a height of 130 m, the 'Diamond Tower' is Italy's highest building with a steel structure and also the country's third highest skyscraper.
Steel details:	For the tower's steel structure high strength steels were used, which, due to their higher yield strength compared to the conventional steel grade S235, permitted a total material cost savings of up to 50%. Since the cost of rolled section sin S460 M is just 10%–15% higher than S235, savings of 30%–40% could be achieved in the material. Further savings could be registered in the workshop: reduction of welding material, reduction of the surface for corrosion protection due to the use of smaller sections and cost savings for transport due to the lightweight structure.
Sustainability:	The use of high strength steel sections contributed to the weight reduction of the whole building, which resulted in reduced costs, less transport, smaller columns and a shallow foundation. Reliance on renewable energy (photovoltaic panels, ground probes, heat pumps, etc.). Highly efficient facades.
Awards:	Green Building certification LEED Gold.

Figure 1 The Torre Diamante. © bauforumstahl.

Figure 2 Steel structure with inclined columns. © bauforumstahl.

Figure 3 Construction phase – the steel structure is already finished. © Oskar Da Riz / Stahlbau Pichler.

Example provided by ArcelorMittal.

DoubleTree Hotel by Hilton

Location: Avcilar, Istanbul, Turkey

Architect: Uras + Dilekçi Architecture

Building description: The project had initially started as a 14-storey full steel and glass transparent auto showroom tower with two floors of basement in concrete. It was converted instead to a 27-floor hotel building.

Steel details: Thanks to the weldability and design flexibility of ArcelorMittal Histar 460 quality steel jumbos, the basement columns were reinforced in situ and the foundations were refurbished. The structure was converted to its new use without sacrificing the existing construction

Sustainablility: The tower was erected in half the time compared to RC, using only the backyard and parking area of the building as a minimal job site. The trucks carrying the ready-to-erect steel parts arrived only during the nighttime through normal traffic, thus minimizing the neighbourhood disturbance.

Awards: The building is now applying for LEED hotel-in-use certification. The 110 m tower is the highest all-steel building in Turkey and received a TUCSA Steel Building award in 2013.

Figure 1 DoubleTree by Hilton, Avcilar, Istanbul. © Uras+Dilekçi Architecture.

Figure 2 Inside view – visible steel structure. © Uras+Dilekçi Architecture.

Example provided by ArcelorMittal.

the original version of all first-generation certificates because the contents and assessment schemes of most certificates orient themselves on it. Even the LEED system in the United States was originally derived from BREEAM. BREEAM was developed at the end of the 1980s and achieved market maturity in 1990. Having started out as a national system for office and residential buildings, the certification scheme is now used worldwide for a wide range of types of use. Over the years, BREEAM has been revised several times, extended to include the widest possible range of building types and internationalised. The currently valid version for the UK dates from 2014. In BRE's opinion, BREEAM can be used to assess any building in the world. BREEAM offers the possibility to also certify buildings that do not conform to one of the predefined types of use or building.

Although the system is currently in use in various countries worldwide, the focus is still mainly on the UK. Certification in the residential building sector in particular has led to the widespread use of BREEAM there and thus to the large number of registrations and certifications. This widespread use in the UK is backed by government requirements, for example, that all residential buildings completed after 1 May 2008 should be certified in accordance with a specific standard, the BREEAM Code for Sustainable Homes. In the absence of this certification, the seller is obliged to hand over a document to the owner or purchaser in which it is expressly stated that the building has been constructed in accordance with the valid standards but does not fulfil the higher standards of the system with regard to, for example, energy and water efficiency. Compared with other systems,

BREEAM has the largest number of buildings that are registered (i.e. submitted for certification) or certified. At present, there are over 40,000 registered projects worldwide and 15,000 that have already been certified. Outside the UK, 674 buildings have been certified in accordance with the BREEAM International scheme in Europe. In addition to a building project in the United States, the certified buildings are mainly in Europe, including Germany, the Netherlands, Spain, Turkey, Luxembourg, Italy, Belgium and France. The weighting, reachable points and categories differ in some countries, for example in the Netherlands [1]. For example, in Germany, the Centrum Galerie in Dresden and the Pollux building in Frankfurt are certified and received the BREEAM 'Excellent' award.

A BREEAM assessment is voluntary, but many clients in the UK, particularly retail property developers and public authorities, have to provide a certification of building within guidelines as a condition for financing. In Germany, the scheme is mainly used for assessing existing building stock (BREEAM In-Use). However, an update of the whole BREEAM scheme and version for new buildings specially tailored to countries should also appear. The latest method of evaluation, BREEAM International New Construction Buildings 2013, will be explained here, and it will be shown one can make use of the environmental performance of steel for the construction of sustainable buildings. A BREEAM assessment considers a building in its context (including, for example, its connection to the public transport network and its negative ecological impact) with the result that situation and location can strongly affect the overall result. Credits are awarded in 10 categories relating to the quality characteristics of the building, as follows (see also Figure 5.3):

1. Management: holistic management strategies, operational and process management;
2. Health and Wellbeing: indoor and outdoor related;
3. Energy: consumption and CO_2 reduction;
4. Transport: transport-related CO_2 emissions and location-related factors;
5. Water: consumption and efficiency;
6. Material: environmental effects and impacts of the construction materials used during the life cycle;
7. Waste: waste produced and efficient avoidance;
8. Land Use and Ecology – of the site: the demand made on the site and the impact on the surroundings with regard to ecological factors;
9. Pollution: the various negative emissions like sound, light or pollutants;
10. Innovation – additional criteria for exceptional achievements: awarding of additional points for exceeding individual defined criteria by means of particular innovation.

The credits for the categories are weighted and added together to give a total score. The building is then rated as Unclassified or on a scale that ranges from Acceptable (one star) to Outstanding (five stars).

Bosch Siemens Experience Center

Location:	Hoofddorp, Netherlands
Architecture/vision:	William McDonough, Michael Braungart
Building description:	A durable, flexible and multiuse office building in the first sustainable full-service office park in the Netherlands.
Steel details:	Material-efficient, lightweight floor system.
Sustainability:	Adaptability and the cradle-to-cradle philosophy were the starting point for this design.
Awards:	BREEAM good (version 2010) ; new projects FOX/AMWB and FIFPRO – same structure as BSH – BREEAM Excellent (BREEAM version 2014).

Figure 1 Bosch Siemens Experience Center, Hoofddorp. © Slimline Buildings.

Figure 2 During the construction phase – the hole pattern in the steel beams of the Slimline floor system for accommodation of all common building installations is still visible. © Slimline Buildings.

Example provided by Slimline Buildings.

Head Office of AGC Glass Europe

Location:	Louvain-la-Neuve, Belgium
Architects and engineers:	Philippe Samyn and Partners architects & engineers
Building description:	12.120 m^2 office building for 575 members of staff. The interior design can be changed through open-plan areas, closed offices or separate wings if necessary. The same flexibility applies to meeting areas and restaurants, where the space is divided up by using movable, acoustic partition walls.
Steel details:	Steel roof in the main atrium in S690. Movable light steel stairs also in S690.
Sustainability:	In terms of energy performance, the aim for the building is to achieve zero energy. Energy saving throughout (natural light, insulation, etc.) with the use of efficient materials (energy-saving circulation, regulation, etc.) and reliance on renewable energy (photovoltaic panels, ground probes, heat pumps, etc.) made it possible to do so.
Awards:	VALIDEO and BREEAM certified.

Figure 1 Head office of AGC Glass Europe. © Simon Schmitt.

Figure 2 Crosswise to the office wings, a gallery (or atrium) provides an obligatory route for everyone in the building. © Marie-Françoise Plissart.

Figure 3 Slipped under the offices following the natural gentle slope (1.5%–2.5%), the parking area is discreetly positioned under the building, preserving the surrounding landscape. © Jean-Michel BYL.

Example provided by Philippe Samyn and Partners.

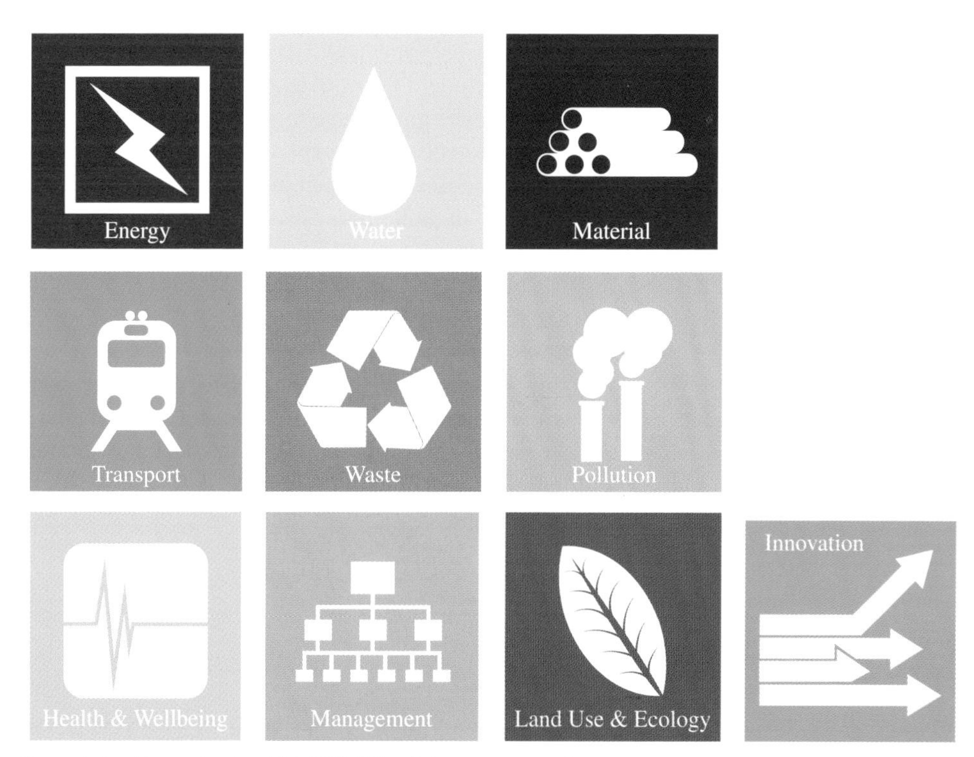

Figure 5.3 Categories in which BREEAM points can be scored.

5.2 EFFECT OF STRUCTURAL DESIGN IN THE CERTIFICATION SCHEMES

5.2.1 Life-cycle assessments and environmental product declarations

A building is evaluated on the basis of the results of an impact assessment for the life cycle. To do this, emissions affecting the air, water and soil are grouped according to impact potential and converted using characterisation factors based on recognised models of environmental effects. The environmental problems resulting from the emissions are represented by the following indicators:

1. climate change: global warming potential (GWP) in kg CO_2 equivalents;
2. destruction of the stratospheric ozone layer: ozone depletion potential (ODP) in kg R11 equivalents;
3. summer smog: photochemical ozone creation potential (POCP) in kg C_2H_4 equivalents;
4. forest and fish mortality: acidification potential (AP) in kg SO_2 equivalents;
5. overfertilisation: eutrophication potential (EP) in kg PO_4 equivalents.

The consumption of primary energy is a further indicator calculated in a life-cycle assessment (LCA), and this is often expressed as

1. nonrenewable primary energy requirement (PE_{ne});
2. renewable primary energy requirement (PE_e);
3. total primary energy requirement (PE_{ges}).

Renewable sources of energy include, for example, solar energy, geothermal energy, hydroelectric power, wind power and biomass.

Environmental product declarations (EPDs), prepared by manufacturers and trade associations or the national databases with their average values, are used as data sources for conducting LCAs of buildings. Several European producers under the lead management of bauforumstahl have prepared an EPD for structural steels [2] and based on this for hot-dip galvanised structural steels [3]. EPDs for several other steel products are available. (See Chapter 3.9.)

In the LCA, it is important that the values for Module D (benefits and impacts beyond the system limit defined by EN 15804 and EN15978 [4], [5]) are always taken into account when using EPDs or values from the national databases. This contains a credit for structural steel because of its recycling potential. (See Table 5.1, more about recycling potential in Chapter 3.9.1.).

5.2.1.1 *DGNB criterion ENV 1.1: life-cycle assessment – emissions-related environmental impacts*

Aspect: Ecological Quality
Group of criteria: Global and Local Environment Impact
Specific impact factor: 7
Proportion of overall evaluation: 7.9%

Table 5.1 Effect of structural design in steel and steel composite construction on the assessment in certification schemes – EPDs and LCA [6]–[8].

EPDs and Life-Cycle Assessments					
	Aspect	Criteria	Importance Factor/Maximum Score	Proportion of Total Score	Comments with Respect to Steel Construction
DGNB	Ecological Quality – Global and Local Environment Impacts	ENV 1.1 Life Cycle Assessment – Emissions-related Environmental Impacts	7	7.9%	Use available EPDs: e.g. for structural steel (EPD-BFS-20130094)[5], or for hot-dip galvanised structural steel (EPD-BFS-20130173)[6].
	Ecological Quality – Resource Consumption and Waste Generation	ENV 2.1 Life Cycle Assessment – Primary Energy	5	5.6%	Take Module D into consideration – recycling potential.
LEED	Materials and Resources (MR)	Building Life Cycle Impact Reduction	3	3.1%	
		Building Product Disclosure and Optimization – Environmental Product Declarations	2	2.1%	
BREEAM	Materials (Mat)	Mat 01 Life cycle impacts	6	6.82%	

Table 5.2 Weighting of the environmental indicators in DGNB.

Global Warming Potential (GWP)	Ozone Depletion Potential (ODP)	Photochemical Ozone Creation Potential (POCP)	Acidification Potential (AP)	Eutrophication Potential (EP)
40%	15%	15%	15%	15%

The ENV1.1 Life-Cycle Assessment – Emissions-Related Environmental Impacts criterion is evaluated in accordance with the results of an LCA of the building. The environmental problems resulting from the emissions include global warming, destruction of the stratospheric ozone layer, summer smog, forest and fish mortality, and overfertilisation of waterbodies and the soil and are represented by the indicators given earlier.

Expressed in simple terms, these indicators are determined in accordance with the mass, materials, volume and building physics related properties of the building and compared with an average reference value that depends on the use profile of the building being evaluated (office building, commercial building, school, etc.). A higher or lower score is given depending on whether the building exceeds or undershoots the reference value. Within the evaluation, the indicators have the weightings shown in Table 5.2.

Construction-related and use-related results are added together in the LCA of a building. Here construction is apportioned as production, maintenance, dismantling and disposal, including any benefits and impacts beyond the system boundary, such as transport and maintenance that are not yet considered in the DGNB scheme. With regard to use, supply and waste management services and repairs during the use phase have to be taken into account. The LCA of the building is calculated in accordance with EN 15978 [5], and this means the modules A1–A3, B2, B4, B6, B7, C3–C4 and D of this standard or EN 15804 [4] are currently included (see Chapter 2).

The above-mentioned environmental indicators are converted to a value per m^2 of net floor space (NFS) that is covered and enclosed to the full height per year of use (kg environmental impact eq./m^2 NFS * a). The life expectancy to be used depends here on the respective building type; for office buildings, a life expectancy of 50 years is used.

For the use-related part, the reference values for the evaluation are determined along the lines of the standard for the current energy saving ordinances and those for the construction-related part are determined using average values, for example, from the German construction ministry and previous DGNB certifications for the same use profile. Where possible, the LCA of the building should be established during the planning phase. It can then serve as an important instrument for optimising the environmental quality of the building. The characteristic values to be used for the individual construction materials should be taken from the respective EPD where possible. EPDs have already been prepared for many products. The characteristic values for construction materials/products for which no EPD is currently available can be taken from national databases (see also Chapter 3.10).

5.2.1.2 *DGNB criterion ENV 2.1: life-cycle assessment – primary energy*

Aspect: Ecological Quality
Group of criteria: Resource Consumption and Waste Generation
Specific impact factor: 5
Proportion of overall evaluation: 5.6%

The ENV 2.1: Life-Cycle Assessment – Primary Energy criterion evaluates the following indicators:

1 nonrenewable primary energy requirement (PE_{ne});
2 total primary energy requirement (PE_{ges});
3 proportion of renewable primary energy.

The aim here is to exceed the statutory regulations in order to benefit the prevention of climate change and the global protection of resources. In an analogous manner to evaluation in accordance with the ENV 1.1 criterion, a comparison is made with a reference value based on the current energy saving ordinances and average values from previous DGNB certifications for the same use profile; there is a credit for exceeding this. Renewable sources of energy include, for example, solar energy, geothermal energy, hydroelectric power, wind power and biomass. The three above-mentioned evaluation indicators each have the same effect on the end result.

5.2.1.3 *LEED Materials and Resources (MR) – life-cycle impact reduction and building product disclosure and optimization*

MR Credit: Building Life-Cycle Impact Reduction
Maximum score: 3
Proportion of total score: 3.1%

MR Credit: Building Product Disclosure and Optimization – Environmental Product Declarations
Maximum score: 2
Proportion of total score 2.1%

The Materials and Resources category aims to minimise the 'grey energy' and environmental impacts associated with the production, processing, transport, maintenance and disposal of construction materials. The requirements are stipulated in such a way that a life-cycle approach that aims to achieve optimised environmental performance and resource efficiency scores more points.

In the Materials and Resources category the new version of LEED demands an LCA of the whole building. The highest score can be achieved if it can be demonstrated that the building under consideration is at least 10% better than a comparable average building in the environmental category GWP and at least two of the other environmental impacts – ODP, POCP, AP, EP, or for nonrenewable primary energy (see Chapter 3.3). Here, the comparative building can be chosen, for example, by the LEED Accredited Professional. European producers under the lead management of bauforumstahl have prepared an EPD for structural steels [2], and based on this for hot-dip galvanised structural steels [3]. EPDs for several other steel products are available (see Chapter 3.11). It is important with LEED, although not yet prescribed bindingly, that the LCA always uses values for benefits and impacts outside the system boundary (Module D in accordance with EN 15978; see Chapter 2.2). Here a credit for steel products is given according to the recycling potential, and this improves the LCA by about 50%.

Other points can be awarded for submitting EPDs for 20 permanently installed building products and proof that over 50% (based on the total cost of the material) of the installed products give better values in at least three of the common environmental categories (see above and Chapter 3.3.1) than the industrial average. All LCA results prepared in accordance with ISO 14044 can be submitted; however, Type III declarations per ISO 14025 and EN 15804 (see Chapter 3.6), which are developed by involving independent third parties and additionally checked independently, give a score that is four times higher. The EPDs for structural steel and hot-dip galvanised structural steel mentioned above and presented in Chapter 3.11 fulfil this requirement.

The use of structural steel also results in points being awarded where the recycled content of the building material is concerned. The recycled content is calculated from the sum of the postconsumer recycled content and a half of the preconsumer recycled content – for European grades of structural steel the calculation is 62% + ½(8%) = 66% [9]. The value to be used is thus very high. In earlier versions of LEED the maximum score was given for 20%. The score is now given for the fraction of recycled material in the total material costs, and it has to be at least 25%. The recycled content is often 100%, especially for producers of rolled sections in Western Europe, so depending on the producer, the individual value can be requested from the producer or distributor.

5.2.1.4 *BREEAM category material – mat 01 life-cycle impact*

Criteria: Mat 01 Life-cycle impacts
Maximum score: 6
Proportion of total score: 6.82%

With BREEAM, the environmental compatibility of a building product is assessed using BRE's *Green Guide to Specification*. The guide provides an assessment of components made from the most important building products and their applications in different types of building. The components are evaluated on a scale from A+ to E. Components with a steel content achieve very good evaluations in the guide and thus lead to high BREEAM credits in this category. An LCA can be used as an alternative verification procedure to the *Green Guide* to evaluate the criteria in the Material category. The use of existing national LCA evaluation systems is permitted. The systems in other European countries are currently not listed, though, so BRE Global first has to give its approval before they are used.

In the LCA, the following boundary conditions must be met as minimum requirements: at least three environmental indicators have to be specified; the calculation takes into account the complete building life cycle including use and recycling; and the LCA has to be carried out in accordance with international standards (ISO 14040, ISO 14044, et al.). The EPDs presented in Chapter 3.11 provide the necessary base data.

5.2.2 Risks to the environment and humans

The effects of pollutants on the local environment are taken into consideration in certification schemes. Substances that pose a risk to the soil, the air, the ground or surface water or the health of human beings, flora and fauna during transport, processing, use or disposal should not be used. These include, for example, halogens, heavy metals, biocides or volatile organic compounds (VOCs).

Steel is a building material that is free from emissions, absolutely odourless and completely resistant to microbiological infestation and may be protected by hot-dip galvanising. Steel construction is a dry method of construction. It does not cause any residual moisture in the finished building and thus does not harbour any risk of mould formation either. This means steel does not contribute to microbiological contamination or loading of the indoor air. The building material steel is also free from emissions even when it comes to the evaluation of the concentration of volatile organic compounds emitted. Manufacturer's information on pollutants should be taken into account when using coatings that may affect the indoor air. Sufficient harmless products are available, such as intumescent paints and other water-based coatings, to obtain most of the desirable properties (see Table 5.3).

5.2.2.1 DGNB criterion ENV 1.2: local environmental impact

Aspect: Ecological Quality
Group of criteria: Global and Local Environment Impact
Specific impact factor: 3
Proportion of overall evaluation: 3.4%

As part of Ecological Quality, the ENV 1.2 Local Environmental Impact criterion takes the effects of pollutants on the local environment into consideration in the LCA in addition to the global environmental effects.

All materials and building products should be recorded in a parts catalogue according to cost groups. The high risk groups of materials and substances are then checked individually in a product-related manner using a criteria matrix specified by DGNB and subsequently evaluated. This matrix is based on the corresponding EU guidelines, technical rules for hazardous substances and sector-related design codes. Four levels of quality are possible, whereby the fourth is the best quality and gives the best score for the evaluation. It is important here that the coatings used on components in all the cost groups are of the same quality because, with some exceptions [6], according to the ENV 1.2 criterion only the lowest level

Table 5.3 Effect of structural design in steel and steel composite construction on the assessment by certification schemes – risks for the environment and humans [6]–[8].

Risks for the Environment and Humans					
	Aspect	Criteria	Importance Factor/ Maximum Score	Proportion of Total Score	Comments
DGNB	Ecological Quality – Global and Local Environment Impacts	ENV 1.2 Local Environmental Impact	3	3.4%	Steel is emission free, as is hot-dip galvanised steel. VOC = 0 [9]. With coatings, observe manufacturer's instructions. Depending on location, choose nonhazardous products. Application over large areas does not take place at building site.
	Socio-cultural and Functional Quality – Health, Comfort and User-friendliness	SOC 1.2 Indoor Air Quality	3	3.2%	
LEED	Materials and Resources (MR)	Low-Emitting Materials	3	3.1%	
BREEAM	Materials (Mat)	Mat 03 Responsible sourcing of materials	3	3.4%	
	Health and Wellbeing (Hea)	Hea 02 Indoor air quality	4	4.3%	

of quality achieved is taken into consideration in the final assessment. Proof that the corresponding levels of quality have been achieved can be provided in the form of technical information from the manufacturer, material safety data sheets (MSDSs), Types I and III environmental declarations, and manufacturers' declarations regarding ingredients and constituents of the formulation.

It should be noted that where it is decisive for the component being considered the ENV 1.2 criterion only evaluates the VOC content of the product and not its VOC emissions in use. The quantitative emissions of volatile substances in the interior are taken into consideration in the SOC 1.2 criterion covering hygiene inside the building.

5.2.2.2 DGNB criterion SOC 1.2: indoor air quality

Aspect: Socio-cultural and Functional Quality
Group of criteria: Health, Comfort and User Satisfaction
Specific impact factor: 3
Proportion of overall evaluation: 3.2%

To achieve the required quality of the indoor air from a hygienic point of view, it is necessary to minimise the concentration of harmful emissions. These include volatile organic compounds and formaldehyde. In addition to release of gases from the materials installed, the air change rate has a major effect on the hygiene of the indoor air. During certification, the quality of the indoor air is evaluated by on-site measurements (at the latest 4 weeks after completion of the building, without furniture) and by calculating the mechanical and natural air change rates.

5.2.2.3 *LEED Indoor Environmental Quality (EQ)*

EQ Credit: Low Emitting Materials
Maximum score: 3
Proportion of total score: 3.1%

This category covers traditional indoor air quality aspects such as ventilation, source control, monitoring for user-determined contaminants as well as emerging design strategies, including a holistic, emissions-based approach. Thermal comfort, requirements for lighting quality and advanced lighting metrics, and acoustics are addressed as well. The aspects of air quality range from low emitting building materials to requirements for the building services engineering.

5.2.2.4 *BREEAM category material – mat 03 responsible sourcing of materials*

Criteria: Mat 03 Responsible sourcing of materials
Maximum score: 3
Proportion of total score: 3.4%

BREEAM also takes into account procuring building products in a responsible manner. In the case of steel, this involves purchasing products from a producer who has an environmental management system. Not only is the steelmaker involved but also the product manufacturer and the steel constructor. In addition to the primary steel construction it also applies to products, such as cold-rolled profiles for purlins, facades and roofs. The latest reports of the Worldsteel Association document that over 90% of steel producers are accredited with a recognised environmental management system. Furthermore, a large number of manufacturers of steel building products are certified in accordance with ISO 14001 [10]. (See also Section 5.2.1.4.)

5.2.2.5 *BREEAM category health and wellbeing*

Criteria: Hea 02 Indoor air quality
Maximum score: 4
Proportion of total score: 4.3%

The environmental section Health and Wellbeing reflects the influences on the human living and working in a building, which includes visual, thermal and acoustic issues, and the quality of air and water. In addition, the effective design measures that promote low risk, safe and secure access to and use of the building are reviewed. The criteria for indoor air quality deals with a healthy internal environment through the specification and installation of appropriate ventilation, equipment, materials and finishes.

5.2.3 Costs during the life cycle

Planning, construction, operation, maintenance and cleaning and dismantling of buildings involves considerable costs, which have to be met during the course of the life cycle by rental income. In order to use financial resources economically, the total costs incurred in the construction, operation and disposal of a building should be kept to a minimum.

The preconception that it would be very difficult to determine the cost of steel construction because of the fluctuations in raw material prices is only true to a limited extent. More influential are the costs for construction plant and equipment, wages and transport. In addition, there are capital costs resulting from financing the building project. The overall economic efficiency of a project will be improved if these costs are minimised by means of good planning and rapid completion.

There are several sources available for obtaining the necessary information to make an initial estimate of the manufacturing costs for steel construction. An initial indication can be obtained from national construction organisations [11] or from the planning aid titled *Steel Price Book* [12] from bauforumstahl. Both sources give typical price ranges for steel construction projects and are suitable as a guide for early cost planning.

When finalising the design of a building project, several quotations from different steel fabricators should be obtained at an early stage. Steel fabricators offer useful advice on the integrated planning and economics. Cellular beams, for example, are often very material efficient because they allow different steel grades and profiles to be combined together in a single beam. Compared with the classical S235 steel, the use of higher strength steels allows considerable material savings to be achieved. The extra costs for S355 steel are about 5% for rolled profiles, but the usable yield strength is 50% higher. With S460 steel, the price increase is about 9%, but the added costs for material procurement are offset by a gain in tensile strength of 90% compared with S235 steel. For a multistorey car park at Düsseldorf Airport, beams made from S460 steel instead of S235 achieved savings of 24% based on steel tonnage and 17% based on building costs (Figure 5.4).

Construction that is suitable for dismantling and recycling pays for itself when it comes to dismantling a building or individual components. The dismantling costs consist mainly of the costs for labour and disposal. The proven collection rate of 99% [2] for structural steel in Europe shows that the dismantled steel components and steel scrap are important commodities for which a comprehensive market has developed over many decades. Accordingly, there are also zero costs for

Figure 5.4 Rental car centre at Düsseldorf Airport. © Deutsche Industrie- und Parkhausbau GmbH.

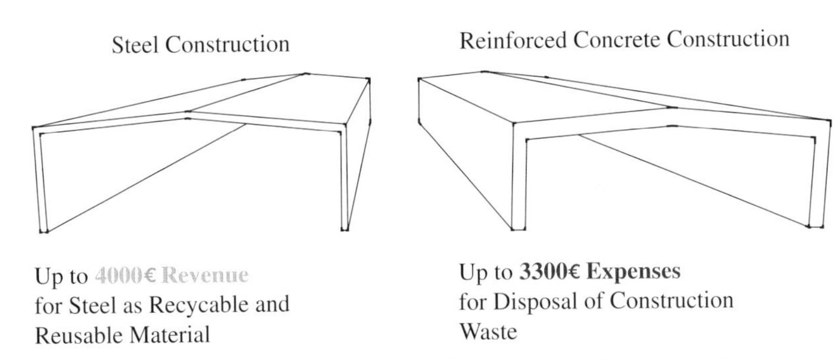

Figure 5.5 Revenue or expenses at the end of life of typical single-storey buildings. Additional information: gross floor area = 900 m² for both variants, 16.8 t steel – scrap price: ~250 €/t, 111.3 t of reinforced concrete – concrete rubble disposal costs, reinforced: ~30 €/t, foundations not considered.

disposal – and what is more, there is remuneration as well. With good dismantlability and subsequent separability of the individual material fractions, the labour costs for dismantling can often be covered by the income from material sales alone.

There are many benefits associated with constructing with steel that can improve the result for the LCA of a building. These include, for example, prefabrication, which is typical for steel construction and makes shorter construction periods possible, which in turn allows earlier use of the building and earlier repayments of debt. If dismantling and disposal costs are also taken into account in the assessment in future, the building material steel will also score here due to its ease of dismantlability and recyclability. The latter makes reusable steel components and steel scrap much sought-after products (see Figure 5.5 and Table 5.4).

Table 5.4 Effect of structural design in steel and steel composite construction on the assessment by certification schemes – costs during life cycle [6]–[8].

Costs during Life Cycle					
	Aspect	Criteria	Importance Factor/ Maximum Score	Proportion of Total Score	Comments
DGNB	Economic Quality – Life Cycle Costs	ECO 1.1 Building-Related Life-Cycle Costs	3	9.6%	Planning aid 'Steel Price Book' (Kosten im Stahlbau). [12] and national sources e.g. [11]. Use of higher strength grades of steel. Ductility and loadbearing reserves. Short building time – early use. Dismantlability.
BREEAM	Materials (Mat)	Mat 05 Designing for Robustness	1	1.1%	
	Management (Man)	Man 05 Life cycle cost and service life planning	3	1.6%	

5.2.3.1 *DGNB criterion ECO 1.1: building-related life-cycle costs*

Aspect: Economic Quality
Group of criteria: Life-Cycle Costs
Specific impact factor: 3
Proportion of overall evaluation: 9.6%

There are considerable costs associated with the planning, construction, operation, maintenance and cleaning and dismantling of property.

In an analogous manner to the environmental criteria, the economic quality over the whole life cycle of a building is also taken into consideration in the sustainability assessment. The time of occurrence has to be considered here – in contrast to the environmental impacts of an LCA, which are independent of time. The calculation is carried out on the basis of the cash-value method. Within the scope of a certification, the period under review is set differently depending on the type of use. For office buildings, it is 50 years. The costs for dismantling and disposal as well as possible income for recyclable products or reuse are currently not taken into consideration in the certification process.

Life-cycle costs of a property in a certification scheme have to be compared with those of other buildings in order to evaluate them. This presupposes that certain aspects are fixed in order to conduct calculations in the same manner for all buildings. These include taking an assumed hourly rate for cleaning of €17/hour, €0.20/kWh for use of electricity and an average interest rate of 5.5%.

5.2.3.2 *BREEAM category material – mat 05 designing for robustness*

Criteria: Mat 05 Designing for robustness
Maximum score: 1
Proportion of total score: 1.1%

Criteria 'Designing for robustness' is classified into life-cycle costs because adequate protection of exposed elements of the building and landscape minimises the frequency of replacement and maximises materials optimisation. (See also Sections 5.2.1.4 and 5.2.2.4.)

5.2.3.3 *BREEAM category management – man 05 life-cycle cost and service life planning*

Criteria: Man 05 Life-cycle cost and service life planning
Maximum score: 3
Proportion of total score: 1.6%

The management category deals with the delivery of a functional and sustainable construction process. The construction site has to be managed in an environmentally and socially considerate, responsible and accountable manner, especially in terms of resource use, energy consumption and pollution. The needs of current and future building users and other stakeholders have to be satisfied, considering life-cycle costing and service life planning in order to improve design, specification and through-life maintenance and operation.

5.2.4 Flexibility of the building

In sustainable construction, it is of major importance to have a high degree of flexibility and convertibility of use coupled with highly efficient use of the area available. Saving floor space – or making better use of existing space – can enhance the return on investment, particularly in areas where land prices and rents are high. From an economic point of view, a highly efficient use of floor space should be achieved. Structural steel allows buildings to be constructed that are slim and make particularly efficient use of materials and floor space. Compared with concrete construction with comparable thermal insulation properties, steel construction allows floor areas to be increased by up to 12% by reducing wall cross sections. Composite columns are smaller and so increase the effective rentable area. The longer spans achievable by means of steel construction eliminate columns and supporting walls in the room. Lightweight partition walls can be used to divide up the floor space, and they are flexible and can be installed or removed quickly and in a dry manner.

In addition to the orientation of the supporting structure, the major services also have to be adaptable in case of changes in use. Steel composite floors allow

Table 5.5 Effect of structural design in steel and steel composite construction on the assessment by certification schemes – flexibility of the building [9]–[11].

Flexibility of the Building					
	Aspect	Criteria	Importance Factor/ Score Achievable	Proportion of Total Score	Comment
DGNB	Economic Quality – Growth in Value	ECO 2.1 Flexibility and Adaptability	3	9.6%	Space-efficient. Slim supporting structure. Large spans and less columns. Capability for adding storeys.
LEED	Materials and Resources (MR)	Design for Flexibility	1	1.0%	
BREEAM	Innovation (Inn)	Inn 01 Innovation	10	10%	

penetrations to be made in the floor slab near to the columns at a later date. Provision for additional openings can be made in composite beams by use of cellular beams. Openings can be introduced later by local stiffening. Although changes are often made to the services and floor plan after a few years, for example, where there is a change of tenant, the choice and arrangement of the supporting structure of a building significantly influences its adaptability. The supporting structure can be easily strengthened at a later date. However, it is often worthwhile to choose higher strength steel at the beginning to ensure flexibility from the start. Furthermore it is possible with a lightweight steel construction to add storeys or introduce mezzanine floors or similar measures to existing buildings to make more efficient use of space (see Table 5.5).

5.2.4.1 *DGNB criterion ECO 2.1: flexibility and adaptability*

Aspect: Economic Quality
Group of criteria: Life-Cycle Costs
Specific impact factor: 3
Proportion of overall evaluation: 9.6%

Seven indicators are taken into account and evaluated to assess the flexibility and adaptability of a building under consideration:

1. Efficient use of floor area: To determine how efficiently floor area is used, the usable area is expressed as a function of the gross floor area for area a (BGFa) for the specific use profile in accordance with DIN 277-1: usable area/BGFa. In order to score lots of points, this factor should be as high as possible.
2. Height between floors: The average floor height is very important. This is taken to be the height from the top edge of the unfinished floor to the lower edge of

the unfinished ceiling. A height between floors in excess of 3.0 m gives the maximum score. No points are given for heights less than this.

3 Building depth: The building depth between both facades has to be available for 70% of the facade length (no interior shafts, stairwells or lifts). The maximum score is given here for a building depth of 12.5–14.5 m. Half the points can be obtained for depths down to 10.0 m or up to 16.5 m. Dimensions outside this range do not score any points.
4 Vertical division: The relationship between the gross floor area of the standard storey and the number of central areas of the development is assessed. The smaller the gross floor area of each central area, the more the building can be compartmentalised and the greater the flexibility with respect to the size of possible units. The maximum score is achieved here for less than 400 m^2 per central area. Next levels are up to 600 m^2 and up to 1200 m^2.
5 Floor plan: When dividing the floor space into smaller units, more toilet units are necessary in accordance with the division. A certain quantity of these should already be available or the connections should be installed so that toilet units can be retrofitted. If escape routes do not pass through another unit, the individual units can be better adapted to the user or the use. This contributes to the building being more usable. For this indicator, one can score half of the points for fulfilling each of these prerequisites.
6 Construction: The structural design is evaluated with respect to certain components whose condition affects the conversion of buildings: (a) interior walls should not be loadbearing, (b) dividing walls can be installed without interfering with the floor and the ceiling, (c) dividing walls can be reused, and (d) floors have reserves for greater payloads for various forms of conversion. These properties each account for a quarter of the points for this indicator and are added to give the total score.
7 Building services: The convertibility of use of building services is considered using ventilation/air conditioning technology, cooling, heating and water as parameters. The possibility of being able to adapt to different room situations and the related constructional measures are evaluated. Points are awarded individually for all the parameters so that building services alone account for 40% of the total points for the ECO 2.1 criterion. The building services that perform best are those whose distribution is arranged in a flexible manner so that they can be adapted to a modified layout without any need for constructional measures.

5.2.4.2 *LEED Materials and Resources (MR) – design for flexibility*

MR Credit: Design for Flexibility
Maximum score: 1
Proportion of total score: 1.0%

The target is to conserve resources associated with the construction and management of buildings by designing for flexibility and ease of future adaptation and for the service life of components and assemblies. (See also Section 5.2.1.3.)

5.2.4.3 *BREEAM category innovation*

Criteria: Inn 01 Innovation
Maximum score: 10
Proportion of total score: 10%

Awarding of additional points for overfulfilment of criteria by using special innovations. The particular advantages of steel construction described above can also lead to bonus points.

5.2.5 Recycling of construction materials, dismantling and demolition capability

Due to its physical properties and the joining techniques commonly used in steel construction, steel can be recycled and preferly reused. Joints that can be unbolted with only a minimum of effort allow materials to be separated easily according to material type. Components recovered in this way can either be reused directly or melted down and turned into new products (see Table 5.6).

Recycling is generally taken to mean any recovery process that enables waste materials to be processed into products, materials or substances, either for the original purpose or for other purposes. This means it also includes, for example, material recycling where rubble is used for road building. There are different levels at which recycling can be achieved as follows:

- In 'true' recycling, a substance is processed in such a way after use that it subsequently exhibits the same quality as it did before.
- In contrast to this, 'upcycling capability' means a building material can be processed in such a way that it is of better quality than previously.
- There is a clear loss in quality with downcycling. Well-known examples of this are in papermaking, where recycled paper is no longer capable of achieving the same whiteness of paper produced via the primary production route or the above-mentioned recycling of rubble for use as a filler in road building.

Recycling-friendly construction assumes that thought is given to recyclability at a later date when planning a building. The designer therefore has a decisive influence during the planning phase on a building's suitability for recycling later. First of all, the choice of building products plays a decisive role. Construction materials that can be reused after dismantling or can be sent for true recycling are preferable to those where only the material can be recycled. Structural steel, for example, is characterised by outstanding recyclability – what is more, after remelting modern thermomechanical rolling processes can even be used to achieve a higher grade than that of the starting material. Besides the basic recyclability of construction materials, the dismantling and separability of the building products used has an effect on the ease of recycling of the whole construction. A dismantling-friendly

Table 5.6 Effect of structural design in steel and steel composite construction on the assessment by certification schemes – recycling of building materials, dismantling and demolition capability [6]–[8].

Recycling of Building Materials, Dismantling and Demolition Capability					
	Aspect	Criteria	Importance Factor/ Maximum Score	Proportion of Total Score	Comments
DGNB	Technical Quality – Quality of Technical Implementation	TEC 1.6 Ease of Dismantling and Demolition	2	4.1%	For structural steel: 99% collection rate with 11% reuse and 88% recycling [5]. Recycling content of structural steel: postconsumer: 62% preconsumer: 8% [7]. Reversible connections. Can be sorted according to material type. Cradle-to-cradle capability. In DGNB structural steel components rank 'A', which is the highest
LEED	Materials and Resources (MR)	Construction and Demolition Waste Management Planning	Minimum requirement		
		Building Product Disclosure and Optimization – Sourcing of Raw Materials	2	2.1%	
		Building Product Disclosure and Optimization – Material Ingredients	2	2.1%	
		Construction and Demolition Waste Management	2	2.1%	
BREEAM	Materials (Mat)	Mat 03 Responsible sourcing of materials	3	3.4%	
	Waste (Wst)	Wst 02 Recycled aggregates	1	1.5%	Blast furnace slag from steelmaking

design allows materials to be sorted according to material type and whole components to even be returned to a product recycling stream. This satisfies a significant criterion for high grade reuse. Increasing the reusability and recyclability also offers monetary benefits for the property owner because the individual material fractions are immediately available sorted and the effort required for dismantling is reduced. These benefits are only apparent, though, when the building is completely demolished. One can also benefit from a dismantling-friendly building concept when carrying out use-related renovation measures.

The following principles should thus be adopted to during planning so that a building is suitable for recycling and reuse:

- The materials used should be recyclable.
- The number of materials used should be kept as small as realistically possible.
- Avoid difficult-to-separate composites.
- The use of products with a long service life and the choice of nondestructive joints enable products to be used again in other buildings or in another part of the building.

All of the properties mentioned above can be obtained optimally using steel construction, as follows:

- The assumed life of steel components is well in excess of 50 years. Depending on the location and the loading coupled with suitable protective measures, for example, hot-dip galvanising, it can be significantly more than 100 years. (See Chapter 4.6.)
- Steel frames can often be dismantled easily and can thus be reused.
- Even steel beams can be separated easily from the concrete slab. As far as possible, the use of a concrete encasement is not encouraged, as it contributes little to the structural efficiency.
- Components that are no longer reusable can also be cut up quickly and effectively using a flame cutter and then melted down again and recycled or even upcycled.
- Steel is the most recycled building material worldwide. About 35% [13] of total steel production is based on recycled steel scrap. This proportion is significantly higher in the case of structural steel [9].

5.2.5.1 *DGNB criterion TEC 1.6: dismantling and ease of demolition*

Aspect: Technical Quality
Group of criteria: Quality of the Technical Execution
Specific impact factor: 2
Proportion of overall evaluation: 4.1%

A key aim of sustainable construction is to reduce eco-unfriendly waste and to return used products to the material loop. Besides the use of recyclable materials, the technical feasibility of dismantling is also of great importance when it comes to the dismantlability and ease of recycling of a building. This results in correct sorting of dismantled items and recovery of high grade recycled material. Certification takes into account the ease of recycling of the building components and products used. The assessment is based on two indicators:

1. Recycling-oriented material selection: All components including the structural system are considered. All structural steel components are classified in highest rank = 'A'. That means 'Re-use or recycling to a comparable product'. This means that the recycled product is in its essential characteristics similar to the characteristics of the primary material. The recycled product is capable to cover the entire spectrum of the primary material or even more. Furthermore, components that are proven reused in the considered building can also be ranked 'A'.
2. Recycling-compatible design: Only components that are replaced more often, such as nonloadbearing walls, windows, floorings and roof coverings, are considered. Steel components can be classified in the highest rank = 'Dismantling friendly construction'. This means a possibility of a nondestructive separation of the component and its layers. A recycling of the materials recovered without limitation is possible.

5.2.5.2 *LEED Materials and Resources (MR) – waste management and sourcing of materials*

MR Credit: Construction and Demolition Waste Management Planning
Minimum requirement

MR Credit: Building Product Disclosure and Optimization – Sourcing of Raw Materials
Maximum score: 2
Proportion of total score: 2.1%

MR Credit: Building Product Disclosure and Optimization – Material Ingredients
Maximum score: 2
Proportion of total score: 2.1%

MR Credit: Construction and Demolition Waste Management
Maximum score: 2
Proportion of total score: 2.1%

The choice of construction material has a significant effect in this category. The recyclability of steel building products is beneficial even when it comes to satisfying the minimum requirement for presenting a waste management plan for building and demolition waste. A concept for sorting and recycling waste during construction and dismantling has to be submitted. With the steel products used, reference to established collection and treatments systems (i.e. the scrap trade) can be made. There are additional points for separating waste into different recycling and recovery streams in accordance with the EU Waste Framework Directive (see Chapter 2.3), or creating a total of less than 12.2 kg of waste per m^2 of floor space that has to be disposed of via landfill or combustion. Steel construction can contribute to a good result here because its reuse rate of 11% and recycling rate of 88% [2] allow these requirements to be fulfilled easily, at least for the structural steel products used. (See also Sections 5.2.1.3 and 5.2.4.2.)

5.2.5.3 *BREEAM category material – mat 03 responsible sourcing of materials*

Criteria: Mat 03 Responsible sourcing of materials
Maximum score: 3
Proportion of total score: 3.4%

BREEAM also takes into account procuring building products in a responsible manner. Criterion Mat 03 is described in Section 5.2.2.4.

5.2.5.4 *BREEAM category waste*

Criteria: Wst 02 Recycled aggregates
Maximum score: 1
Proportion of total score: 1.5%

This criteria recognises and encourages the use of recycled and secondary aggregates, thereby it aims for the reduction of virgin material and optimising material efficiency in construction. Structural steel itself has no influence on this criteria, but blast furnace slags from the steelmaking process are useful as aggregates for concrete production.

5.2.6 Execution of construction work and building site

Any negative impacts of the building phase on the environment, for example, due to noise, dust, waste or contamination of the soil, should be minimised. This is important for the successful completion of a project to maintain a good relationship with local residents and the local authorities. The minimum statutory requirements in many European countries with respect to avoiding waste, providing protection against noise and dust, and protecting the soil and the environment are relatively stringent.

Prefabrication in steel construction workshops complies with environmental and occupational health and safety standards, and furthermore is independent of the weather. The high degree of prefabrication and the manner in which the processing is carried out at the building site means that steel construction is clean, quiet, low in vibrations, safe and precise. In addition, the high degree of prefabrication enables the construction work to progress speedily. On average only 35% of the work on a steel framed structure is carried out at the building site. (For comparison: with other forms of construction 80% of the work is often carried out on the building site.) The factory-based manufacture of components, including any coatings required, reduces the burden on the environment at the building site and at the same time improves the quality of the work carried out. The prefabricated parts can then be assembled at the building site, saving time and to a large extent using a dry method of construction, whereby waste and emissions of noise and dust at the building site are reduced significantly. In addition, prefabricating the supporting structure away from the building site saves time because a large part can be fabricated while the site is being prepared and foundations are being placed. The reduction of the construction time also brings economic benefits, because the building is ready for occupancy sooner and capital repayments (e.g. via rental income) can then be made earlier.

Manufacturing in the factory also offers the workers benefits: they have a permanent workplace that is independent of weather conditions and where they are familiar with work sequences, have a high standard of occupational health and safety, and have working conditions that are more comfortable than those on the building site. To achieve the maximum score in certification schemes, it is necessary to exceed the legal requirements by means of additional controls and concepts. Although these involve additional effort, steel construction allows them to be implemented easily. The measures required often exist already and only need to be documented accordingly (see Table 5.7).

5.2.6.1 DGNB criterion PRO 2.1: construction site/construction process

Aspect: Technical Quality
Group of criteria: Quality of the Technical Execution
Specific impact factor: 2
Proportion of overall evaluation: 1.0%

As noted earlier, negative impacts of the building phase on the environment, such as those due to noise, dust, waste or contamination of the soil, should be minimised. The evaluation is based on satisfying and exceeding statutory requirements.

5.2.6.2 DGNB criterion PRO 2.2: quality assurance of construction execution

Aspect: Technical Quality
Group of criteria: Quality of the Technical Execution
Specific impact factor: 3
Proportion of overall evaluation: 1.4%

Quality management is very important to ensure the required levels of quality for sustainable construction are also achieved. Documentation for quality assurance

Table 5.7 Effect of structural design in steel and steel composite constructions on the assessment by certification schemes – execution of construction work and building site [6]–[8].

Execution of Construction Work and Building Site					
	Aspect	Criteria	Importance Factor/ Maximum Score	Proportion of Total Score	Comments
DGNB	Process Quality – Quality of the Building Construction	PRO 2.1 Construction Site/Construction Process	2	1.0%	High degree of prefabrication, 65% manufactured in factory. Clear planning of construction sequences as a result of just-in-time deliveries. Offcuts already recirculated in factory. Low waste construction site, easy sorting of metal fractions. Steel construction is clean and low in noise and vibration.
		PRO 2.2: Quality Assurance of Construction Execution	3	1.4%	
LEED	Sustainable Sites (SS)	Construction Activity Pollution Prevention	Minimum requirement		
	Materials and Resources (MR)	Construction and Demolition Waste Management Planning	2	2.1%	
		Construction and Demolition Waste Management	2	2.1%	
BREEAM	Management (Man)	Man 01 Sustainable procurement	9	4.7%	
		Man 03 Construction site impacts	5	2.1%	
	Waste (Wst)	Wst 01 Construction waste management	3	4.5%	

is the central component of good risk management and can serve as a quality characteristic for the constructed building. The evaluation of the quality assurance of the building construction is carried out by documenting the materials and consumables used, use of safety data sheets and conducting some typical measurements to check that the energy-related and building-acoustic-related quality are achieved.

5.2.6.3 *LEED Sustainable Sites (SS) – construction activity pollution prevention*

SS Credit: Construction Activity Pollution Prevention
Minimum requirement

The category of sustainable construction depends on the impact of the building on its environment. In this case, this is the construction site, and also the building itself. Important issues are dust and noise emissions, rainwater management, 'light pollution' occurring, heat islands and the preservation of 'green' areas.

5.2.6.4 *LEED Materials and Resources (MR) – waste management*

MR Credit: Construction and Demolition Waste Management Planning
Minimum requirement

MR Credit: Construction and Demolition Waste Management
Maximum score: 2
Proportion of total score: 2.1%

A concept for the sorting and recycling of waste during construction and dismantling has to be submitted. With the steel products used, reference to established collection and treatments systems (i.e. the scrap trade) can be made. (See Section 5.2.5.2.)

5.2.6.5 *BREEAM category management – sustainable procurement and construction site*

Criteria: Man 01 Sustainable procurement
Maximum score: 9
Proportion of total score: 4.7%

Criteria: Man 03 Construction site impacts
Maximum score: 5
Proportion of total score: 2.6%

The construction site has to be managed in an environmentally and socially considerate, responsible and accountable manner, especially in terms of resource use, energy consumption and pollution. (See also Section 5.2.3.3.)

5.2.6.6 BREEAM category waste

Criteria: Wst 01 Construction waste management
Maximum score: 3
Proportion of total score: 4.5%

The most challenging waste criterion addresses site waste management. Credits are achieved by exceeding good and best practice benchmarks. All steel construction products are manufactured off site and therefore site waste is minimal and, for many cases, for example, structural frame components, is zero, making it easy to achieve these credits using steel construction. (See also Section 5.2.6.)

REFERENCES

[1] Dutch Green Building Council. (2014) BREEAM-NL Nieuwbouw en Renovatie Keurmerk voor duurzame vastgoedobjecten Beoordelingsrichtlijn september 2014 versie 1.01. Rotterdam.
[2] EPD-BFS-20130094. (2013) *Environmental Product Declaration: Structural Steel: Open Rolled Profiles and Heavy Plates*. Berlin: Institute Construction and Environment (IBU).
[3] EPD-BFS-20130173. (2013) *Environmental Product Declaration: Hot-Dip Galvanized Structural Steel: Open Rolled Profiles and Heavy Plates*. Berlin: Institute Construction and Environment (IBU).
[4] EN 15804. (2012) *Sustainability of Construction Works – Environmental Product Declarations – Core Rules for the Product Category of Construction Products*. Brussels: European Committee for Standardization.
[5] EN 15978. (2011) *Sustainability of Construction Works – Assessment of Environmental Performance of Buildings – Calculation Method*. Brussels: European Committee for Standardization.
[6] German Sustainable Building Council – DGNB e.V. (2012) *Neubau Büro- und Verwaltungsgebäude – DGNB Handbuch für nachhaltiges Bauen*. Version 2012. Stuttgart.
[7] BRE Global Ltd. (2011) *BREEAM New Construction – Non-Domestic Buildings*. Technical Manual SD5073 – 2.0:2011. Watford.
[8] U.S. Green Building Council. (2013) *LEED Reference Guide for Building Design and Construction* v4. Washington, D.C.
[9] bauforumstahl e.V. (2014) *EPD Structural Steel: LEED Supplementary Sheet*. Düsseldorf.
[10] ISO 14001. (2015) *Environmental Management Systems – Requirements with Guidance for Use*.
[11] BCSA TATA Steel SCI. (2015) The free encyclopedia for UK steel construction information: 'Cost of structural steelwork'. Available at www.steelconstruction.info/Cost_of_structural_steelwork. Accessed 29 October 2015.
[12] bauforumstahl e.V. (2015) *Kosten im Stahlbau 2015 – Basisinformationen zur Kalkulation*. Düsseldorf.
[13] Bureau of International Recycling (BIR). (2015) Ferrous report 2010–2014, *World Steel Recycling in Figures*. Brussels.

Chapter 6

Case studies and life-cycle assessment comparisons

6.1 LCA COMPARISON OF SINGLE-STOREY BUILDINGS

Markus Kuhnhenne, Dominik Pyschny and Raban Siebers

With the introduction of green building labels such as LEED, DGNB and BREEAM (see Chapter 5), life-cycle assessment (LCA) has become an integral part of the sustainability assessment of buildings. The evaluation and DGNB labeling of industrial buildings started in 2009 and requires LCA evaluations for these types of buildings. To gain knowledge about the environmental impacts of different construction methods for single-storey industrial buildings, different forms of constructions are considered. The focus is on the structural frame and the associated construction products. In addition, LCAs of several types of building envelopes were compared.

National databases (see Chapter 3.10) or environmental product declarations (EPDs) (see Chapter 3.6) employ reference units such as 1 kg or 1 m^3. However, it is necessary to compare functional units, for example, whole structures or a basic module, according to the specific situation, to achieve meaningful results (see Chapter 2.2). By calculating different quantities depending on the structural concepts for a comparable function, and the data per unit as mentioned above, realistic results are achieved. For a common single-storey building (Figure 6.1) the simple frame structure can be regarded as a functional unit.

6.1.1 Structural systems

The structural system of a single-storey building can be accomplished using different static systems. Depending on the chosen design, the required material quantities may vary for a given building size. Also, depending on the construction material, a particular structural system may be found as the optimum solution.

Sustainable Steel Buildings: A Practical Guide for Structures and Envelopes, First Edition.
Edited by Bernhard Hauke, Markus Kuhnhenne, Mark Lawson and Milan Veljkovic.

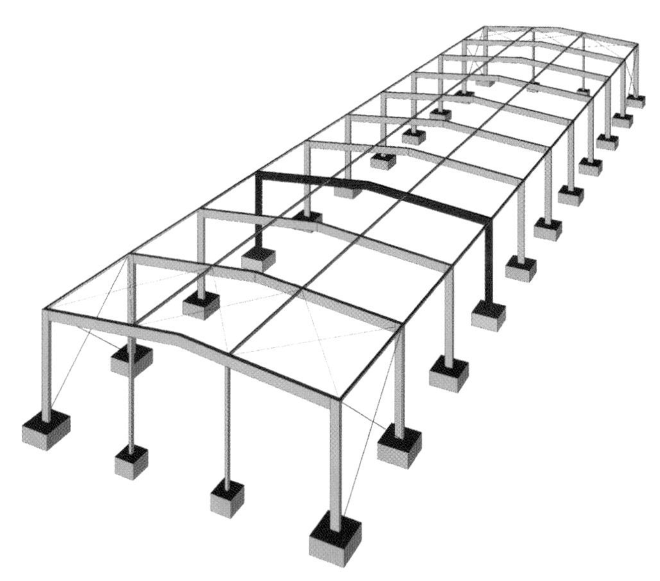

Figure 6.1 Isometric view of a single-storey building's structural system with the regarded functional unit, a basic frame structure.

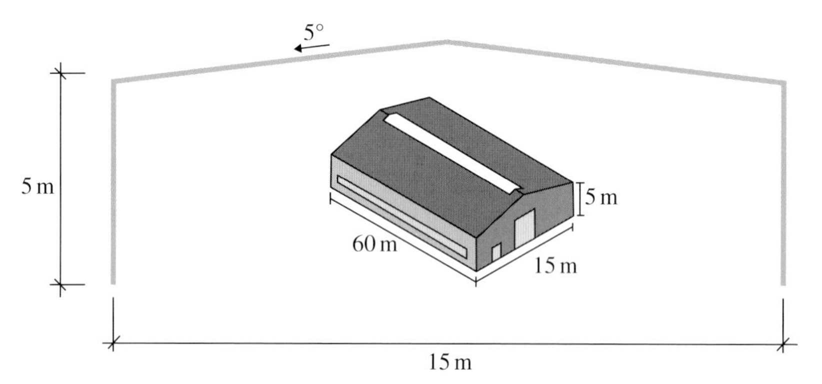

Figure 6.2 Dimensions of the single-storey building and its structural frame.

The following comparison deals with the main structural system of a typical single-storey building with span 15 m, 5 m eaves height, roof pitch 5°, frame spacing 6 m, wind load and a snow load of 75 kg/m^2 (Figure 6.2). Two different structural systems with different construction materials are considered (see Table 6.1).

Design details of the different structural systems are presented below. The associated quantities are the basis for the following LCA. In addition to the complete structural frame as a functional unit, columns and beams are looked at individually. For research purposes only, the comparison is made on the individual member level.

6.1.1.1 *Pinned-base portal frame, pad foundations*

See Figures 6.3 and 6.4 and Table 6.2.

Table 6.1 Static systems and construction possibilities.

Structural System	Materials
1. Pinned-base portal frame, pad foundations	Structural steel Frame: Grade S 355 Foundation: C 25/30
2. Rigid-base columns, pinned beam, sleeve foundation	Reinforced concrete Columns, girder: strength class C30/37 Foundation: C 25/30 Reinforced concrete, timber Columns: strength class C30/37 Glue-laminated timber girders: BS 16 Foundation: C 25/30

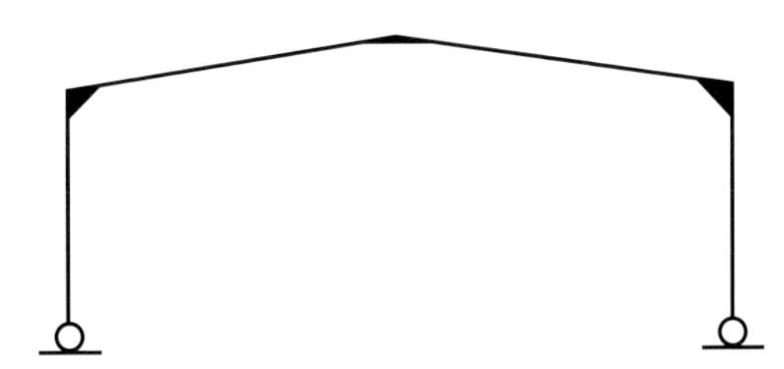

Figure 6.3 Structural system: pinned-base portal frame.

Figure 6.4 Single-storey building with a steel frame, symbols for steel frame in grade S 355. © Kerschgens Stahl & Mehr GmbH.

Table 6.2 Structural steel frame, steel grades S 355.

Steel Frame	Grade S 355	Reinforcement BSt 500 (Yield Strength 500 MPa)
Columns	IPE 400	–
Girder	IPE 360	–
Pad foundation C 25/30	1.5 m × 1.5 m ×0.4 m	19.9 kg/m^3

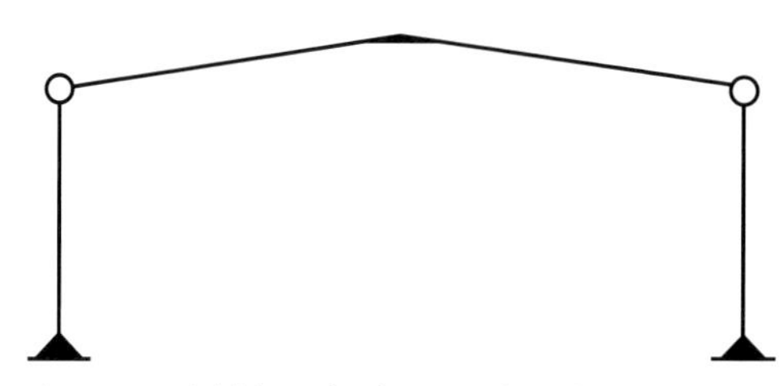

Figure 6.5 Structural system: rigid-based columns, pinned beams.

Figure 6.6 Single-storey building with a precast RC frame, symbol for RC frame. © Dr. Gerhard Köhler, Werksvertretungen.

6.1.1.2 *Rigid-based columns, pinned girder, sleeve foundations*

See Figures 6.5 and 6.6 and Table 6.3.

The design of the reinforced concrete (RC) foundations concrete class C25/30, reinforcement BSt 500 [yield strength 500 MPa]) depends on the different superstructures. Therefore, the foundations are included in the comparison. Minor components, such as screws, rods, starter bars and so forth, are not considered. All four different structural systems provide the same functionality of the single-storey building. The design of the steel construction for easy material ordering, fabrication, manufacturing and construction is carried out prior to the optimisation of component sizes. Hot-rolled steel sections are used in common sizes (see Table 6.4 and Figure 6.7).

Table 6.3 Reinforced concrete frame.

Reinforced Concrete Frame		Reinforcement BSt 500 (Yield Strength 500 MPa)
Columns C30/37	40 cm × 40 cm	108.1 kg/m^3
RC girder C30/37	Precast concrete unit T 80	202.5 kg/m^3
Sleeve foundation C25/30	185 × 185 × 26 cm sleeve height 80 cm	48.1 kg/m^3

Table 6.4 Reinforced concrete timber frame, RC columns and glue-laminated timber girders.

Reinforced Concrete Timber Frame			Reinforcement BSt 500 (Yield Strength 500 MPa)
Columns C30/37		40 × 40 cm	108.1 kg/m^3
Timber (GL) girder BS 16	b = 14 cm, hs = 71 cm, hap = 101 cm, rin = 80 m, lc = 13.94 m	–	–
Sleeve foundation C 25/30	–	191 × 191 × 24 cm, sleeve height 60 cm	53.2 kg/m^3

Figure 6.7 Single-storey building with RC/timber frame, symbol for RC/timber frame. © Michael Fassold, Sägewerk und Holzhandels Gmbh.

6.1.2 LCA information

Steel components permit easy recycling, and reuse is the only usual and acceptable way of treatment (see Chapter 3.11.1). For wooden components, incineration is the best way to avoid landfill and regaining energy. The ÖKOBAUDAT includes the fitting dataset. According to the EPDs for concrete, it is assumed that 96% of

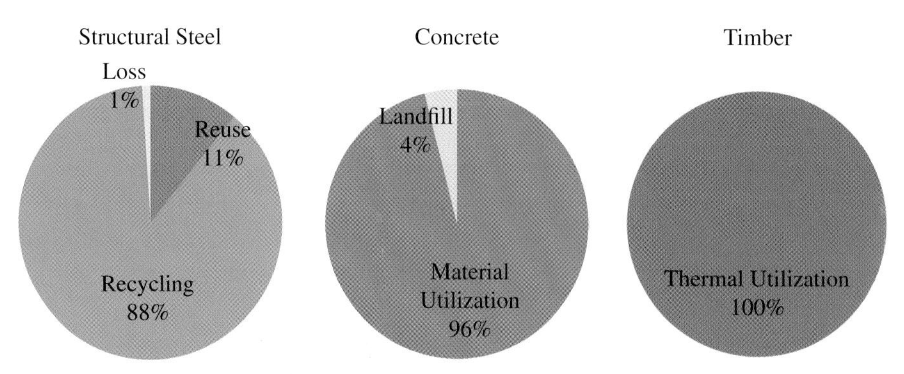

Figure 6.8 End-of-life scenarios for different materials [1], [3], [4].

the concrete is utilized and 4% is landfilled (Figure 6.8). In the case of utilization, the concrete is processed to a substitute for gravel from primary materials. This leads to a reduction of the required energy and is included in the LCA calculation. Table 6.5 shows the environmental data for the relevant materials.

6.1.3 Frame and foundations – structural system

Comparing the various building materials (see Figure 6.9), structural steel leads to lightweight structures because of its high strength. This leads not only to less material for a specific component, in this case the frame structure, but often, for example, fewer columns, smaller foundations or less material transport to the site. Figure 6.10 shows this for the considered single-storey building in steel or reinforced concrete construction.

The steel frame in S 355 (steel) is compared with a reinforced concrete frame (RC) and an RC-timber frame (RC/timber). The foundations are sized in accordance with the different frame systems and are taken into consideration. Benefits and loads at the end of life of the product are first displayed separately and then summed up for evaluation purpose with the values for the product stage. In EN 15978, this separate display of the individual modules is requested, but a common evaluation is allowed. Thus, the entire life cycle of a building material, including recycling or disposal, is presented as one total value. For better comparison with different structures and buildings the values per frame are converted to values per m^2 gross floor area (GFA). Figures 6.11–6.16 show the environmental indicators as mentioned in Chapter 3.3.1, including the benefits or loads for the end-of-life scenarios – recycling (steel), incineration (timber) or downcycling (concrete) – per m^2 GFA.

Looking at all environmental indicators, none of the structural systems has a clear advantage. However, for even higher strength steels, the environmental performance is increased (see Section 6.4). In GWP, ODP, AP and especially in EP, the steel construction performs very well. In this comparison of structural systems, the recyclability of structural steel without losses in material properties plays an important role. Moreover, because of its high strength, structural steel achieve slender and material-efficient structures.

Table 6.5 Environmental data for construction products from EPDs [2]–[4] and ÖKOBAUDAT [1].

Material	Source and Life-Cycle Stage According to EN 15978 [6]	Reference Unit (RU)	Primary Energy, Not Renewable	Primary Energy, Renewable	Total Primary Energy	Global Warming Potential (GWP)	Ozone Depletion Potential (ODP)	Acidification Potential (AP)	Eutrophication Potential (EP)	Photochemical Ozone Creation Potential (POCP)
			[MJ/BE]	[MJ/BE]	[MJ/BE]	[kg CO_2–Äqv./BE]	[kg CFC11–Äqv./BE]	[kg SO_2–Äqv./BE]	[kg PO_4–Äqv./BE]	[kg Ethen Äqv./BE]
Structural steel	EPD-BFS-20130094	kg	10.59	0.93	11.52	0.78	1.45E-10	1.79E-03	1.58E-04	2.98E-04
Production	A1–A3	Kg	17.80	0.84	18.64	1.74	1.39E-10	3.47E-03	2.89E-04	7.55E-04
Benefits & loads	D (11% Reuse, 88% Recycling)	kg	−7.21	0.09	−7.12	−0.96	6.29E-12	−1.68E-03	−1.31E-04	−4.57E-04
Concrete C25/30	EPD-IZB-2013411	kg	0.25	0.01	0.26	0.08	2.49E-10	1.12E-04	1.84E-05	1.41E-05
Production	A1–A3	2365 kg/m³	909	77.10	986	211.1	6.94E-07	4.26E-01	6.04E-02	4.36E-02
		kg	0.38	0.03	0.41	0.09	2.89E-10	1.24E-04	1.97E-05	1.50E-05
Benefits & loads	C3	kg	0.01	0.00	0.01	0.00	3.10E-14	5.42E-06	1.17E-06	7.08E-07
	D	kg	−0.13	−0.02	−0.15	−0.01	−3.99E-11	−1.71E-05	−2.46E-06	−1.69E-06
	C3 + D (96% Utilisation, 4% Landfill)	kg	−0.12	−0.02	−0.14	−0.01	−3.98E-11	−1.17E-05	−1.30E-06	−9.79E-07
Concrete C30/37	EPD-IZB-2013431	kg	0.29	0.02	0.31	0.09	2.66E-10	1.23E-04	2.01E-05	1.54E-05
Production	A1–A3	2365 kg/m³	984	82.70	1067	231.90	7.35E-07	3.23E-01	5.13E-02	3.93E-02
		kg	0.41	0.03	0.57	0.11	2.93E-09	1.94E-04	2.73E-05	1.99E-05
Benefits & loads	C3	kg	0.01	0.00	0.01	0.00	3.10E-14	5.42E-06	1.17E-06	7.08E-07
	D	kg	−0.13	−0.02	−0.15	−0.01	−3.99E-11	−1.71E-05	−2.46E-06	−1.69E-06
	C3 + D (96% Utilisation, 4% Landfill)	kg	−0.12	−0.02	−0.14	−0.01	−3.98E-11	−1.17E-05	−1.30E-06	−9.79E-07
Rebars	ÖKOBAUDAT 2015	kg	10.21	1.88	12.10	0.75	4.95E-11	1.78E-03	1.78E-04	1.74E-04
Production	4.1.02 Rebar steel A1–A3	kg	10.21	1.88	12.10	0.75	4.95E-11	1.78E-03	1.78E-04	1.74E-04
Benefits & loads	No recycling potential	–	0.00	0.00	0.00	0.00	0.00E + 00	0.00E + 00	0.00E + 00	0.00E + 00
Glue-laminated timber	Ökobau.dat 2015	kg	0.60	19.80	20.40	−0.13	4.44E-11	1.18E-03	2.31E-04	4.75E-05
Production	3.1.03 Glue-laminated timber A1–A3	515 kg/m³	2034	19070	21104	−696.20	7.47E-09	4.48E-01	1.03E-01	8.72E-02
		kg	3.95	37.03	40.98	−1.35	1.45E-11	8,69E-04	2.01-04	1.69E-04
Benefits & loads	3.4.03 Wooden composites in incineration plant	kg	−4.50	35.96	31.46	−0.15	−1.39E-11	2.16E-03	2.51E-04	1.35E-04
	C3	kg	0.51	0.05	0.56	1.80	1.40E-12	5.86E-04	1.47E-04	4.01E-05
	D	kg	−8.96	−1.12	−10.08	−0.60	−2.98E-11	7.09E-04	−9.71E-05	−7.40E-05
	C3 + D	kg	−8.45	−1.07	−9.52	1.20	−2.84E-11	1.29E-03	4.99E-05	−3.39E-05

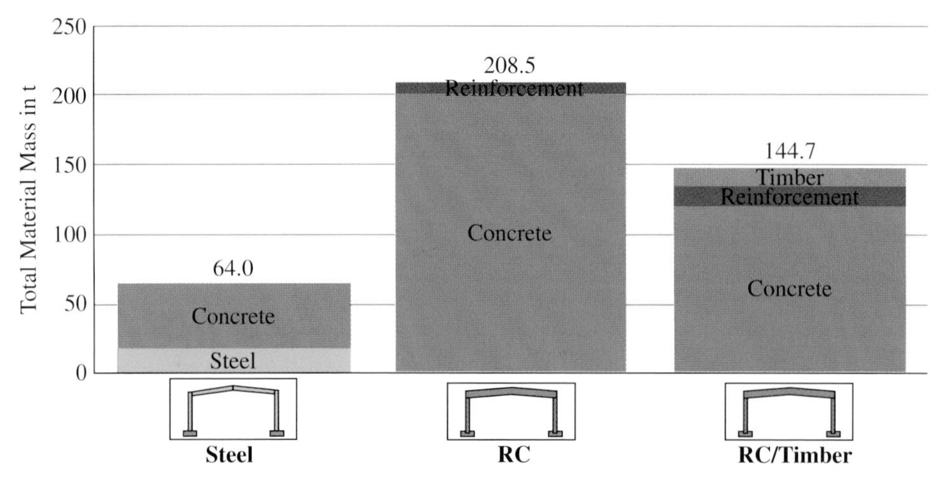

Figure 6.9 Quantities for the structural systems: frames and foundations, in tonnes.

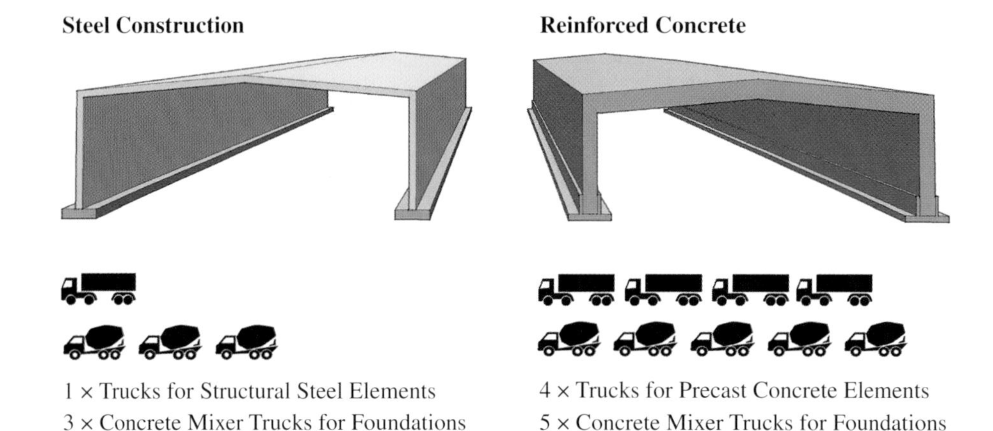

Figure 6.10 Influence of construction methods on the number of transports to the site.

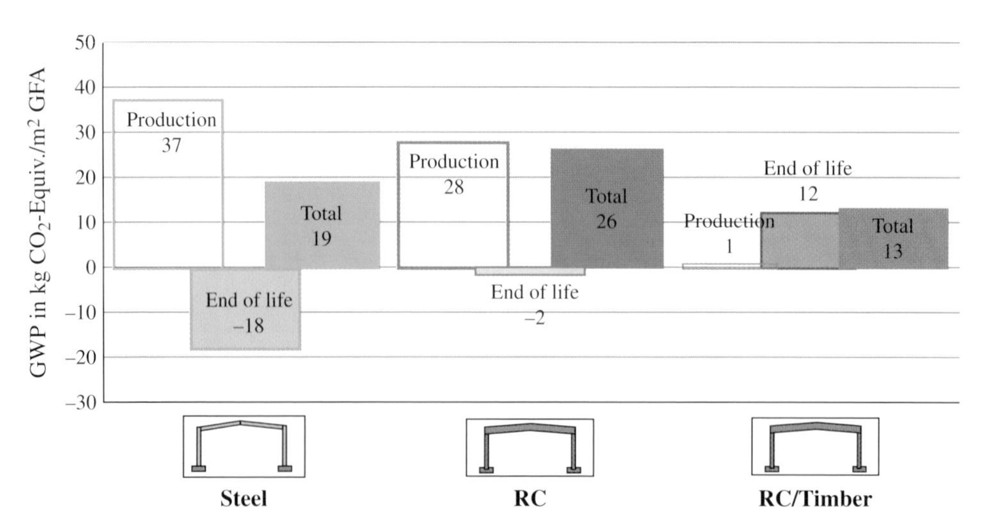

Figure 6.11 Global warming potential (GWP) in kg CO_2-equiv. per m^2 GFA.

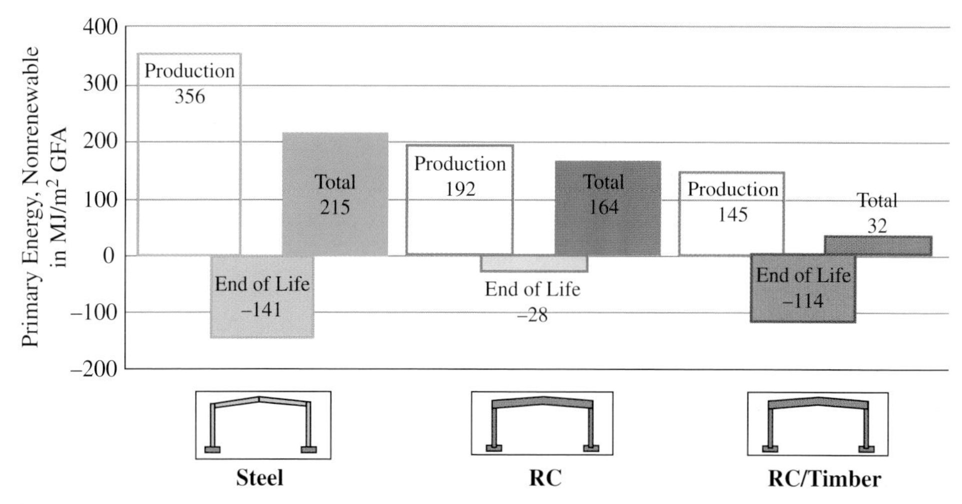

Figure 6.12 Primary energy, nonrenewable in MJ per m^2 GFA.

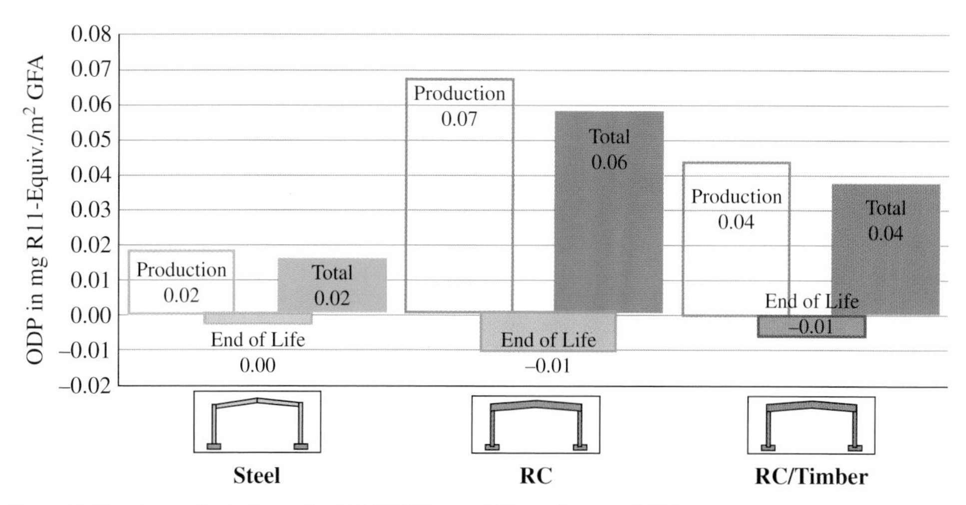

Figure 6.13 Ozone depletion potential (ODP) in mg R11-equiv. per m^2 GFA.

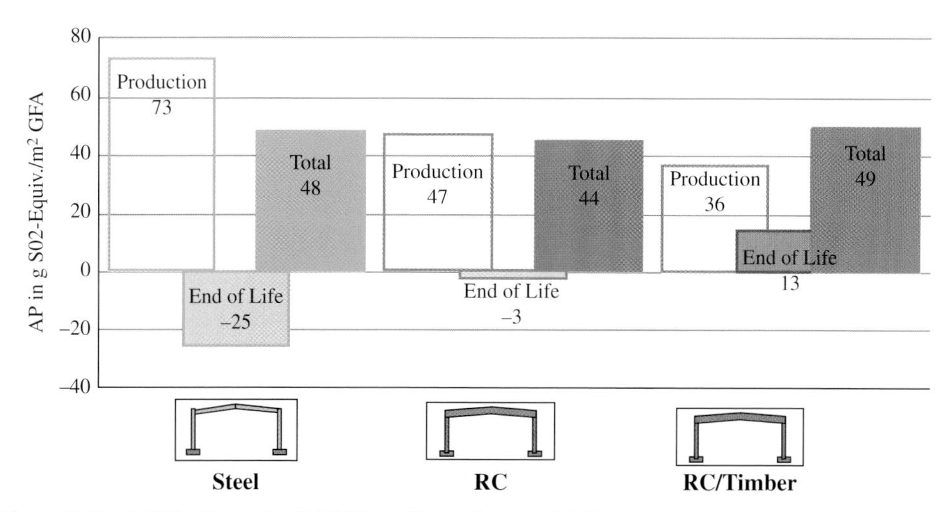

Figure 6.14 Acidification potential (AP) in g SO_2-equiv. per m^2 GFA.

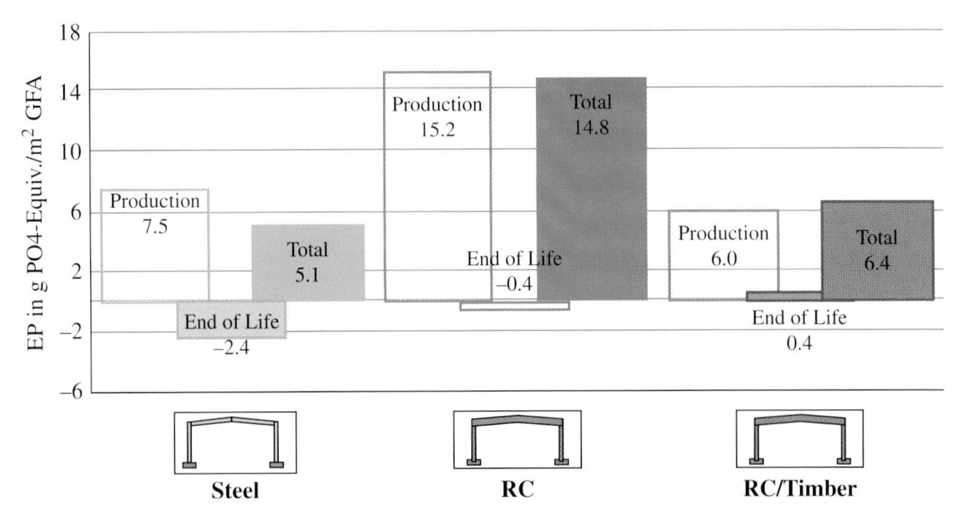

Figure 6.15 Eutrophication potential (EP) in g PO_4-equiv. per m² GFA.

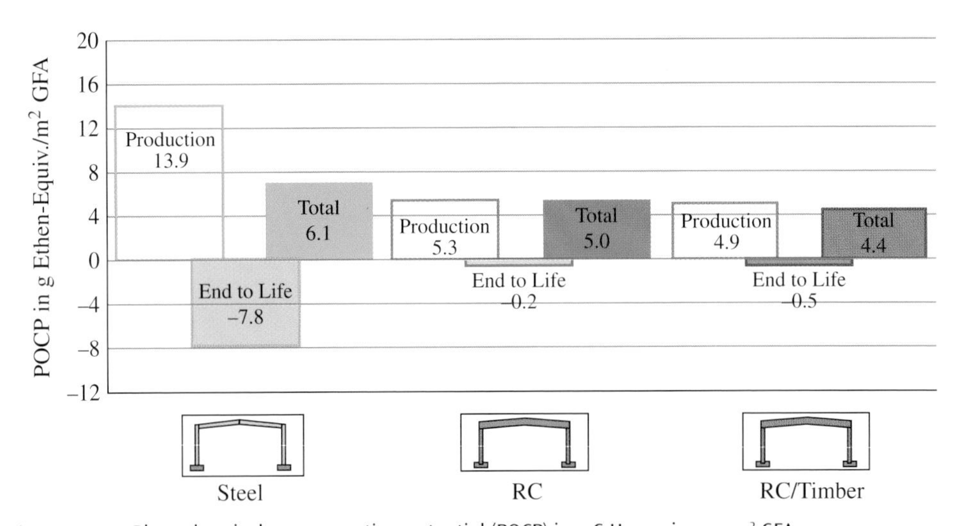

Figure 6.16 Photochemical ozone creation potential (POCP) in g C_2H_4-equiv. per m² GFA.

6.1.4 Column without foundation – single structural member

For the columns (combined compression and bending member), the steel column as compared to the reinforced concrete column achieves much lower masses and better results for GWP. For total primary energy demand, the reinforced concrete column superficially has the advantage. However, the foundations, which are not considered here, are larger for the RC columns. A true conclusion can therefore only be established in relation to the overall structural system (see Figures 6.17–6.19).

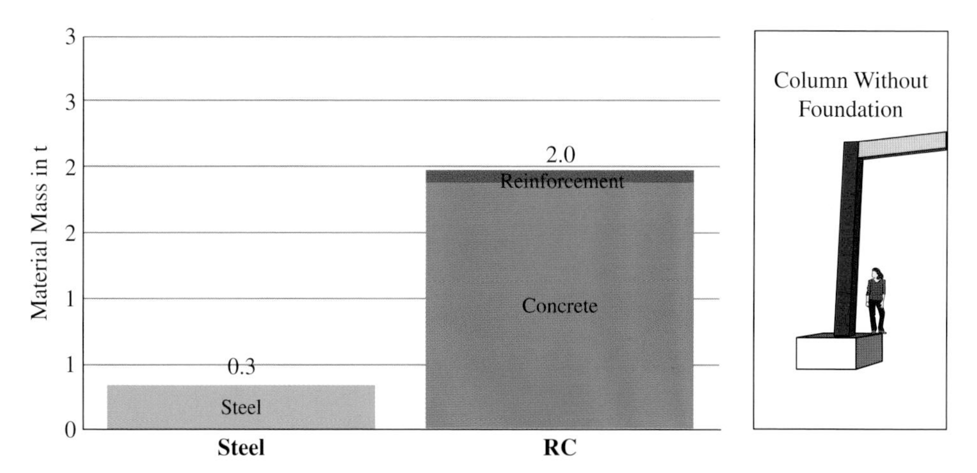

Figure 6.17 Quantities for columns without foundation, in tonnes.

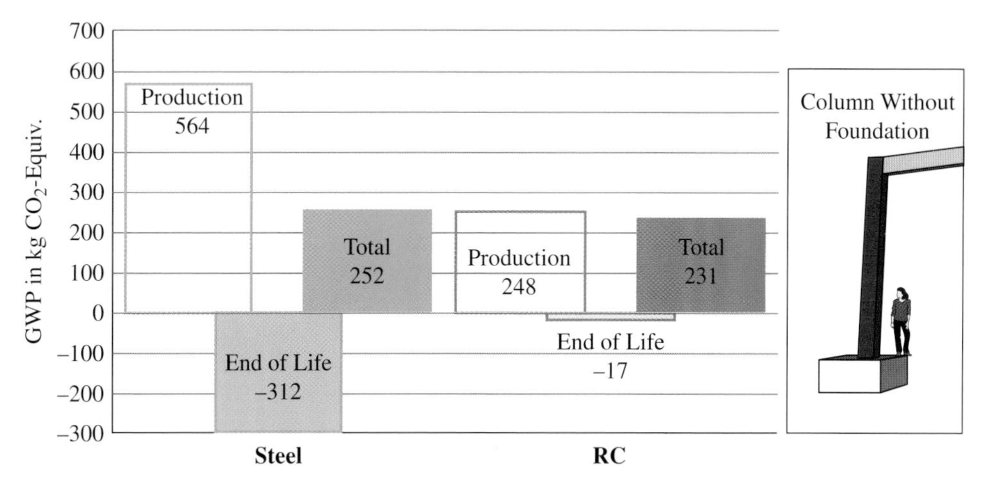

Figure 6.18 GWP in kg CO_2-equiv. per column.

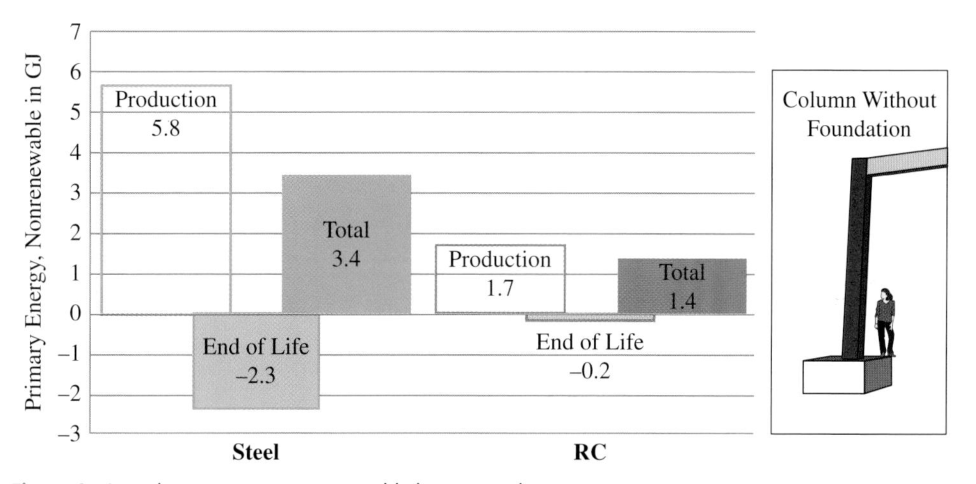

Figure 6.19 Primary energy, nonrenewable in GJ per column.

6.1.5 Girder – single structural member

For girders (bending member), the large material quantities of the reinforced concrete system and the good performance of the glue-laminated timber truss are apparent. The use of a section with high steel grade is apparent for the bending member. It is evident that if only single members are considered the results may be distorted (see Figures 6.20–6.22).

6.1.6 Building envelope

Different possibilities for the building envelope are compared for an otherwise identical single-storey building: an uninsulated building, three equivalently insulated buildings and a 'super'-insulated building. In Table 6.6 the various building envelopes and their building physical properties are listed.

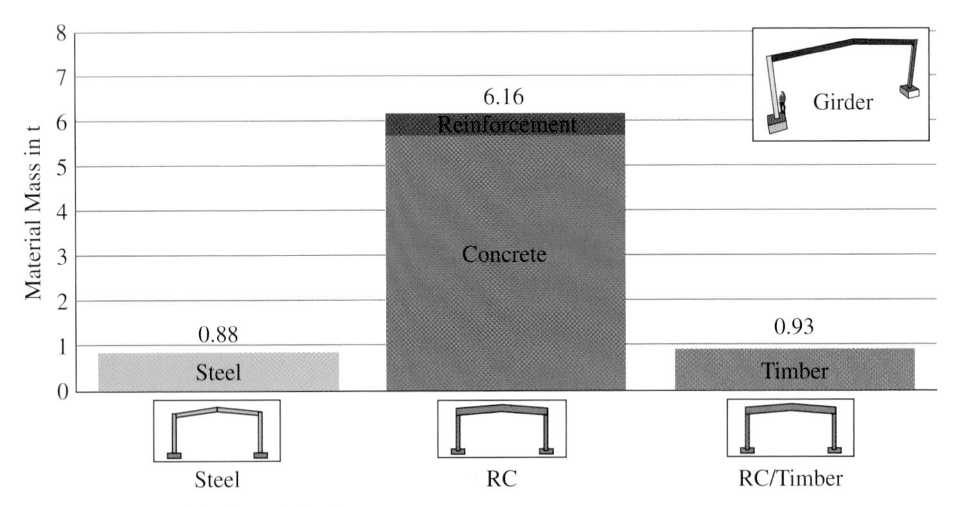

Figure 6.20 Quantities for a girder, in tonnes.

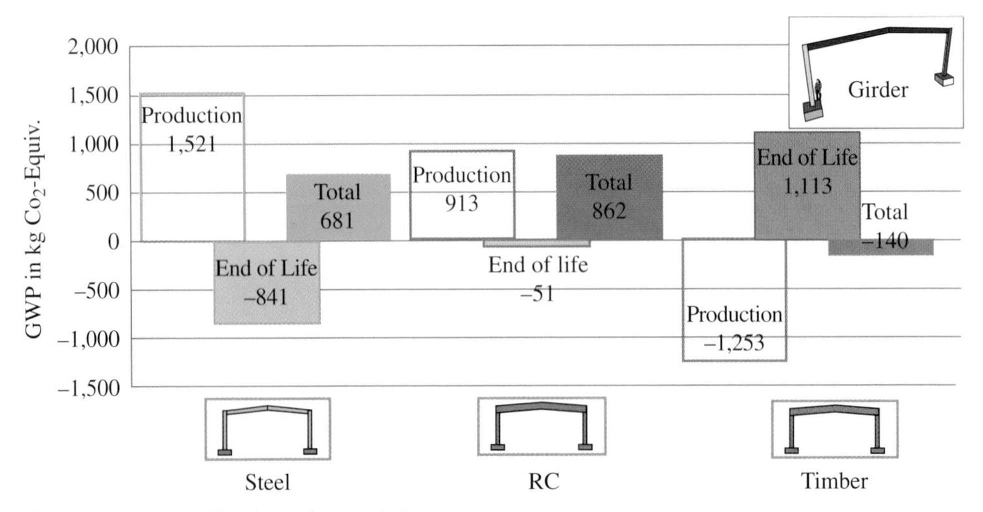

Figure 6.21 GWP in kg CO_2-equiv. per girder.

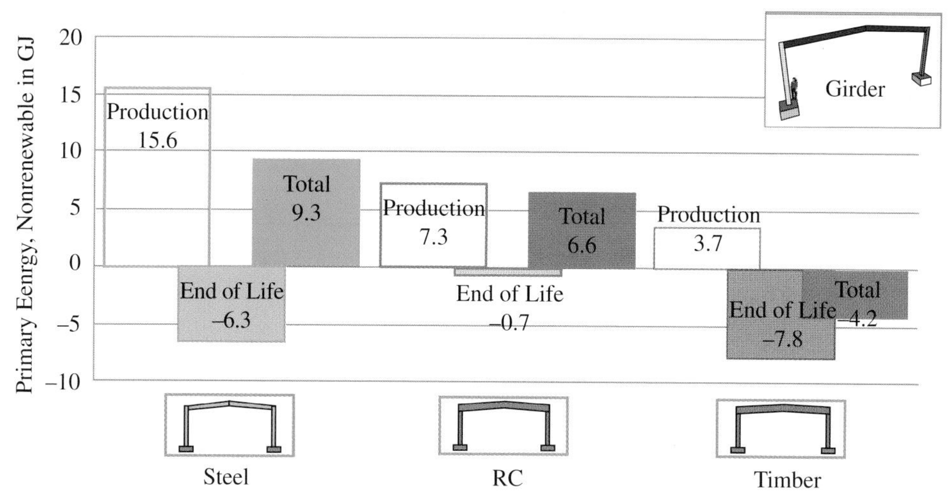

Figure 6.22 Primary energy, nonrenewable in GJ per girder.

As expected, the environmental impact for the uninsulated building for the product stage summed up with benefits and loads is the lowest (Figures 6.23 and 6.24). The equivalently insulated envelopes of the 'warm' buildings show similar results. The superinsulated building with 200 mm polyurethane sandwich panels gives the highest value. However, it is noted that the increase of GWP and total primary energy for the superinsulated building is relatively moderate compared to the increase of insulation, which is more than a factor of two. It is significant that by using sandwich elements, especially in comparison to the aerated concrete, a more eco-efficient insulation with a thinner panel can be achieved.

For the next step, the two polyurethane sandwich panel variants are evaluated for the operational phase of the building. This is used to calculate how long it takes for the superinsulated building to pay off in terms of the primary energy demand during the building's operation.

6.1.7 Comparison in the operational phase

The non renewable primary energy, which is required for the product stage and end of life of the building envelope, is converted from GJ to MWh. The assumed average annual energy demand during the operational stage can be compared as in Table 6.7. The assumed annual non renewable primary energy demand can be summed over time linearly, with a presumed utilization period of 20 years. For simplicity, the non renewable primary energy for the product stage and end of life of the building envelope is graphically displayed as an initial offset (Figure 6.25).

In terms of primary energy demand, the superinsulated building pays off after about 4.5 years compared to the insulated building 1 (see Figure 6.25). The additional energy for the product stage and end of life compared to a standard insulated building is compensated by the lower demand for operational energy. Figure 6.26 illustrates this finding and shows a real energy saving after about 4.5 years. This comparison of the entire energy demand for a typical single-storey

Table 6.6 Baseline data of the comparison of different building envelopes.

	Uninsulated Building	Insulated Building 1	Insulated Building 2	Insulated Building 3	Superinsulated Building
Symbol					
External walls	Trapezoidal profiles (sheets), cold, U = 5.88	Steel-PUR-sandwich panels, 80 mm, U = 0.33	Aerated concrete, 300 mm, U = 0.31	Cassette wall (linear tray) 145 + 40 mm, U = 0.29	Steel-PUR-sandwich panels, 200 mm, U = 0.13
Roof	Trapezoidal profiles U = 7.14 W/($m^2 \cdot K$)	Foil roof, 140 mm MW, U = 0.28 W/($m^2 \cdot K$)	Foil roof, 140 mm MW, U = 0.28 W/($m^2 \cdot K$)	Foil roof, 140 mm MW, U = 0.28 W/($m^2 \cdot K$)	Foil roof, 320 mm MW, U = 0.12 W/($m^2 \cdot K$)
Skylight	2.4 W/($m^2 \cdot K$)	2.4 W/($m^2 \cdot K$)	2.4 W/($m^2 \cdot K$)	2.4 W/($m^2 \cdot K$)	2.4 W/($m^2 \cdot K$)
Windows	1.3 W/($m^2 \cdot K$)	1.3 W/($m^2 \cdot K$)	1.3 W/($m^2 \cdot K$)	1.3 W/($m^2 \cdot K$)	1.3 W/($m^2 \cdot K$)
Doors	4.0 W/($m^2 \cdot K$)	4.0 W/($m^2 \cdot K$)	4.0 W/($m^2 \cdot K$)	4.0 W/($m^2 \cdot K$)	4.0 W/($m^2 \cdot K$)
Gates	2.9 W/($m^2 \cdot K$)	2.9 W/($m^2 \cdot K$)	2.9 W/($m^2 \cdot K$)	2.9 W/($m^2 \cdot K$)	2.9 W/($m^2 \cdot K$)
Structural system	Pinned-base portal frame in steel grade S355				
Foundations	Pad foundations				
Base plate	Not insulated, U = 0.44 W/($m^2 \cdot K$)	Insulated, U = 0.24 W/($m^2 \cdot K$)			

MW = mineral wool.

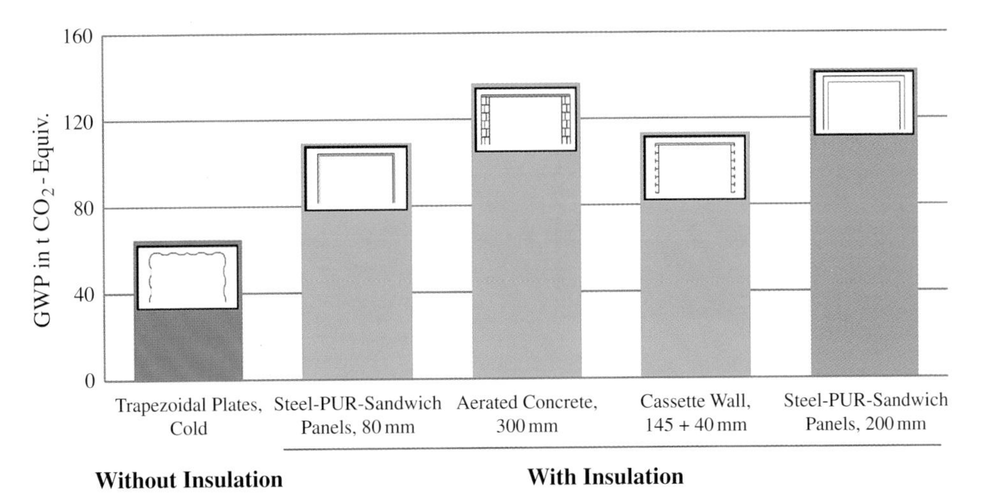

Figure 6.23 GWP for product stage (A1–A3) summed with end of life (D) for external walls, in t CO_2 equivalent.

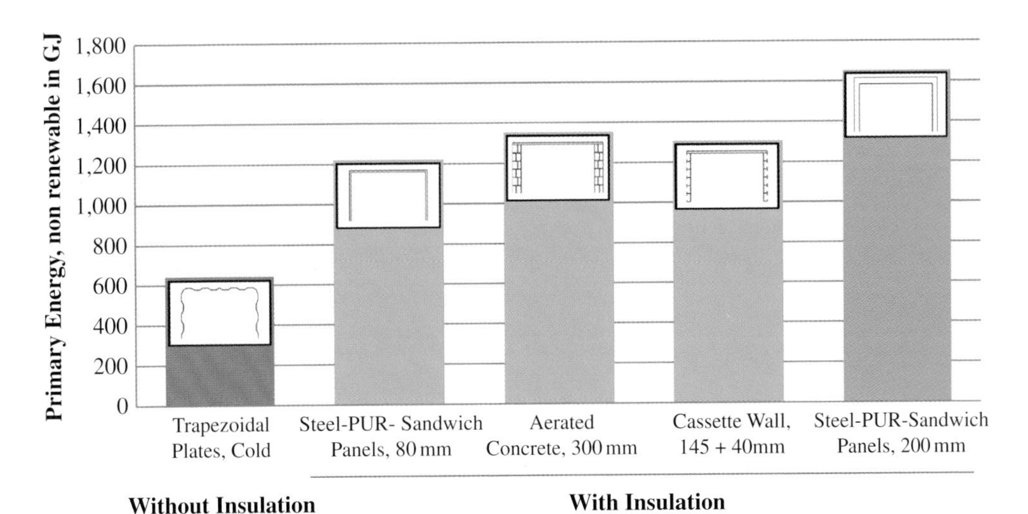

Figure 6.24 Non renewable primary energy for product stage (A1–A3) summed with end of life (D) for external walls, in GJ.

building with different building envelopes is for demonstration purposes. The comparison of building envelopes with different insulation properties shows the importance of considering the entire life cycle. Buildings are designed for a long period of use, and so the decisions made during planning and construction may have long-term consequences that must be considered.

6.1.8 Conclusions for single-storey buildings

With the comparison of the environmental performance of different structural systems and materials but the same functionality, it is evident that the efficient design of steel structures is advantageous. It is not only the reduced material

Table 6.7 Primary energy demand for warm building 1 and superbuilding considering product stage, end of life and operational energy use.

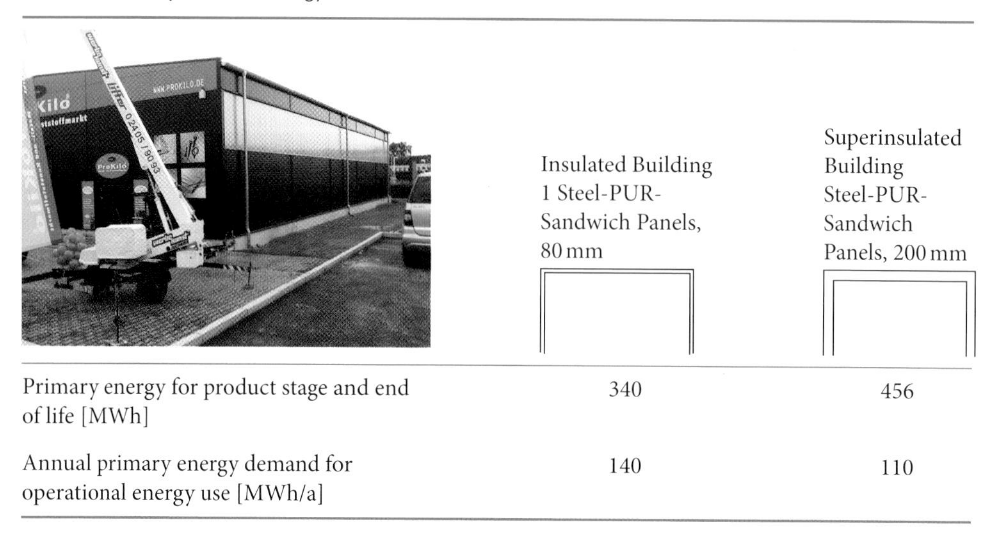

	Insulated Building 1 Steel-PUR-Sandwich Panels, 80 mm	Superinsulated Building Steel-PUR-Sandwich Panels, 200 mm
Primary energy for product stage and end of life [MWh]	340	456
Annual primary energy demand for operational energy use [MWh/a]	140	110

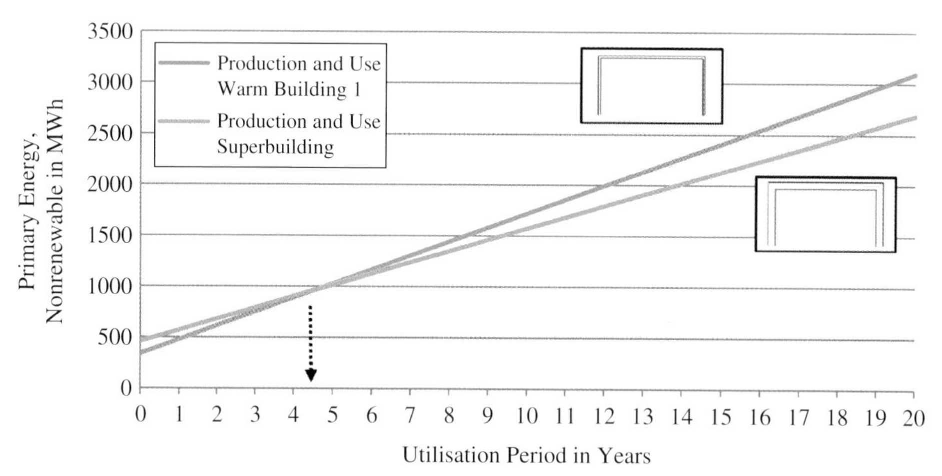

Figure 6.25 Comparison of nonrenewable primary energy demand for the building envelope of insulated building 1 and superinsulated building over a utilization period of 20 years including the production stage and end of life as well as operational energy use.

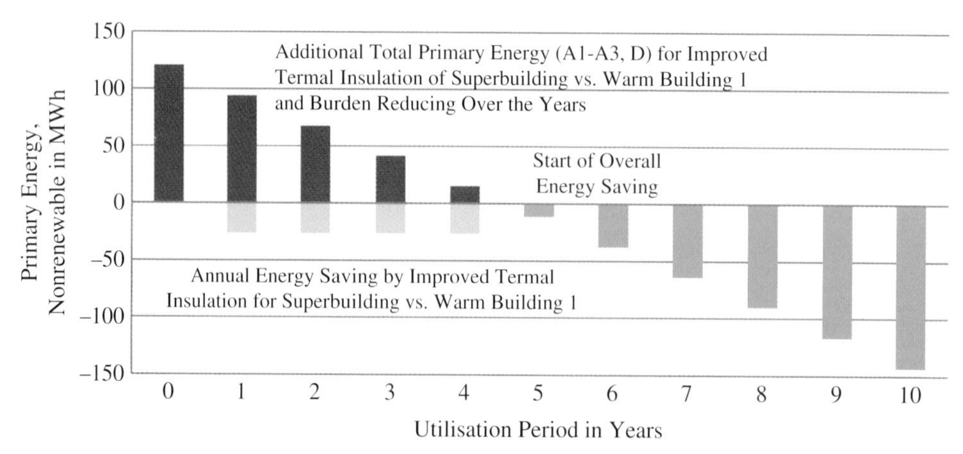

Figure 6.26 Amortisation of improved thermal insulation due to energy savings over the utilisation period.

quantities for a certain structural element – here the frame of a single-storey building – but also the holistic view of the reduced number of columns, smaller foundations and less transport to the construction site. Another advantage of steel is its 'cradle-to-cradle' property: after the dismantling of a building the steel members can be directly reused or recycled, saving natural resources. By using high strength steel, especially for tension and bending members, the LCA can be improved further. It is evident that the level of comparison – for example, material, member or functional unit - has significant influence on the results. When comparing the environmental performance of construction materials, a representative structure must be chosen. The holistic concept of building LCA requires that benefits and loads, which appear at the end of life of a building, should be considered. The comparison of construction materials at the required level of a functional unit showed that structural steel is very competitive. Depending on the specific situation or aim, a complete functional unit – a structural system or its major structural elements – must be compared.

The comparison of building envelopes with different thermal properties shows also the importance of considering the entire life cycle and not just the production stage. Buildings are usually designed for a long period of use, so the decisions made during the planning and construction phase should be taken into account. The total energy demand for a typical single-storey building with various building envelopes is a good example of this holistic approach (see also [7]).

6.2 LCA COMPARISON OF LOW RISE OFFICE BUILDINGS

Raban Siebers

The evaluation of office buildings in terms of fire protection is often carried out in accordance with the respective national building regulations. Three-storey office buildings constitute the largest share of buildings completed. An evaluation of a single building from a fire-protection point of view is generally not carried out because the project is too small and the method of construction is simple. The preparation and comparison of design variants is also not carried out for cost reasons. It will be shown here that low rise office buildings using steel and composite construction are economical including a fire-protection methods.

6.2.1 The low rise model building

A model building that can be used very flexibly throughout its whole life cycle was designed to carry out the investigation (Figure 6.27).

A structural system that matches the respective methods of construction was chosen and a preliminary design was prepared. To ensure comparability between the systems with their different methods of construction, detailed solutions were developed for the complete forms of construction (e.g. including the facade, secondary elements, etc.).

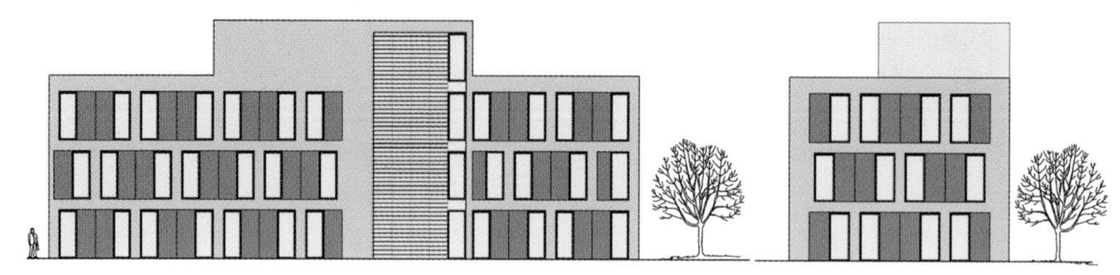

Figure 6.27 Front and side view of the considered building (exemplary design) [8].

The following basic principles of design and construction were implemented for the model building using the methods of construction compared (see also Figure 6.28 and Table 6.8):

- A high degree of flexibility was provided for the user.
- The supporting structure was a frame construction in all cases. Building components were only envisaged within the grid pattern.
- A building core was used to stabilise the building.
- The construction depth was approximately 13 m, and the long span structure made it possible to have offices with double orientation and natural lighting.
- Office space was provided on the three storeys. In addition, an upper floor was envisaged for the building services.
- All building services could be adapted to meet the respective use.
- The evaluation of the fire performance of the building was conducted in accordance with the Model Building Code in Germany. This provides planners and building owners with a high degree of reliability.
- For simplicity, foundations were assumed similar for the steel and the RC building.

Calculations of the structural design for both methods of construction were prepared by engineering consultants, each specialised in the respective method of construction. In order to ensure that both variants were comparable, the following boundary conditions were defined beforehand:

1. Arrangement/location of supporting structure:
   - Supporting and stiffening components were to be arranged within the strip grid.
   - The design of the supporting structure was carried out using the usual design for the method of construction. Special designs were not considered.
2. Location of building:
   - The model building was located in the Frankfurt area of Germany.
   - Height above sea level: 180 m.
3. Assumed loads:
   - Additional load: $g = 2.00\ kN/m^2$ (raised floor, suspended ceiling)
   - Exterior facade: $g = 5.00\ kN/m$
   - Imposed load: $p = 2.00\ kN/m^2$ (category B1 as per DIN 1055-3:2006-03)
   - Storage space: $p = 6.00\ kN/m^2$ (category E2 as per DIN 1055-3:2006-03)

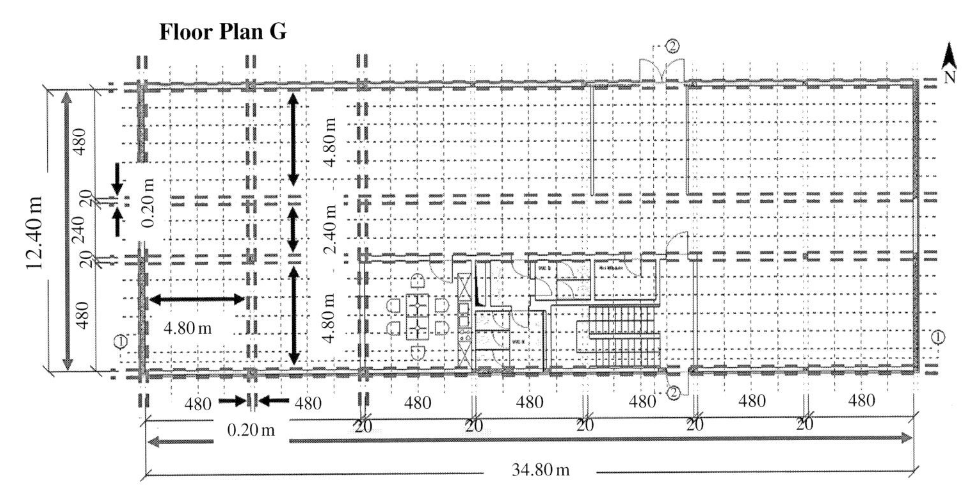

Figure 6.28 Floor plan of the model building [8].

Table 6.8 Summary of building dimensions.

Building Data	
Dimensions	34.8 × 12.4 m, plus facade
Clear floor height	3.46 m
Height of upper edge of finished floor of second upper storey	6.93 m above ground level
Height of balustrade	Max. 1.00 m above upper edge of finished floor
Floor area	431.5 m^2
Size of units	Evaluation of fire performance < 400 m^2
Grid dimensions	Strip grid 4.8 × 4.8 m (office unit) plus 0.2 m wide strip, or 2.4 × 4.8 m (corridor area)

- Curtain walls: $p = 0.80$ kN/m^2 (dead weight $g = 2.00$ kN/m)
- Plant loads: $p = 7.50$ kN/m^2

4 Other loads to be considered:
- Snow load, wind load.
- Self-weight of components.

The important components of the office building are shown in Table 6.9.

6.2.2 LCA comparison of the structural system

For the model building described above, a comparison of the LCAs for the supporting structure was made for the two methods of construction chosen, reinforced concrete and composite steel construction with plasterboard. After determining the material masses involved (Figure 6.29), the environmental

Table 6.9 Summary of the key components for the construction methods considered.

	Composite Construction	Reinforced Concrete Construction
Roof and slab design	Roof over top floor: non ventilated roof construction Direction of span of standard storeys: uniaxial in longitudinal direction of building; circulation system (span 5.00 m) Execution: Floor slabs with screed	Roof over top floor: non ventilated roof construction Direction of span of standard storeys: uniaxial in longitudinal direction of building; circulation system (span 5.00 m) Execution: Floor slabs with screed
Beam supports	Composite beams made from rolled steel sections Direction of span: transverse direction of building	T-beams with an effective width of 1 m and a thickness of 200 mm Direction of span: transverse direction of building
Column supports	Continuous supports with rolled steel sections	Pinned reinforced concrete supports, dimension of edge support 20/20 cm (B/H), centre support of larger dimensions depending on number of floors (up to 30/30 cm)
Stabilisation	Bracing in the closed walls of the central access and sanitary core	Walls in the area of the stairwell and central access in reinforced concrete
Facade	Sandwich panels with ribbon window, internal cladding of balustrade using dry construction	Composite thermal insulation system with exterior rendering
Plant room	Technical plant enclosed in steel construction with no fire protection	Enclosed using limestone walls
Structural fire protection	Design measures such as intumescent coatings to steel profiles or plasterboard	No fire-protection measures required, execution in normal concrete of strength C25/30 and with reinforcing steel BSt 500 (yield strength 500 MPa)

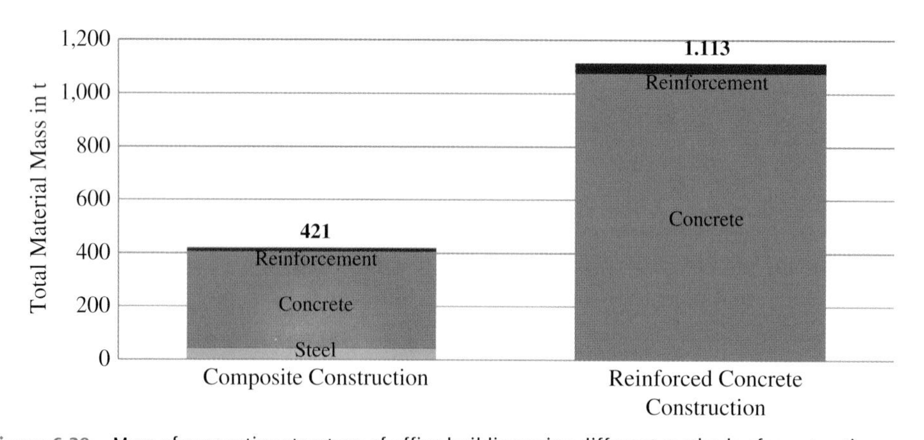

Figure 6.29 Mass of supporting structure of office building using different methods of construction.

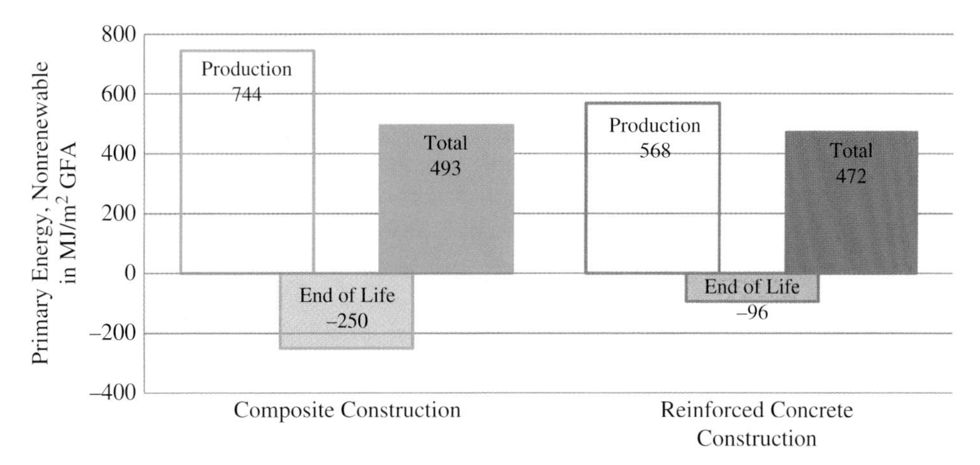

Figure 6.30 Nonrenewable primary energy consumption for supporting structure of office building using different methods of construction.

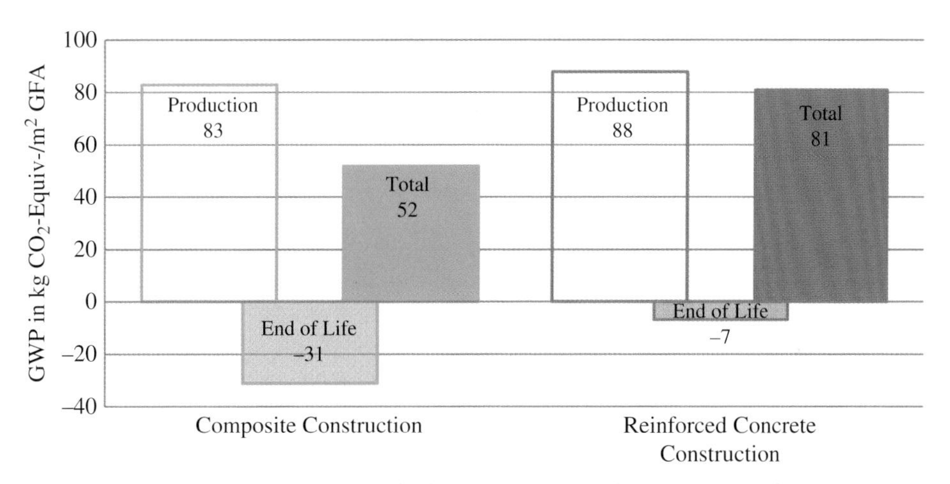

Figure 6.31 GWP for supporting structure of office building using different methods of construction.

impacts for the two variants could be determined using the EPDs [2]–[4] and data from ÖKOBAUDAT 2015 [1] for the remaining construction materials (see Table 6.5). The values for production and for recycling, recovery and disposal were listed separately and a balance sheet subsequently drawn up. Only the renewable primary energy and the GWP for both supporting structure variants are shown here as examples of all the values evaluated for the complete LCA (Figures 6.30 and 6.31).

The benefits of the recyclability and reusability of structural steel become apparent when the credits or impacts resulting from the reuse and recycling of structural steel or from the recovery and dumping of concrete are taken into account in the energy balance. Environmental impacts are reduced and resources are used more efficiently. The use of broken concrete as a substitute for ballast also manifests itself as a benefit in the energy balance. With regard to the nonrenewable primary energy consumption, both supporting structure variants are almost the same. However, the variant using the steel-composite method of construction scores more highly as a result of the recyclability of the steel components.

These benefits are clearer when it comes to the GWP. The relatively high CO_2 emissions during the manufacture of the cement used in the concrete cannot be compensated for by simple recovery. The recovery (sorting, crushing, screening and mixing) and the dumping of the concrete components only leads to a slight improvement in greenhouse gas emissions (see Figure 6.31).

6.3 LCA COMPARISON OF OFFICE BUILDINGS

Markus Kuhnhenne, Dominik Pyschny and Raban Siebers

In order to gain knowledge about the environmental impacts of different types of structural systems for office buildings, six different building structures are considered here: four steel-composite and two reinforced concrete structures. Since a comparison of the environmental performance of different structural types is only useful and meaningful within the building context, all types have the same functionality and dimensions. They are suitable for different levels of building services paired with various cladding systems.

The following basics for design and structural layout are valid for all six building structures. The steel-composite structures and the reinforced concrete structures are based on identical structural conditions:

- High user flexibility: The frequently reoccurring grid dimension for in interior planning, 5.40 × 5.50 m, is included. Thus, several different floor plans including individual, group and open-plan offices are possible at low effort.
- The building width of 13.7 m provides good ventilation and lighting conditions for office use.
- Six full floors for office use are considered.
- The base plate as foundation (thickness 40 cm) was assumed to be the same for all designs.

Building data:

- Dimensions: 32.4 × 13.7 m
- Floor height: 3.5 m
- Floor area: 448 m^2
- Grid: 5.4 × 7.45 m with central columns, 5.4 × 12.7 m without central columns

The following design loads were defined for the structures:

- Life load: $p = 5.0$ kN/m^2
- Additional load: $g = 1.5$ kN/m^2

For simplification plasterboards and concrete are assumed for the structural fire protection. Table 6.10 shows the dimensions for the slabs and columns of the

Table 6.10 Dimensions of the six building systems.

Option	Structure		Sketch
Concrete 1	Reinforced concrete flat slab	Thickness: 0.30 m Reinforcement: 150 kg/m^3	12.70 m
	Reinforced concrete wall elements	Thickness: 0.20 m Reinforcement: 100 kg/m^3	
	Reinforced concrete columns	Central columns: Ø 0.5 m Reinforcement: 350 kg/m^3	
Concrete 2	Reinforced concrete flat slab	Thickness: 0.27 m Reinforcement: 150 kg/m^3	7.45 m 5.25 m
	Reinforced concrete columns	Edge columns: Ø 0.3 m Central columns: Ø 0.4 m Reinforcement: 350 kg/m Column grid: 5.40 × 7.45 m	
Steel 1	RC slab Composite beam	Thickness: 0.20 m Reinforcement: 75 kg/m^3 Beam: IPE 500, S355	12.7 m
	Steel columns	HEB 220, S355 Column grid: 5.40 × 12.7 m	
Steel 2	RC slab Composite beam	Thickness: 0.20 m Reinforcement: 75 kg/m^3 Beam: IPE 500, S355	12.7 m
	Composite columns	HEB 180, S355 Column grid: 5.4 × 12.7 m	
Steel 3	RC slab Composite beam	Thickness: 0.20 m, Reinforcement: 75 kg/m^3 Beam: IPE 360, S355	7.45 m 5.25 m
	Composite columns	Edge columns: HEB 120, S355 Central columns: HEB 180, S355 Column grid: 5.40 × 7.45 m	
Steel 4	Composite slab	Trapezoidal decking: 135/310 Thickness: 0.20 m Reinforcement: 27 kg/m^3 Beam: IPE 500, S355	12.7 m
	Composite columns	HEB 180, S355 Column grid: 5.4 × 12.7 m	

different structural systems. 'Concrete 1' is a concrete wall construction with openings for windows. The structural systems 'Concrete 2' and 'Steel 1–4' are framed constructions that can be combined with common facade system, from all types of glass facades to classical facades with regular windows.

6.3.1 LCA information

Databases for this comparison are the available EPDs according to DIN EN 15804 [5] and the ÖKOBAUDAT [1] (see Chapter 3.10). In this study, the environmental indicator 'primary energy not-renewable' is considered. It includes mainly the use of natural gas, petroleum, coal and nuclear power. The used data is listed in Table 6.11.

6.3.2 Results of the LCA for the building systems

Figure 6.32 left shows the masses of the materials used for the six different building systems (slabs and columns) per m² GFA. Concrete dominates more than 90% of the masses. At the same time, it can be observed that the steel-composite solutions have an average weight of about 550 kg/m² compared to the concrete variants with 700–850 kg/m². Figure 6.32 right represents the associated primary energy demand (nonrenewable). Compared to the ratio of the masses, steel has more influence on the primary energy demand (reinforcing steel as well as structural steel and steel decking). Nevertheless, the steel-intensive solutions 'Steel 1–3' have a lower overall primary energy demand than the concrete constructions because of the material efficiency of steel.

Figure 6.33 shows the primary energy demand (non renewable) of the six different structural systems (slabs and columns) per m² GFA split into the different life-cycle stages: the product stage, the end-of-life stage including benefits from recycling and the sum of these two values. It can be seen that the benefits at the end of life of the steel-intensive solutions are much higher than these for the concrete variants. The steel solutions have in total the lowest energy demands and are more competitive than the concrete solutions.

6.3.3 Results of the LCA for a reference building

In a second step, the five open variants 'Concrete 2' and 'Steel 1–4' are augmented with a ground slab, a facade (curtain wall facade of steel and stainless steel with transparent and opaque filling) and a roof system, in order to show the environmental impact of the different building components on the building level.

Figure 6.34 left shows the masses of the components for the whole building per m² GFA split by building components. The slabs clearly dominate and are responsible for about 70% of the masses. At the same time, it can be observed that the steel-composite solutions have an average weight of about 800 kg/m² in contrast to the concrete variant with almost 1000 kg/m². Figure 6.34 right shows the associated primary energy demand (non renewable). Compared to the ratio of the masses, the facade has a quite big influence on the primary energy demand of

Table 6.11 Used data base for different construction products [1]–[16].

Material/Source	Module According to EN 15804	Reference Unit (RU)	Primary Energy, Nonrenewable [MJ/RU]	Comment
Structural Steel EPD-BFS-20130094 [4]	Total	t	10630	
	A1–A3	t	17900	
	D	t	–7270	11% reuse, 88% recycling
Concrete C 20/25, EPD-IZB-2013411 [2]	Total	m^3	546.2	
	A1–A3	m^3	846	
	C3	m^3	19.2	Building rubble processing
	D	m^3	–319	96% material utilisation, 4% landfill
Concrete C 30/37, EPD-IZB-2013431 [3]	Total	m^3	684.2	
	A1–A3	m^3	984	
	C3	m^3	19.2	Building rubble processing
	D	m^3	–319	96% material utilisation, 4% landfill
Reinforcement, ÖKOBAUDAT 2013, process 4.1.02 [1]	Total	kg	11.2	
	A1–A3	kg	11.2	
	D	kg	–	No recycling potential
Trapezoidal sheet, EPD-IFBS-2013211 [12]	Total	m^2	193	
	A1–A3	m^2	373	
	C4	m^2	0	10% landfill
	D	m^2	–180	90% recycling
Gypsum plaster fire-protection board, EPD gypsum poducts [13]	Total	kg	3.45	
	A1–A3	kg	3.35	
	C3	kg	0.1	Gypsum waste processing
Facade, M-EPD-SFA-000003 [14]	Total	m^2	1049.94	
	A1–A3	m^2	1859.19	
	C3	m^2	24.48	Dismantling, recovery and thermal utilisation
	D	m^2	–833.73	Dismantling, recovery and thermal utilisation

(*Continued*)

Table 6.11 (Continued)

Material/Source	Module According to EN 15804	Reference Unit (RU)	Primary Energy, Nonrenewable [MJ/RU]	Comment
Roof insulation, EPD-DRW-2012131 [15]	Total	m^3	1857.16	
	A1–A3	m^3	1933.68	
	C4	m^3	29.46	100% landfill
	D	m^3	–105.98	Thermal utilisation of packaging
Perimeter insulation base plate, EPD-FPX-2010111-D [16]	Total	0.1 m^3	241.219	
	A1–A3	0.1 m^3	343.752	
	D	0.1 m^3	–102.533	90% thermal utilization

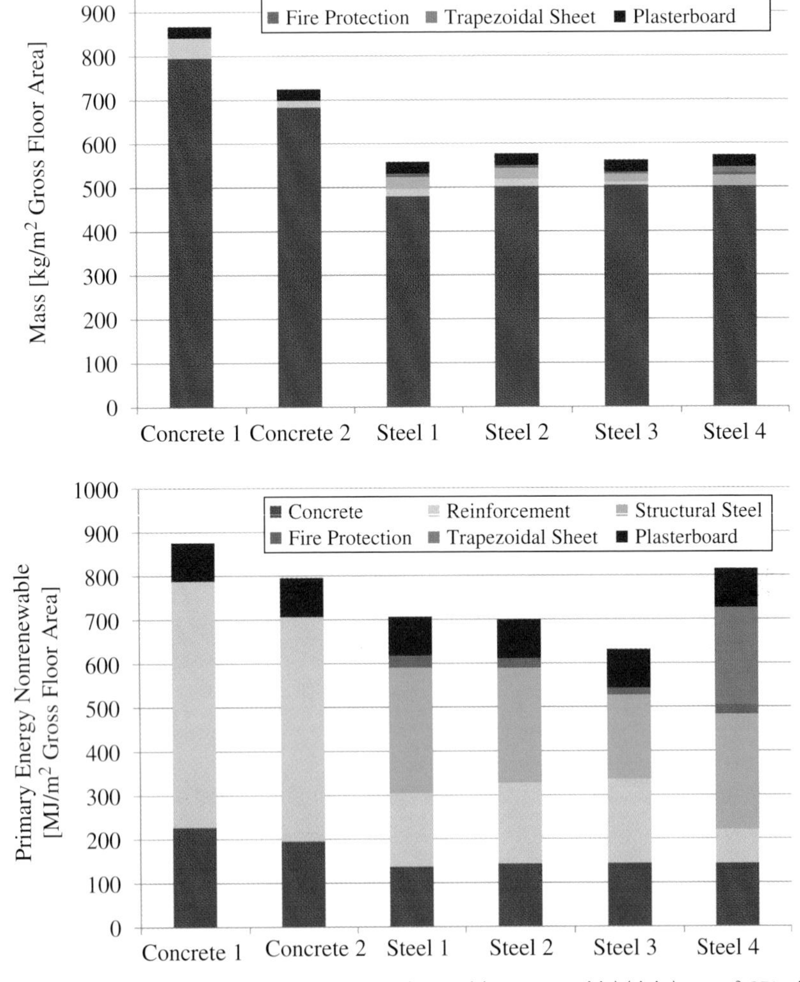

Figure 6.32 Masses (left) and primary energy demand (nonrenewable) (right) per m^2 GFA of one floor (slabs and columns) split by materials.

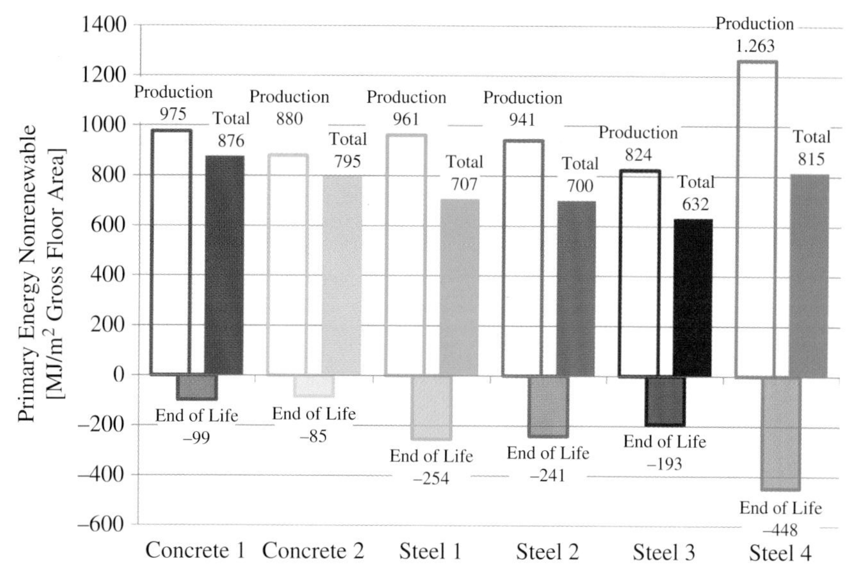

Figure 6.33 Primary energy demand (nonrenewable) per m² GFA of the structural system (slabs and columns) split by life-cycle stages.

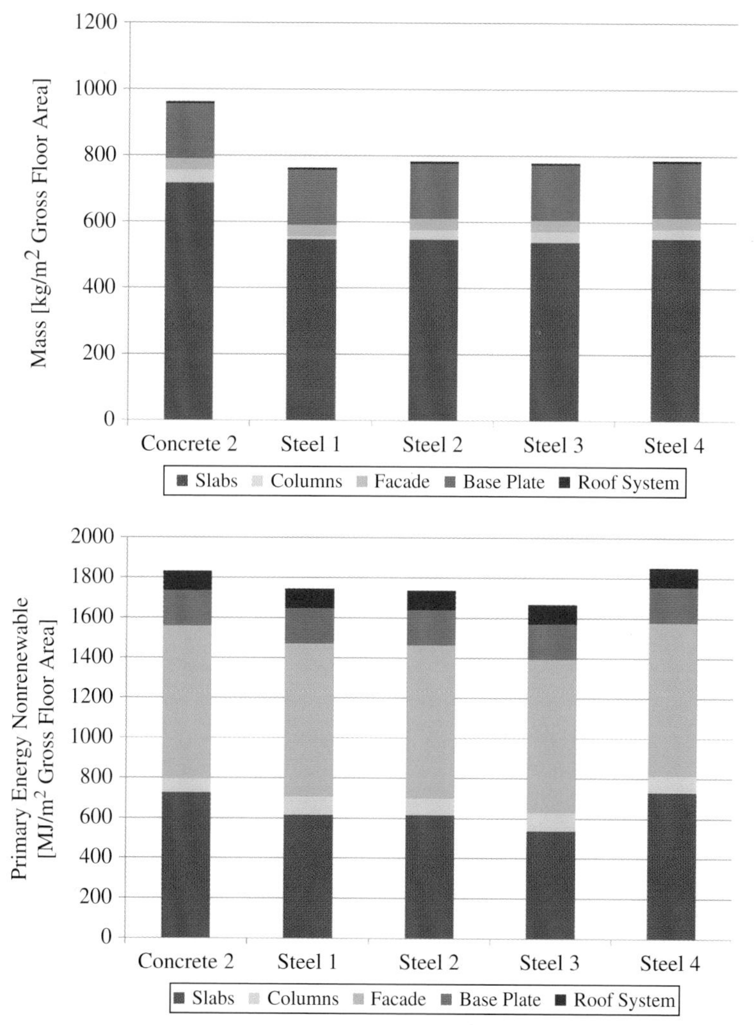

Figure 6.34 Masses (left) and nonrenewable primary energy demand (right) per m² GFA of the whole building split by building components.

Table 6.12 Primary energy demand for the office building "Steel 3" considering product stage, end of life and operational energy use for three different office types.

	Primary Energy Demand, Nonrenewable		
Office type	Product stage (A1–A3) and end of life (D) [MJ]	Heating [MJ/a]	Electricity (lighting, hot water, ventilation) [MJ/a]
Open-plan office		125	400
Individual offices	1669	115	303
Group offices		119	268

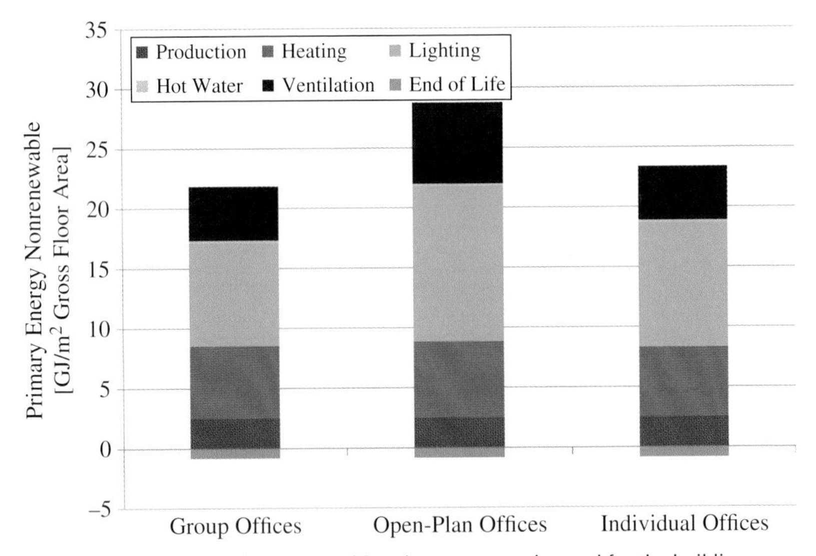

Figure 6.35 Comparison of nonrenewable primary energy demand for the building over a utilisation period of 50 years including production stage, end of life as well as operational energy use.

the building. Thus, overall, the differences between the various structural systems reduce and the variants tend to the same level.

The comparison of the environmental impact of different structural systems for office buildings shows that mass of the elements of an office building are not automatically indicative of the associated primary energy demand.

The option 'Steel 3' (see Table 6.10) is evaluated for the use phase of the building. For this purpose, three different types of office uses are considered:

- open-plan offices;
- individual offices;
- group offices.

The annual operational energy demand has been determined for the three different office types using a software package according to DIN V 18599 (Table 6.12) [17]:

Figure 6.35 shows the comparison of the non renewable primary energy demand for the building over a utilisation period of 50 years including production stage, end of life as well as operational energy use. The operational energy is composed of the energy demands for heating, lighting, hot water and ventilation, as seen in Figure 6.35.

The study of the use stage shows the importance of considering the entire life cycle. Office buildings are designed for a long period of use (50 years or longer), so the energy demand during the use stage dominates compared to the production stage without reference to the type of office use. Furthermore, it can be observed that the open-plan offices have the highest energy use due to the energy demand for lighting and ventilation compared to the group and individual offices.

6.4 MATERIAL EFFICIENCY

Richard Stroetmann

The selection of building materials may significantly influence compliance with sustainability criteria. Life-cycle costs (LCCs), resource efficiency, recyclability and environmental impact are of particular relevance. Increased use of high strength steel not only improves environmental performance and profitability but also the competitiveness of steel structures and the technological advantage of the companies involved in the value-added process.

By the choice of material, dimensioning of components and design features, influence can be exerted on how particular sustainability criteria are met. This significantly influences LCCs, utilisation of resources, environmental impact, recyclability and waste accumulation.

Energy and material efficiency are of special significance due to energy consumption and large mass flows in the construction industry. Optimisation in using steel products is a way to improve resource efficiency and to reduce negative environmental impacts.

6.4.1 Effective application of high strength steels

The application of high strength steels is advantageous if it enables steel consumption to be reduced. Figure 6.36 shows the relative price comparison for heavy plate in various steel grades. The comparison is based on average prices provided by different producers for the German market [18], [19]. Compared to the increase in strength, there is only a moderate increase in price that may be compensated by appropriate weight savings.

Mass reduction requires that the resistance of a structural element is not strongly influenced by buckling or fatigue resistance. Furthermore, serviceability criteria – such as limits for deflections and for dynamic effects (e.g. minimal vibration frequency) – should not control the dimensions of the structural elements that are required.

The necessary dimensions of tension elements are significantly affected by material strength, as cross-sectional areas can be reduced inversely in proportion

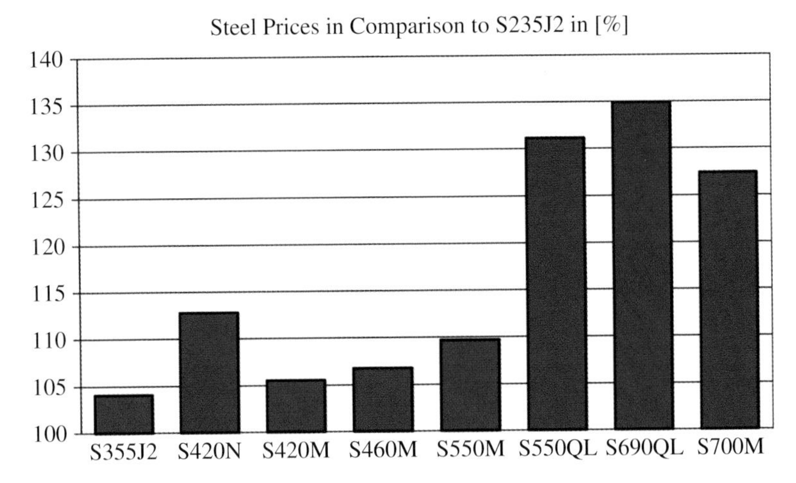

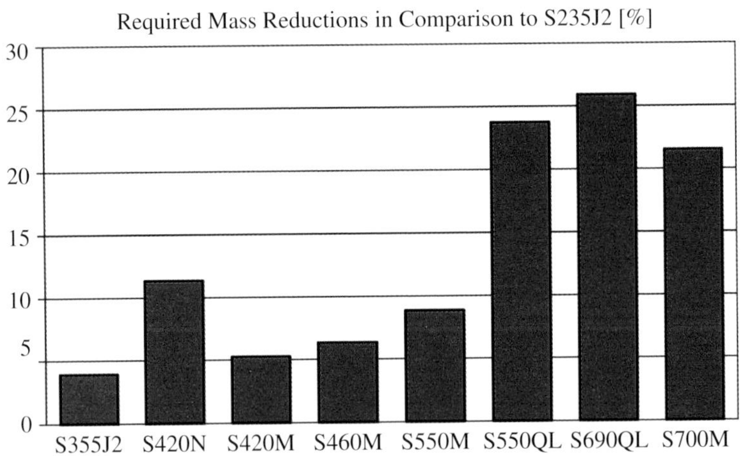

Figure 6.36 Economic efficiency of heavy plate if only the material price is taken into account.

to the yield strength (Figure 6.37). Also for butt welds with a constant included angle, the weld volume increases with the square of the plate thickness. For this reason, a reduction in plate thickness has positive effects on the welding, although the weld should be of compatible strength to the plate.

For the design of class 4 cross-sections affected by local buckling due to compressive stresses resulting from bending and/or normal forces, Eurocode 3-1-5 offers two approaches for calculations: the effective width method and the reduced stress method. For rolled I-sections according to DIN 1025, the web may be in class 4. In this case, the effective width method can be applied beneficially because only the webs have to be reduced and the flanges are fully effective.

Figures 6.38 and 6.39 show the cross-section resistance of HEA and square hollow sections (SHS) under compression as a function of the yield strength. In Europe, the yield strength of I-sections is presently limited to 460 N/mm^2. However, yield strengths up to 690 N/mm^2 are considered in the diagram to show the dependencies. It can be clearly seen that, independent of the local buckling

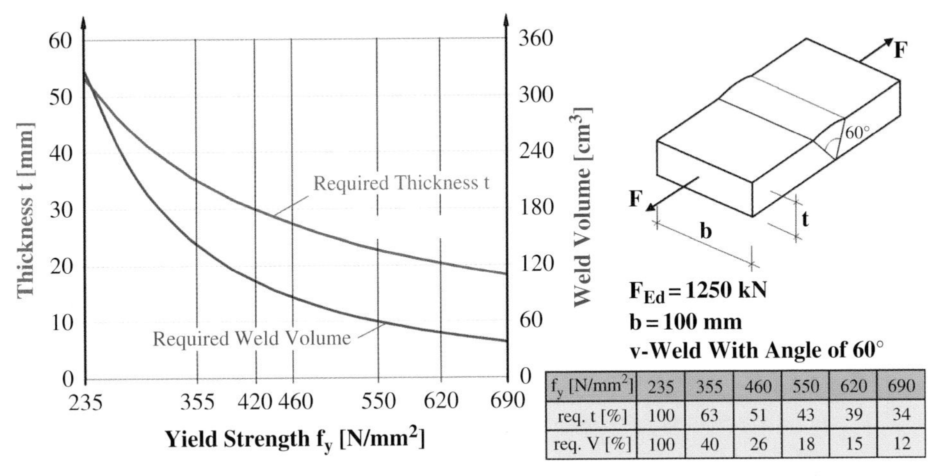

f_y [N/mm²]	235	355	460	550	620	690
req. t [%]	100	63	51	43	39	34
req. V [%]	100	40	26	18	15	12

Figure 6.37 Example for dimensions of tension bars with butt welds and various steel grades.

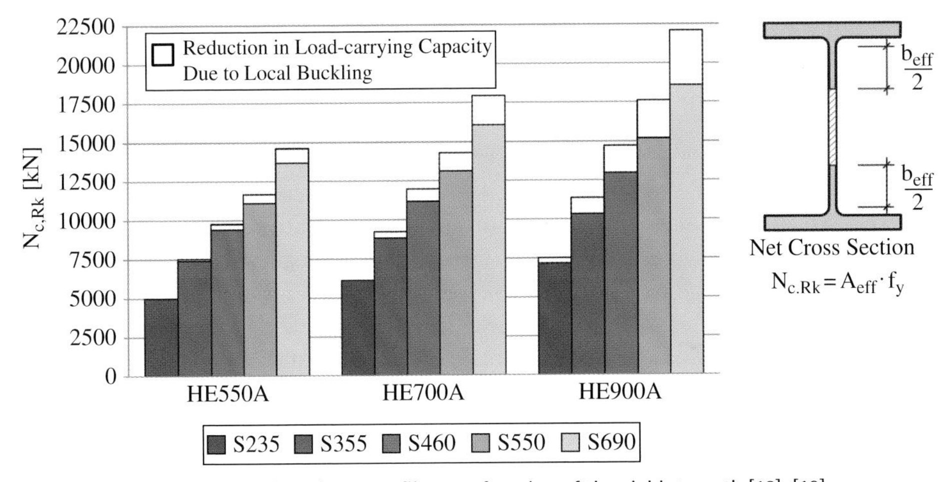

Figure 6.38 Resistance of selected HEA profiles as a function of the yield strength [18], [19].

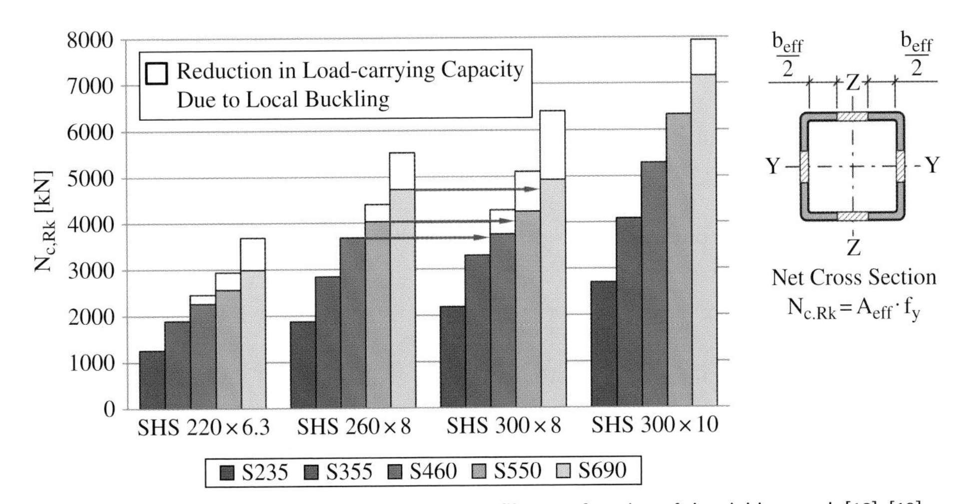

Figure 6.39 Resistance of selected square hollow profiles as a function of the yield strength [18], [19].

effect, the gain of cross-section resistance is significant. Figure 6.39 shows that increasing the side width of square hollow profiles while retaining the plate thickness only slightly increases the resistance if local buckling is important.

The diagrams (a) and (b) in Figure 6.40 show the reduction factor ρ for the buckling of internal and outstand plates; the diagrams (c) and (d) show the reduced stresses $\rho \cdot f_y$ as a function of the b/t-ratio for different steel grades. In addition, the safety factor γ_M (γ_{M0} or γ_{M1}) has to be considered in accordance with the relevant part of Eurocode 3. If the goal is to utilise the yield strength most efficiently, the properties of the cross-section must be designed to be compact. However, it is also possible to increase the resistance of the cross section by using a higher yield strength even for more slender cross sections [see diagrams (c) and (d) in Figure 6.40].

Therefore, high strength has a positive effect on steel consumption if stability-sensitive members are in a range of compact and medium slenderness (see also [22]). Figure 6.41 shows the buckling resistance of a column made of SHSs (300 × 16) for different buckling lengths and steel grades. It shows that the differences decrease when the buckling length increases. In conventional multi-storey buildings with buckling lengths of columns between 3 and 5 m, use of high strength steel has a positive effect on compression resistance. The converse is also true: smaller cross sections may be used for the same compression resistance.

Figure 6.42 shows the load-carrying capacity of steel and composite columns with SHSs in three different sizes and different strength classes as a function of the buckling length. In the compact and medium slenderness regions, high strength has a positive effect on material consumption. From the diagrams, it can be seen that the spacing between the lines for the groups of load-carrying capacities (grey, blue and black lines) for the different cross sections and strength classes becomes smaller with increasing buckling length and decreasing column cross-section. The effect of the buckling length manifests itself more strongly with composite columns than for steel columns. The design of columns as composite sections has little effect at small cross sections because of the smaller radii of gyration and the buckling curves that have to be used, which are less favourable. This offsets the increased cross-section resistance due to less favourable reduction factors for flexural buckling.

The effect of lateral torsional buckling is similar when it comes to torsionally flexible supports with open cross sections, such as I-beams. With increasing span, the bending moments increase and usually so does the risk of lateral torsional buckling. The beam slenderness depends on the span, the geometry of the cross section and the loading (moment distribution, point of application of transverse loads, etc.), and especially on the stabilisation due to connected components. These can be purlins, bracings, roof and wall cladding, concrete slabs or similar (see [23], [24]).

For comparison, the moment capacity of a beam with an IPE 300 section is shown as a function of span for different yield stresses (Figure 6.43). Three cases of intermediate lateral restraint attached to the top flange were investigated: (a) no restraint, (b) with one restraint and (c) with three restraints. The evaluation was carried out in accordance with EN 1993-1-1 using buckling curve b (see [25], Table 6.5, h/b = 2). The moment distribution was taken into account using the

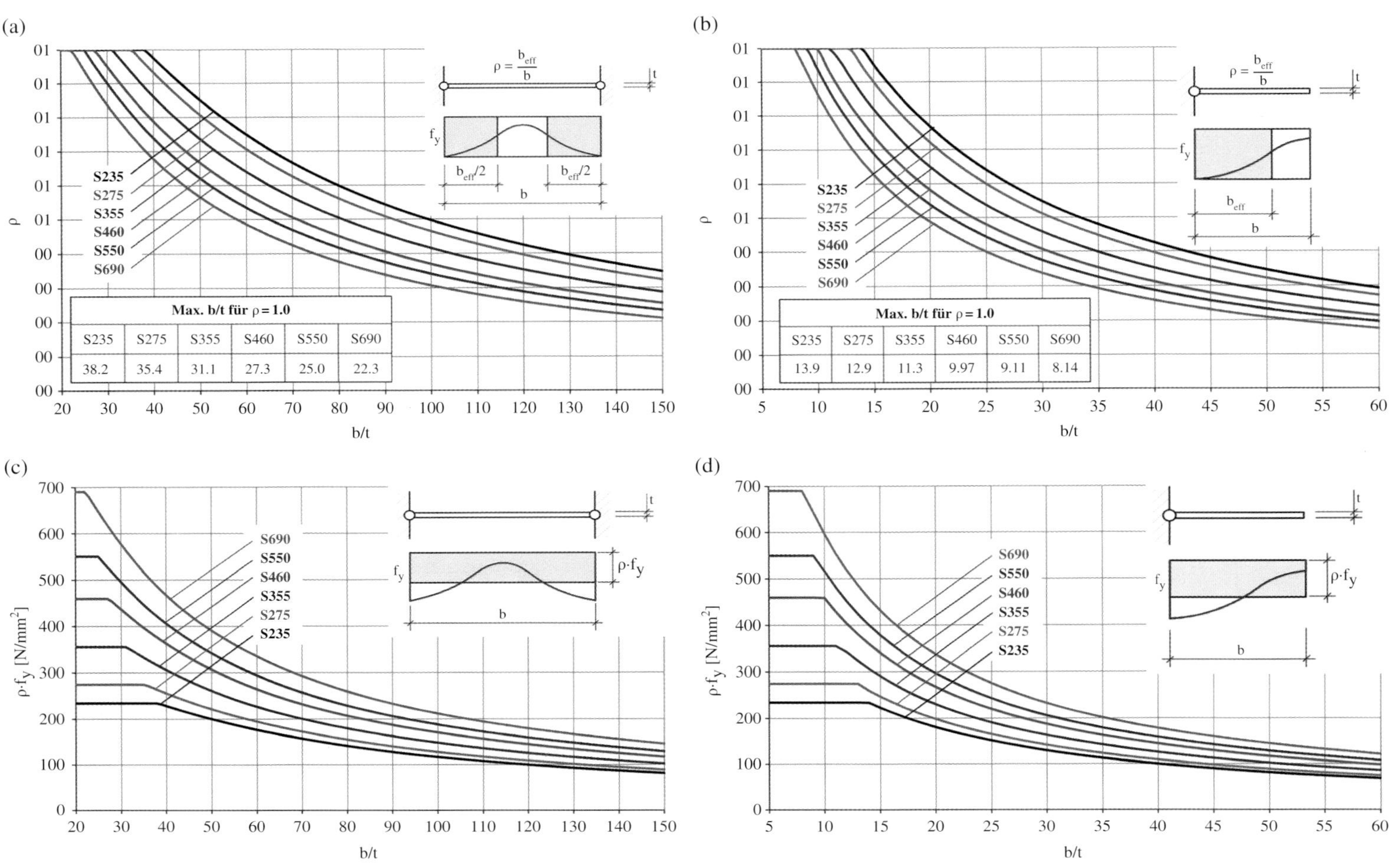

Figure 6.40 Reduction factors ρ and reduced stresses ρ·fy for buckling of internal and outstand compression elements (σx, ψ = 1.0) as a function of yield strength [20], [21].

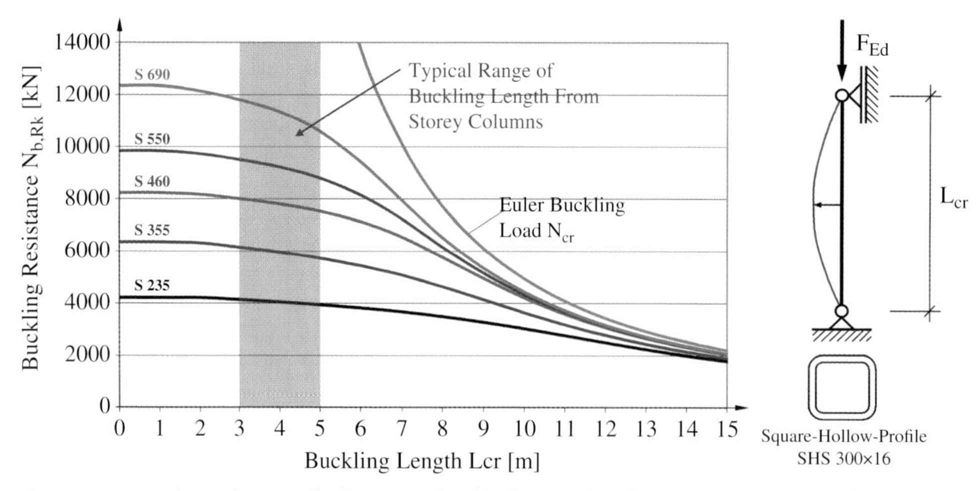

Figure 6.41 Design resistance of columns made of hollow sections for different steel grades [20], [21].

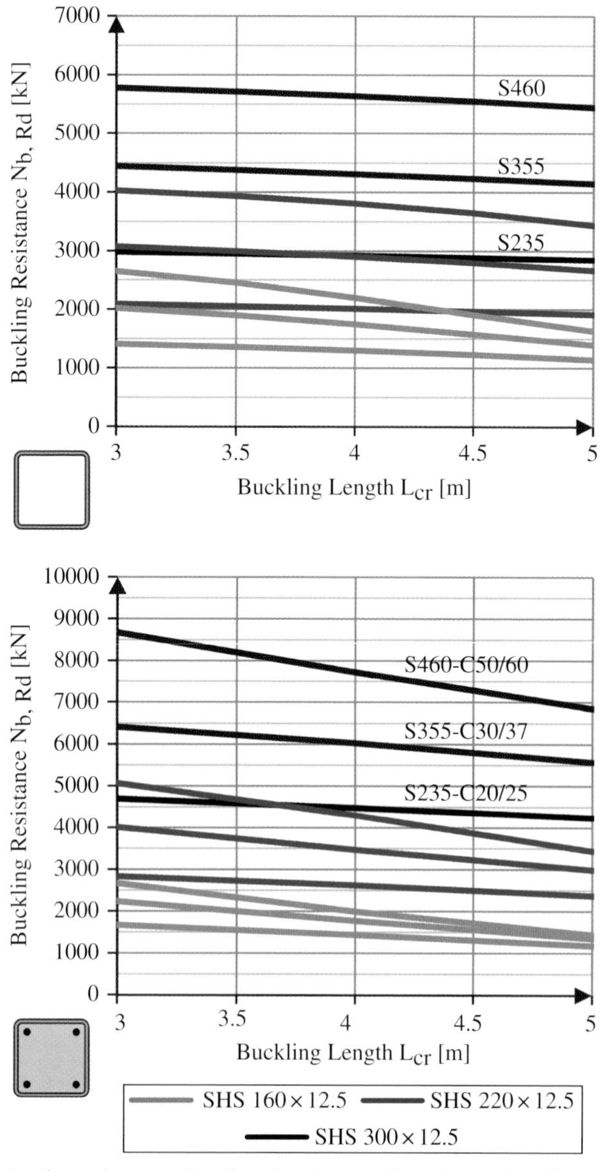

Figure 6.42 Load-carrying capacity of steel and composite columns in various strength categories as a function of buckling length [22].

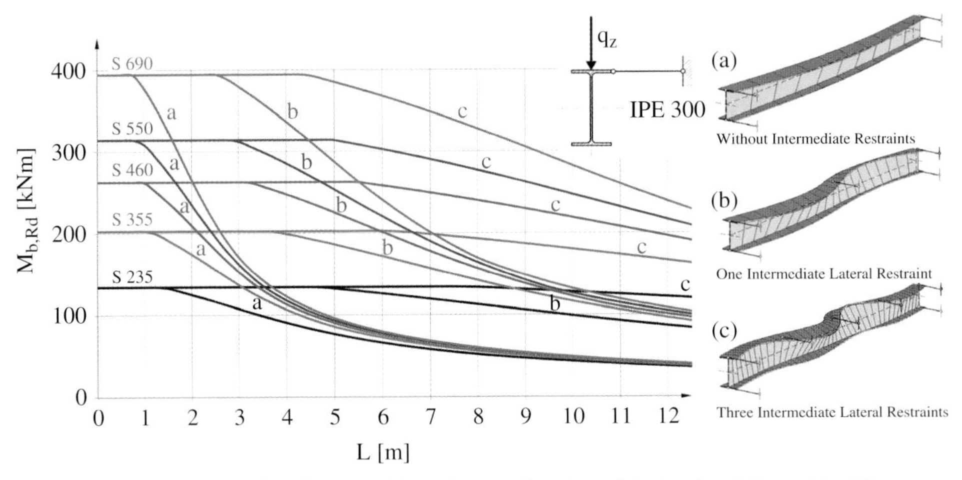

Figure 6.43 Moment capacity of an IPE 300 section as a function of the torsional-flexural buckling slenderness for different yield stresses.

reduction factor $\chi_{LT,mod}$ for lateral torsional buckling [see [25], Section 6.3.2.3(2)]. As can be seen from the shape of the load-carrying capacity curves, high yield stresses have a positive effect in the compact and medium slenderness regions. In order to ensure that this is the case, adequate stabilising measures or use of more torsionally rigid sections (e.g. HEA and HEB sections), or a combination of both measures, should be considered for longer spans (see also [23], [24]).

For structures, where the weight has a significant influence on the structural design, mass reduction is doubly positive: the action effects of the elements are reduced and thus further savings of materials are possible. Examples are large-span structures, such as bridges, stadium roofs, exhibition halls, storage and production halls.

6.5 SUSTAINABLE OFFICE DESIGNER

Li Huang and Martin Mensinger

When it comes to attaining more sustainable design in building projects, the decisions made in early planning phases are of vital significance. At this early stage, it is important to quickly estimate and compare the sustainability qualities of various design proposals. Doing so, however, can prove to be a difficult task due to the lack of detailed design information inherent in early phases. To resolve this the 'Sustainable Office Designer' (SOD) [26] has been developed. It can generate preliminary structural solutions based on early phase architectural designs for composite office and administration buildings, optimising environmental and financial aspects alike. SOD makes it easy to assess and compare sustainability qualities based on the structural solutions it generates.

SOD has been developed as a plugin for SketchUp [27], a free, widely used 3D design and modeling software tool for the early building sketches and designs. The SOD plugin is freely available [28]. Users can quickly create a 3D conceptual design for office buildings featuring floor layout and parameters. Floor layout

Table 6.13 Parameters for configuration.

Parameter	Description	Example Value
Max. room size	Defines the room height of a given floor level.	2.5 m
Flexibility	Defines the height available in a celling for installing HVAC and other mechanical systems.	Low (height 200 mm)
Max. construction height	Defines the height provided for structural elements including slabs and beams.	600 mm
Number of stories	Defines the number of floor levels for each of the building's bars.	5, 3, 4
Length, width	The dimensions for each rectangular bar.	–
Number of inner column rows	In addition to a row of columns on each longitudinal side, there are inner columns for each rectangular bar. The number of inner rows can be zero, one or two.	2, 1, 1
Total column range	Defines the range for possible distance between neighboring columns within one row.	5–15 m
Facade width	Defines the grid on which columns may be positioned.	1 m
Column spacing	Differs from 'Total column range' in that this defines the actual distances possible between neighboring columns within one row. The length of the rectangle is determined based on the options available. For example: with a length of 30 m and a facade width of 1.5 m, potential values for 'Column spacing' are 3 × 10 m, 6 × 5 m, 7.5 × 4 m and 15 × 2 m. The range for spacing in this example will be 2–15 m.	–

defines the shape of a building's ground plan. Currently, SOD works with shapes comprising one or multiple rectangles, called bars. Floor layouts are broken down into bars. Table 6.13 explains the parameters for configuration.

The user interface of SOD and SketchUp is shown in Figure 6.44, illustrating the example of a three-bar building. The designer uses SketchUp to create a volumetric architectural design, featuring information on the positioning range for column spacings. The best solutions are displayed directly on the user interface during and after optimization. Following optimisation, the user is able to visualize the structure with steel columns and beams. A detailed report is also available, which presents the structural solutions selected and dimensions for all structural elements.

SOD employs a genetic algorithm [29] for structural optimisation, a heuristic optimisation method. It searches for the minimum objective value, which is calculated using the object function. Various objective functions can be defined, such as total CO_2 emissions and total material cost. To eliminate infeasible solutions, structural verification is executed in line with Eurocode 4 [30] for composite structures. Penalty values are added to the objective value when structural design rules are violated. This helps to filter the feasible structural solutions from nonfeasible ones. Optimisation is performed on the parametric models obtained for the rectangular 'wings' or zones of the building model.

Parameters for design model, the genetic algorithm, and construction type and objective function can be defined in corresponding subwindows. Two types

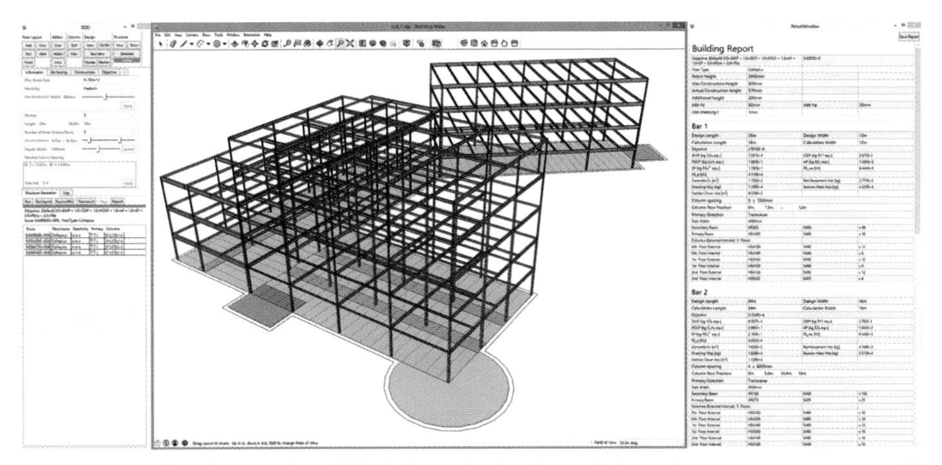

Figure 6.44 User interface: In the middle is the main SketchUp window, where users can create, modify and view the 3D design model. On the left is the main SOD window, where users can configure and select parameters, start optimization and select results for visualization. On the right is the report window, which displays results.

of objective functions are available. One is the weighted sum of life-cycle impact assessment (LCIA) indicators: GWP, ODP, POCP, AP, EP, PEne and PEe, while the other is the weighted sum of different material consumption volume of concrete (Cv), mass of steel reinforcement (Msr), mass of steel profiled sheeting (Msp), mass of steel sections (Mss) and coating area of steel sections (Ass). The user is able to choose either function and define the weighting factors. The default objective function is the weighted LCIA indicators with the factors suggested by DGNB [31].

6.5.1 Database

Databases for SOD are the available EPDs (see Chapter 3.11) and the ÖKOBAUDAT (see Chapter 3.10). The use of construction machinery is also considered (see Table 6.14). The working hours of structural components are obtained from [32].

6.5.2 Example using sustainable office designer

This example involves a building with three rectangular 'wings'. The dimensions are shown in Figure 6.45, while the 3D volumetric design model is shown in Figure 6.46. Facade width is 1 m for all bars. Floor layouts of rectangle bars are colored orange. Possible positions of the inner column rows are defined as red zones. The maximum number of possible rows of inner columns is two for the first bar and one for the other two bars. The color purple indicates building areas that are not part of the rectangles, which are not included in the calculation. The brown color is used for building cores, which are not calculated either. They are presented in the 3D model for visualisation purpose only.

In addition to the values shown in Table 6.13, this example uses a primary beams in both the transverse and longitudinal directions. After optimisation,

Table 6.14 Environmental data for construction products from EPDs and ÖKOBAUDAT 2014 used in SOD.

Material	Comment	Reference Unit (RU)	Primary Energy Not Renewable	Primary Energy, Renewable	Total Primary Energy	Global Warming Potential (GWP)	Ozone Depletion Potential (ODP)	Acidification Potential (AP)	Europhication Potential (EP)	Photochemical Ozone Creation Potential (POCP)
			[MJ/BE]	[MJ/BE]	[MJ/BE]	[kg CO_2–eqv./BE]	[kg CFC11–eqv./BE]	[kg SO_2–eqv./BE]	[kg PO_4–eqv./BE]	[kg Ethen-eqv./BE]
Structural steel	EPD-BFS-20130094	kg	10.59	0.93	11.52	0.78	1.45E-10	1.79E-03	1.58E-04	2.98E-04
Production	A1–A3	Kg	17.80	0.84	18.64	1.74	1.39E-10	3.47E-03	2.89E-04	7.55E-04
Benefits & loads	D	kg	–7.21	0.09	–7.12	–0.96	6.29E-12	–1.68E-03	–1.31E-04	–4.57E-04
Concrete C30/37	EPD-IZB-2013431	kg	0.29	0.02	0.31	0.09	2.66E-10	1.23E-04	2.01E-05	1.54E-05
Production	A1–A3	2365 kg/m³	984	82.70	1067	231.90	7.35E-07	3.23E-01	5.13E-02	3.93E-02
		kg	0.41	0.03	0.57	0.11	2.93E-09	1.94E-04	2.73E-05	1.99E-05
Benefits & loads	C3	kg	0.01	0.00	0.01	0.00	3.10E-14	5.42E-06	1.17E-06	7.08E-07
	D	kg	–0.13	–0.02	–0.15	–0.01	–3.99E-11	–1.71E-05	–2.46E-06	–1.69E-06
Reinforcing steel	ÖKOBAUDAT 2014	kg	11.20	1.24	12.44	0.76	2.53E-09	1.80E-03	1.51E-04	1.78E-04
Production	A1–A3 4.1.02 Reinforcing steel	kg	11.20	1.24	12.44	0.76	2.53E-09	1.80E-03	1.51E-04	1.78E-04
Benefits & loads	No recycling potential	–	0.00	0.00	0.00	0.00	0.00E+00	0.00E + 00	0.00E + 00	0.00E + 00
Profiled sheeting	EPD-IFBS-2013211	kg	17.08	1.38	18.46	1.00	5.41E-09	3.14E-03	3.12E-04	2.55E-04
Production	A1–A3	11.3 kg/m²	373	14	387	27.1	5.65E-08	0.095	8.29E-03	1.19E-02
		kg	33.01	1.24	34.25	2.40	5.00E-09	8.41E-03	7.34E-04	1.05E-03
Benefits & loads	C3	11.3 kg/m²	0	0	0	0.02	1.55E-11	0	1.32E-05	1.02E-05
		kg	0	0	0	1.77E-03	1.37E-12	0	1.17E-06	9.03E-07
	D	11.3 kg/m²	–180	1.6	–178.4	–15.8	4.61E-09	–0.0595	–4.78E-03	–9.03E-03
		kg	–15.93	0.14	–15.79	–1.40	4.08E-10	–5.27E-03	–4.23E-04	–7.99E-04

Work		Comment		Reference Unit (RU)	Primary Energy, Not Renewable	Primary Energy, Renewable	Total Primary Energy	Global Warming Potential (GWP)	Ozone Depletion Potential (ODP)	Acidification Potential (AP)	Eutrophication Potential (EP)	Photochemical Ozone Creation Potential (POCP)
					[MJ/BE]	[MJ/BE]	[MJ/BE]	[kg CO_2–eqv./BE]	[kg CFC11–eqv./BE]	[kg SO_2–eqv./BE]	[kg PO_4–eqv./BE]	[kg Ethen-eqv./BE]
Pump		ÖKOBAUDAT 2014		kg	6.54E-03	1.11E-03	7.65E-03	4.65E-04	2.17E-12	7.81E-07	7.52E-08	5.74E-08
Use		9.1.02 Concrete		2365 kg/m³	15.47	2.62	18.09	1.10	5.14E-09	1.85E-03	1.78E-04	1.36E-04
				kg	6.54E-03	1.11E-03	7.65E-03	4.65E-04	2.17E-12	7.81E-07	7.52E-08	5.74E-08
Crane		ÖKOBAUDAT 2014		-	-	-	-	-	-	-	-	-
Use		9.2.05 Electricity mix		kwh	8.78	1.49	10.27	6.23E-01	3.07E-09	1.03E-03	9.92E-05	7.62E-05
Crane Work for:	Reinforcement	Crane: 40 kw	0.25 h/t	kg	8.78E-02	1.49E-02	1.03E-01	6.23E-03	3.07E-11	1.03E-05	9.92E-07	7.62E-07
	Formwork		0.06 h/m²	m²	21.07	3.58	24.65	1.50	7.37E-09	2.47E-03	2.38E-04	1.83E-04
	Structural steel		0.6 h/piece	piece	210.72	35.76	246.48	14.95	7.37E-08	2.47E-02	2.38E-03	1.83E-03
	Profiled sheeting		0.03 h/m²	m²	10.54	1.79	12.32	0.75	3.68E-09	1.24E-03	1.19E-04	9.14E-05
	Precast		0.2 h/m²	m²	70.24	11.92	82.16	4.98	2.46E-08	8.24E-03	7.94E-04	6.10E-04

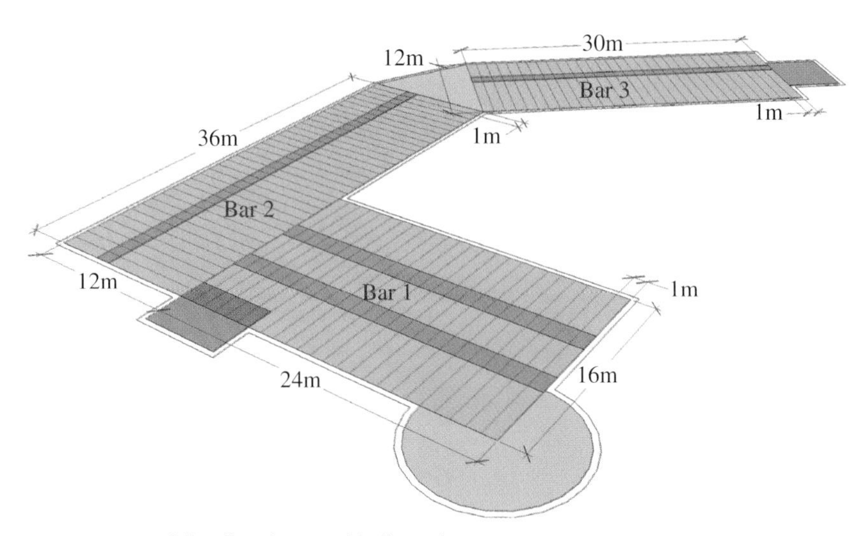

Figure 6.45 Building floor layout with dimensions.

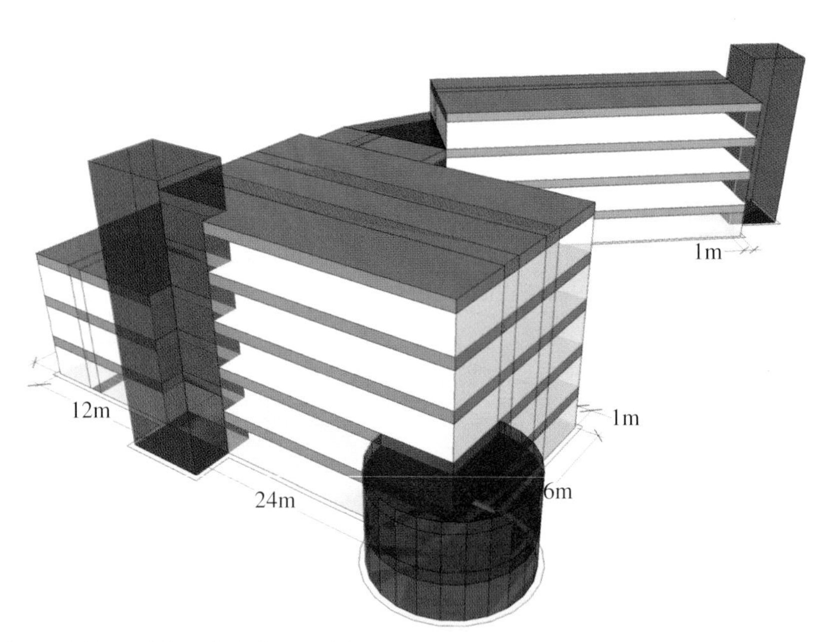

Figure 6.46 Building 3D design model.

which only takes a short time, a structural design is output with an object value of 1.0197E+7 using the default object function: $3.0 \times GWP + 1.0 \times ODP + 1.0 \times POCP + 1.0 \times AP + 1.0 \times EP + 3.0 \times PEne + 2.0 \times PEe$. The optimised structure uses composite slab 'Cofraplus 60' with a thickness of 120 mm. More structural information is shown in Table 6.15.

Table 6.15 Optimised structure design.

Building Report			
Objective: (Default) $3.0 \times GWP + 1.0 \times ODP + 1.0 \times POCP + 1.0 \times AP + 1.0 \times EP + 3.0 \times PEne + 2.0 \times PEe = 8.1537E+6$			
Floor type	Cofraplus	Room height	2.5 m
Max construction height	600 mm	Actual construction height	520 mm
Additional height	200 mm	Slab hc	62 mm
Slab hp	58 mm	Slab sheeting t	1 mm
'Bar' 1			
Length	24 m	Width	16 m
Objective		2.8116E+6	
GWP [kg CO_2 eqv.]	5.091E+4	ODP [kg R11 eqv.]	1.058E-4
POCP [kg C_2H_4 eqv.]	1.684E+1	AP [kg SO_2 eqv.]	1.308E+2
EP [kg PO_4^{3-} eqv.]	1.208E+1	PE_ne [MJ]	7.520E+5
PE_e [MJ]	6.569E+4		
Concrete Cv [m^3]	1.632E+2	Reinforcement Msr [kg]	4.109E+3
Sheeting Msp [kg]	1.638E+4	Section mass Mss [kg]	4.000E+4
Section cover Ass [m^2]	1.362E+3		
Column spacing		4 × 6 m	
Column row position		0 m, 5.6 m, 10.4 m, 16 m	
Primary direction		Longitudinal	
Slab width:		2000 mm	
Secondary beam	IPE160	S355	× 195
Primary beam	IPE220	S355	× 20
Columns (external/internal) 5 floors			
0th floor external	HEA180	S355	× 10
0th floor internal	HEA240	S355	× 10
1st floor external	HEA180	S355	× 10
1st floor internal	HEA220	S355	× 10
2nd floor external	HEA160	S355	× 10
2nd floor internal	HEA200	S355	× 10
3rd floor external	HEB120	S355	× 10
3rd floor internal	HEA160	S355	× 10
4th floor external	HEB100	S355	× 10
4th floor internal	HEB120	S355	× 10
'Bar' 2			
Length	36 m	Width	12 m
Objective		2.3104E+6	
GWP [kg CO_2 eqv.]	4.326E+4	ODP [kg R11 eqv.]	7.305E−5
POCP [kg C_2H_4 eqv.]	1.476E+1	AP [kg SO_2 eqv.]	1.087E+2
EP [kg PO_4^{3-} eqv.]	9.955E+0	PE_ne [MJ]	6.283E+5
PE_e [MJ]	5.495E+4		

(Continued)

Table 6.15 (Continued)

Concrete Cv [m^3]	1.102E+2	Reinforcement Msr [kg]	2.773E+3
Sheeting Msp [kg]	1.105E+4	Section mass Mss [kg]	3.840E+4
Section cover Ass [m^2]	1.132E+3		
Column spacing		6×6000 mm	
Column row position		0 m, 7.2 m, –, 12 m	
Primary direction		Longitudional	
Slab width:		2000 mm	
Secondary beam	IPE200	S355	× 114
Primary beam	IPE330	S355	× 9
Columns (external/internal) 3 floors			
0th floor external	HEA200	S355	× 14
0th floor internal	HEA240	S355	× 7
1st floor external	HEA160	S355	× 14
1st floor internal	HEA200	S355	× 7
2nd floor external	HEB120	S355	× 14
2nd floor internal	HEA160	S355	× 7
'Bar' 3			
Length	30 m	Width	12 m
Objective		2.4976E+6	
GWP [kg CO_2 eqv.]	4.675E+4	ODP [kg R11 eqv.]	8.092E−5
POCP [kg C_2H_4 eqv.]	1.590E+1	AP [kg SO_2 eqv.]	1.177E+2
EP [kg PO_4^{3-} eqv.]	1.079E+1	PE_ne [MJ]	6.803E+5
PE_e [MJ]	5.948E+4		
Concrete Cv [m^3]	1.224E+2	Reinforcement Msr [kg]	3.082E+3
Sheeting Msp [kg]	1.228E+4	Section mass Mss [kg]	4.098E+4
Section cover ass [m^2]	1.100E+3		
Column spacing		6×5000 mm	
Column row position		0 m, 7.2 m, –, 12 m	
Primary direction		Transverse	
Slab width:		4000 mm	
Secondary beam	IPE200	S355	× 96
Primary beam	IPE400	S355	× 28
Columns (external/internal) 4 floors			
0th floor external	HEA180	S355	× 14
0th floor internal	HEA200	S355	× 7
1st floor external	HEA180	S355	× 14
1st floor internal	HEA180	S355	× 7
2nd floor external	HEA140	S355	× 14
2nd floor internal	HEA160	S355	× 7
3rd floor external	HEB100	S355	× 14
3rd floor internal	HEB120	S355	× 7

6.6 SUSTAINABILITY COMPARISON OF HIGHWAY BRIDGES

Tim Zinke, Thomas Ummenhofer and Helena Gervasio

In Germany and most European countries, about two-thirds of the bridges are classified as small and medium span bridges in the range of 5–50 m. Therefore, the analysis focuses on this type of bridge. Since highways are important for economy, bridges that cross a highway with six lanes are considered in this study.

All results presented refer to the three bridges shown in Figure 6.47. All of them have been adapted so that they are of the same length and have the same bridge deck area. Consequently, the bridges can be compared directly because they all can fulfill the identical function within the road network. All results shown in the next sections are based on analyses and findings in [38].

6.6.1 Calculation of LCC for highway bridges

The LCC calculations are carried out with a scenario-based model of the individual bridge components. Additionally, the interval and main inspections (every 3 and 6 years) and the costs for traffic routing (temporal crash barriers, traffic signs, etc.) are considered as discrete events. A discount rate of 2% is used as default value. A lump sum calculation (i.e. a general value for renewal costs of 0.8% with regard to the overall erection costs) is not expedient, since this approach does not allow comparison of different maintenance strategies and does not enable an integral observation of LCCs and external costs. The considered cost elements according to [38] are summarized in Equation (1):

$$NPV_{LCC} = C_{Con} + \sum_{c=1}^{n} \sum_{t=1}^{T} (1+i)^{-t} \cdot \left[C_{(Renew,c)} + C_{(Rehab,c)} + C_{Ins} + C_{TR} \right] + (1+i)^{-T} \cdot C_{Dis}, \quad (1)$$

Bridge A1
1-Span Integral Steel Composite Bridge with 4 VFT Steel Girders

Bridge A2-V
2-Span Steel Composite Bridge with 4 continuous Steel Girders

Bridge A2-B
2-Span Reinforced Concrete Bridge with 4 continuous Girders,
Built with in situ Concrete

Figure 6.47 Bridge types analysed for a two-lane main road on top and a six-lane highway underneath the bridge.

where:

NPV_{LCC} = net present value of the LCCs
C_{Con} = construction costs
$C_{Renew,c}$ = renewal costs of the different components c
$C_{Rehab,c}$ = rehabilitation costs of the different components c
C_{Ins} = inspection costs
C_{TR} = costs for traffic routing
C_{Dis} = costs for dismantling at the end of life
i = discount rate of 2%
t = year when the single measures are carried out
T = service life of 100 years.

The average costs of the single renewal and rehabilitation measures have been determined on the basis of literature studies, assessed projects and manufacturer's data. Uncertainties in prices are incorporated in the analysis and are shown for the individual bridges. For a comparison with existing German regulations, the calculations according to the guideline for the execution of economic efficiency analysis for rehabilitation and renewal with regard to road bridges (Ri-Wi-Brü) [39] are partly included in the analysis. This guideline uses a lump sum calculation and therefore generates a different curve shape of the LCCs over 100 years.

In Table 6.16, input data for bridge A1 and the condition-based maintenance strategy are summarized. The year-specific measures can be identified. Additionally, in Table 6.16, the duration of the single measures and the associated traffic route are specified. These data are necessary for external cost calculations presented in Section 6.6.2. It can be seen that all calculations are based on the same parameters and that an input data variation leads to a change of the results in the different calculation methods.

6.6.1.1 *Results of installation costs*

The comparison of the installation costs is a standard procedure within the tender of a bridge construction project. Since prices are chosen individually by each bidder, a difference of the overall installation costs can have many reasons. In this calculation for all bridges, the same input prices are used so that cost differences result only from different bridge masses and different construction procedures.

In Figure 6.48, the cost distribution for main bridge components can be seen. The bridges A2-V and A2-B have nearly the same substructure and therefore lead to the same costs, whereas bridge A1 is designed as an integral bridge and carries additional bending moments. For A1, the superstructure is most expensive because it consists of a single-span steel composite girder without columns. All three bridges are provided with the same features. The difference of the time-related costs results from different installation periods (A1: 22 weeks, A2: 28 weeks, A3: 39 weeks). Analysing the overall costs with A1 as benchmark, the costs of bridge A2-V are 2% higher and the costs of A2-B are 8% lower.

Table 6.16 Example of component-based maintenance and rehabilitation measures according to the condition-based maintenance strategy shown in Figure 4.94 for bridge A1; additionally, input data for the modelling of the traffic routing has been added and utilized for external cost calculations [38].

Year	Maintenance Measure	Duration [Days]	Time Parallel Activities [Weeks]	Traffic Routing Highway (Below)	Traffic Routing Main Road (Top)
0	Erection	–	22	3 + 3	1n + 1n
17	Top layer road surface	2	0,5	3n + 3n	1 + 0
33	Traffic routing	2	9	3n + 3n	1 + 0
	Renewal caps	26			
	Renewal safety equipment				
	Renewal road surface and basis layer	12			
	Renewal drainage	4,5			
	Complete renewal corrosion protection	15	5	5 s + 1	1n + 1n
	Traffic routing	10			
50	Top layer road surface	2	0,5	3n + 3n	1 + 0
67	Traffic routing	2	10	3n + 3n	1 + 0
	Renewal caps	26			
	Renewal safety equipment				
	Renewal road surface and basis layer	16			
	Renewal drainage	4,5			
	Traffic routing	10	7	5 s + 1	1n + 1n
	Concrete repair superstructure	10			
	Complete renewal corrosion protection	15			
	Concrete repair abutment	12	3	3 + 3	1n + 1n
83	Top layer road surface	2	0,5	3n + 3n	1 + 0
100	Dismantling end of life	10	2	3 + 3#	0 + 0

#= additional closure for one weekend, n = no obstruction of one travel direction, s = utilization of the hard shoulder.

6.6.1.2 *Results of LCCs*

The LCCs combine the installation costs and all costs occurring within the complete service life. For bridge A1, the LCCs for the three different maintenance strategies are shown in Figure 6.49. It is evident that until the year 33 in the lifetime of a bridge, all three strategies lead to similar costs, and afterward the curves vary. For the strategy permitted deterioration, only inspections and the safety-relevant rehabilitation measures are performed. Thus, the superstructure has to be replaced after 50 years. When applying preventive maintenance, many different measures are carried out over the complete life-cycle. Therefore, a lot of small changes of the cost development can be recognized. Overall, the preventive maintenance leads to €1.63 million, the condition-based maintenance to

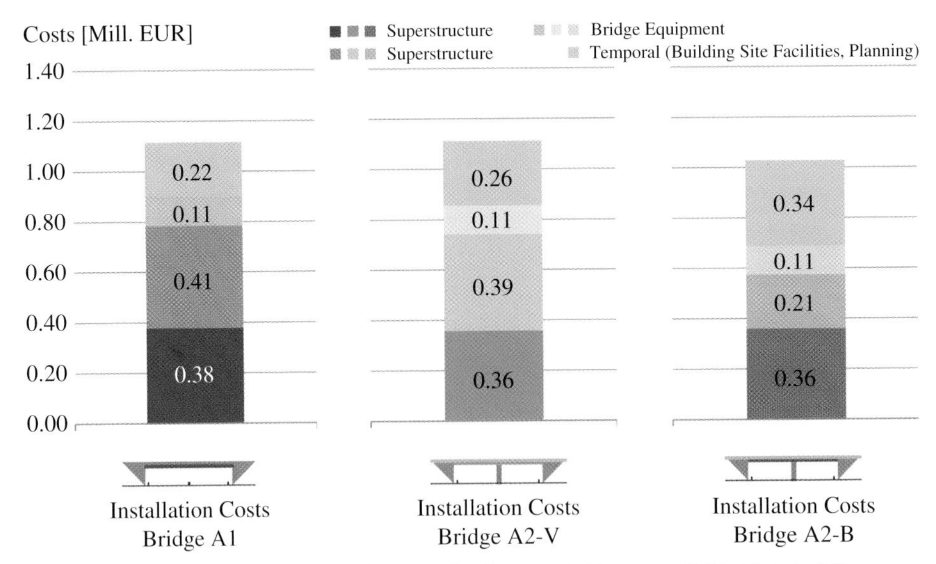

Figure 6.48 Comparison of the Installation costs for the three bridge types divided into building component groups [38].

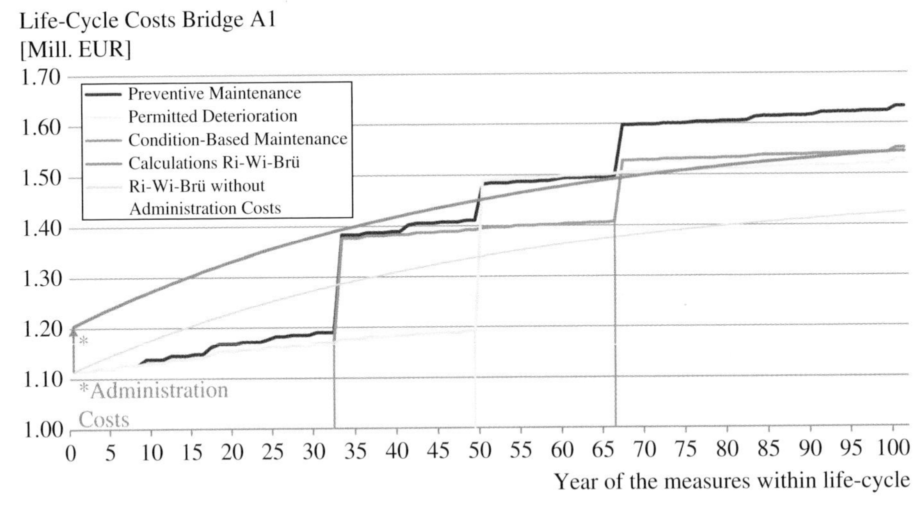

Figure 6.49 LCCs for bridge A1 with regard to different maintenance strategies; additionally the results of [39], which uses a lump sum calculation, are shown, and a discount rate of 2% is used uniformly [38].

€1.55 million and the permitted deterioration reaches an LCC of €1.53 million. It has to be emphasized that all strategies suppose different condition indexes at the end of life. The theoretically 'very good' and 'good' condition indexes after 100 years belonging to the strategies preventive maintenance and condition-based service life, but this is not possible for a permitted deterioration.

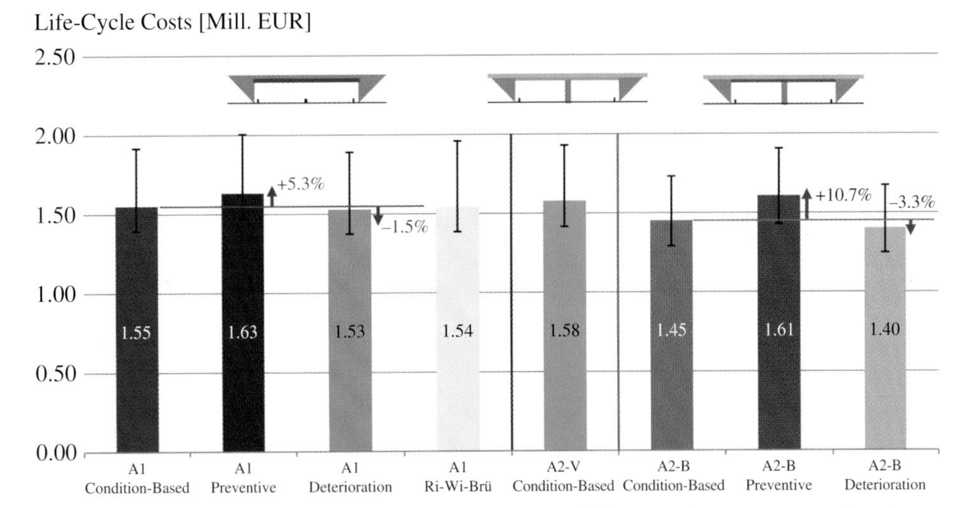

Figure 6.50 LCCs of the three bridge types and with regard to different maintenance strategies. The uncertainties depending on a variation of the input value 'price' are marked by the whiskers [38].

In Figure 6.49, calculations according to the German guidelines Ri-Wi-Brü [39] are also shown. At the first view, the calculated costs according to condition-based maintenance and by applying the guidelines match very well. Unfortunately, the guidelines include administration costs, so that the two curves cannot be compared directly. A correct comparison is conducted by using the light orange line. Here, a strong underestimation of the real costs takes place so that in this case, the Ri-Wi-Brü [39] is not suitable for an application within a sustainability assessment.

The comparison of LCCs for different strategies and different bridges is presented in Figure 6.50. As already shown for bridge A1, also for bridge A2-B, the preventive maintenance strategy causes the highest costs (+10.7%). Costs for the strategy permitted deterioration are 3.3% lower compared to a condition-based maintenance. The LCCs of all three bridges are in the same range, but compared to bridge A1, the costs of the reinforced concrete bridge A2-B are about €0.1 million lower, and those of the steel composite bridge A2-V are about €0.03 million higher.

In addition, the uncertainty of the results subjected to a variation of the input prices used for the calculations are presented. For all bridges, the uncertainties show the same effect. The upward deviation is higher than the downward one. However, the difference of all results calculated with medium input price is smaller than the variation of the results with high or rather low input prices. Therefore, no clearly dominating solution can be identified.

6.6.2 Calculation of external cost for highway bridges

The methodology applied for the calculation of external effects and the transformation into external costs is outlined in the flow chart in Figure 6.51. The calculations are based on the same maintenance strategies as for LCC. The construction time and traffic routing are adjusted to the single rehabilitation measure, and an example

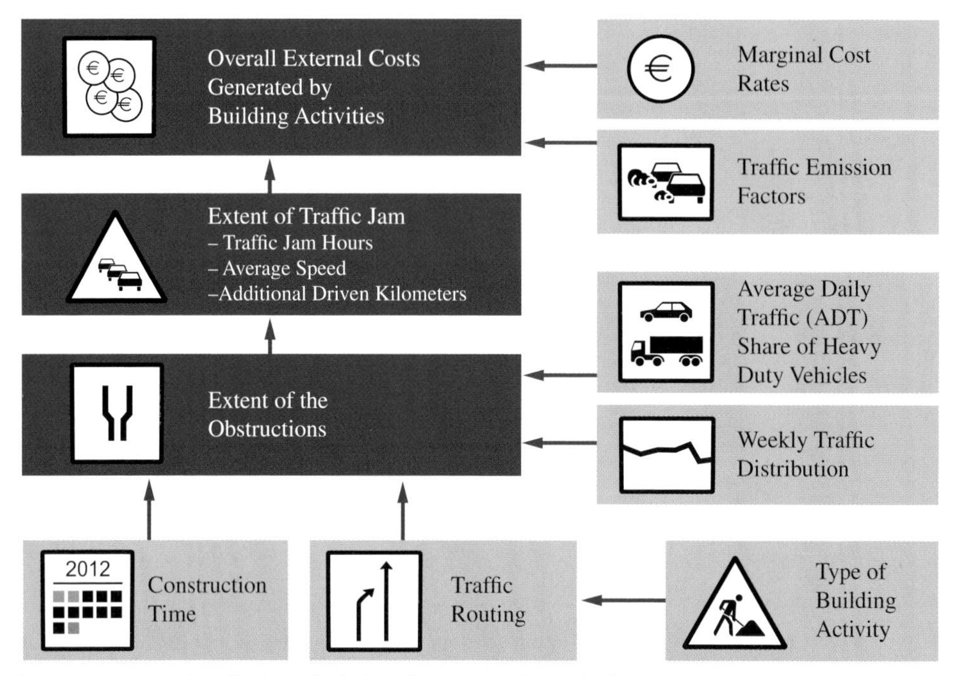

Figure 6.51 Flow chart for the calculation of the external costs [38].

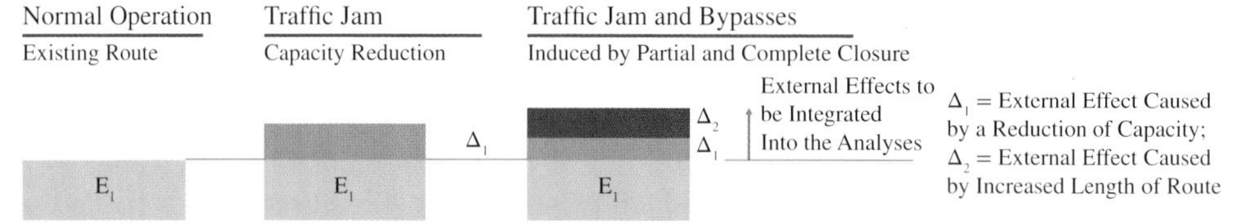

Figure 6.52 External effects from traffic obstruction that are integrated into the calculations. The rectangle E_1 characterizes external effects resulting from normal operation [42].

for the condition-based maintenance strategy for bridge A1 (single-span integral steel composite - see also Figure 6.47) can be found in Table 6.16. The calculation model for traffic emissions and time delay is explained in [40]. All calculations are based on a macroscopic deterministic traffic model, originally proposed in [41].

A fixed average daily traffic (ADT) with a value of 70,000 vehicles per day for the highway and 6,000 vehicles per day for the main road is used for the calculations. A medium loaded highway and an average main road are modelled. Only external effects caused by construction activities are included, and effects occurring during normal operation are neglected; see Figure 6.52.

6.6.2.1 *Results of external costs*

The results of external cost calculations can be categorised according to different aspects. Here, costs arising on the highway (below) and the main road (above) are shown separately. In Figure 6.53, the overall external costs for bridge A1 can

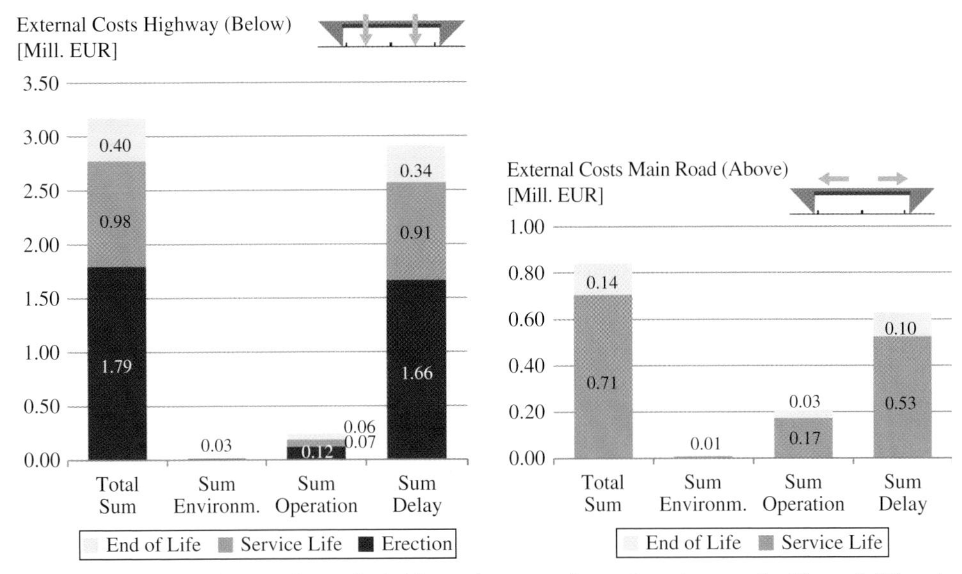

Figure 6.53 Results of external costs for bridge A1 (one-span integral steel composite [Figure 6.47]) and condition-based maintenance strategy for the highway (left) and main road (right) [38].

be seen. The impact on the highway occurring during the construction of the bridge results in more than 50% of the overall costs (€1.8 million). The service life with about €1 million is still very important, whereas the end of life has a minor effect. For the main road (Figure 6.53, right) no external costs arise during construction, because it is assumed that during this time, an existing bridge is still in use. The impact within the service life is in the same range compared to the one of the highway.

For both roads, the impacts caused by induced delays (arising from speed limitation, traffic jam and longer travel distances if roads are closed) are responsible for most of the external costs. For this effect, the following monetisation rates (marginal cost rates) are used: time delay cars, €18.64/h and trucks, €40/h. Monetised environmental impacts play a minor role. Also, operation costs resulting from additional travel kilometres, longer vehicle utilization and increased fuel consumption are less important compared to the delay costs. For the operation, monetization rates of €0.28/km for cars and €0.3/km for trucks have been applied [38].

The summed external costs from Figure 6.53 can also be displayed according to their occurrence during the life-cycle; see Figure 6.54. The curves show the same shape as the LCCs in Figure 6.49. This is because external costs have also been calculated on the basis of individual maintenance measures.

In Figure 6.54, the three maintenance strategies are compared. For both the highway and the main road, preventative maintenance induces the highest external cost, followed by permitted deterioration. The condition-based strategy causes the lowest external costs. It can be concluded that a grouping of maintenance measures results in a recognizable minimization of external costs. At the same time, the construction period is responsible for most of the impacts. Although the

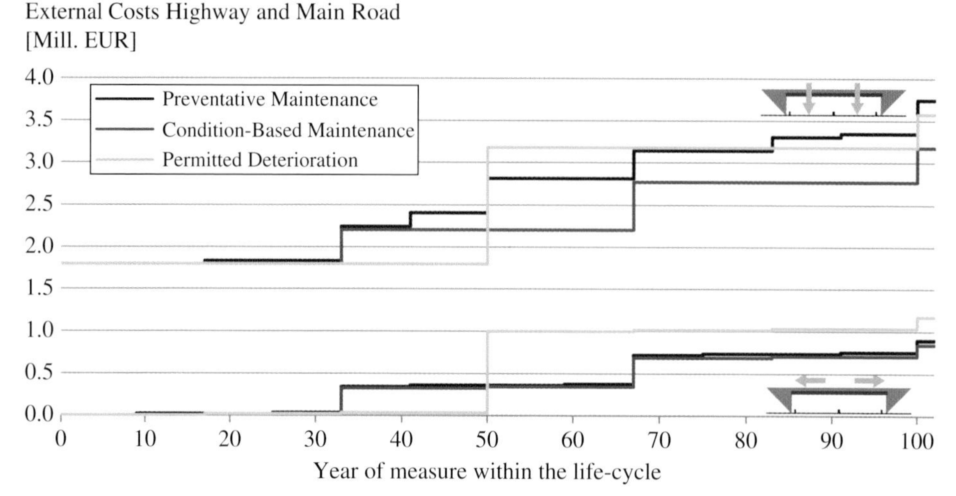

Figure 6.54 Development of external costs for bridge A1 (one-span integral steel composite [Figure 6.47]) during the life-cycle for the highway and main road [38].

average daily traffic is much higher on the highway, impacts resulting from a (partial) closure of the main road on top of the bridge also induce significant external costs.

Figure 6.55 compares the results for the three different bridge types taking the condition-based maintenance strategy as a basis. As mentioned, the construction phase is most important for the total external costs. In Figure 6.55, bridge A1 with the shortest construction period causes minimum impacts. During the life-cycle and at end of life, the impacts are nearly identical. All external costs are not discounted because they mainly characterise goods and preferences, which cannot be traded on a market.

Overall, the external costs are 2.5–5 times higher than the LCCs, but the results are subject to large uncertainties, which are not shown here. The presented calculations on the basis of average monetisation rates prove that a negligence of external effects can lead to wrong decisions. For the bridges analysed, bridge A2-B has the lowest LCCs but causes highest external costs. Of course, within a sustainability assessment system, these aspects have to be complemented with additional aspects, such as environmental and further social impacts. Nevertheless, LCC and external cost calculation are two powerful methods to quantify impacts during the construction and service life of bridges and address parameter variations and uncertainties.

6.6.3 Calculation of LCA for highway bridges

Also for LCAs, the same input data as for LCC and external cost calculations are used. The calculation process is shown in Figure 6.56 and follows with the requirement of EN ISO 14040 [43]. The system boundaries for the construction-related

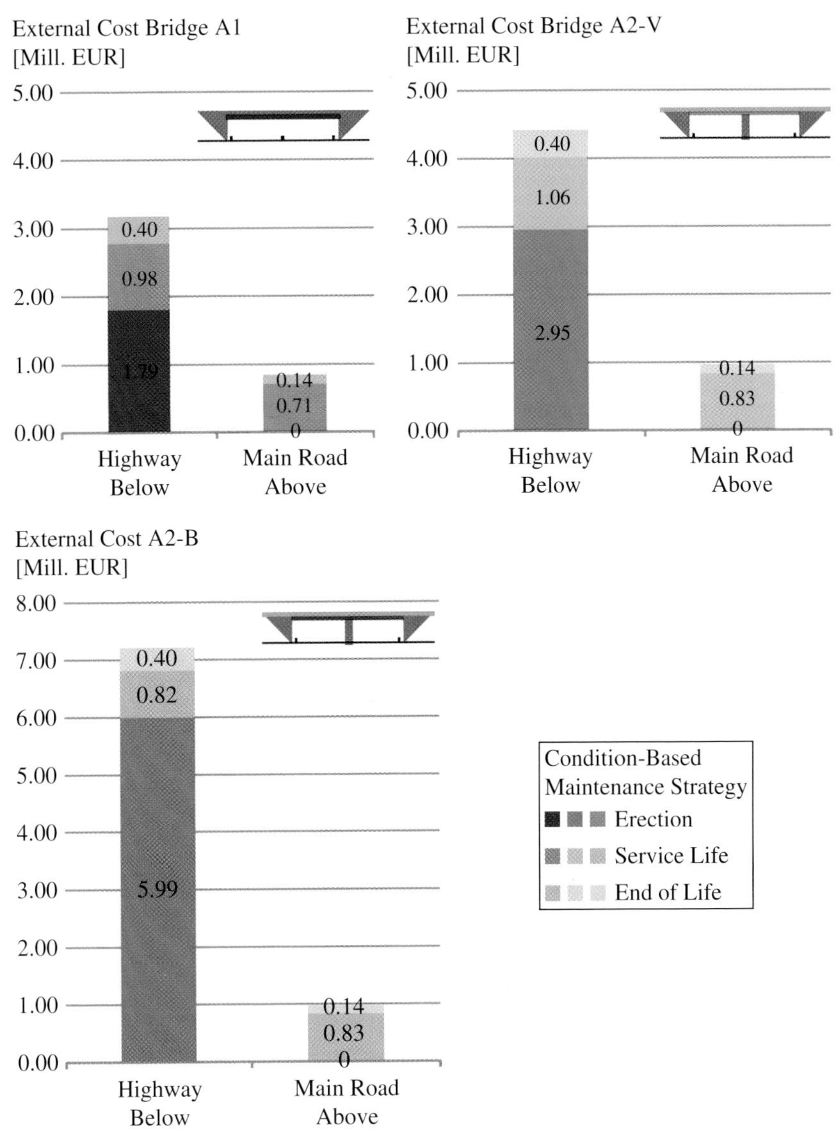

Figure 6.55 Comparison of external costs for the three bridge types (see Figure 6.47) distinguishing between highway and main road [38].

impacts are identical with those on the LCC calculations, and the pollutant emissions are calculated with the same model as for the external costs.

Since construction-related impacts and external environmental effects both are incorporated into the analyses and both have different result uncertainties, these impacts are displayed separately. External environmental impacts result from a changed traffic speed and the arising exhaust emissions during construction, maintenance, rehabilitation and dismantling. For a further explanation of the effects from traffic, see Figure 6.52.

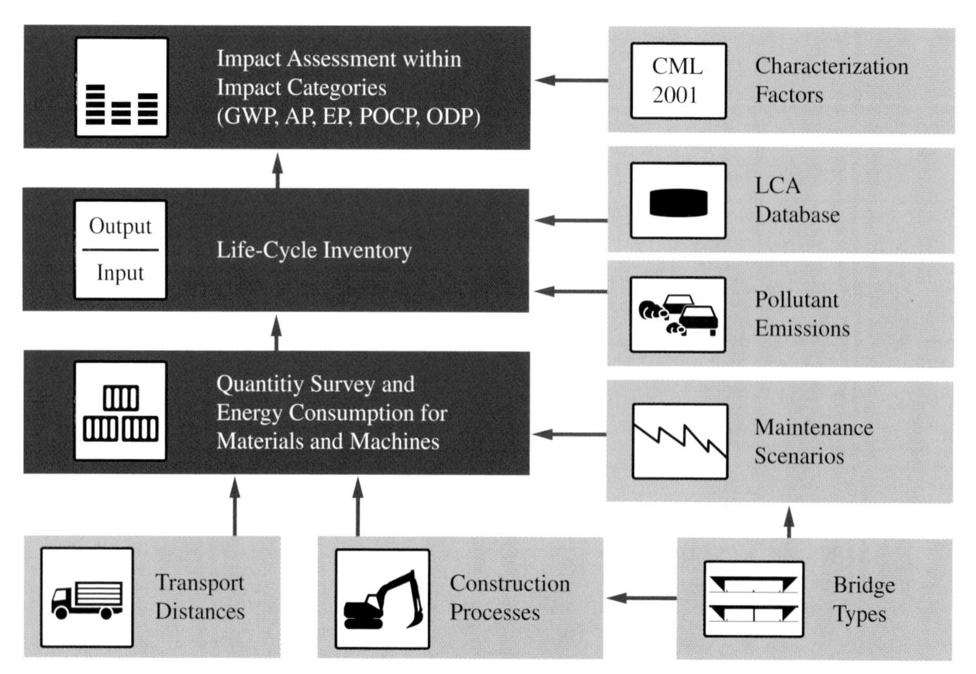

Figure 6.56 Flow chart for the calculation of the external costs.

6.6.3.1 *Results of impact categories GWP and POCP*

Two impact categories are presented that are important for the assessment of the result and show diverging effects. The GWP is an established impact category to describe global effects and is a political key indicator. As seen in Figure 6.57, the impacts resulting from the analysed constructions only vary within the range of −3% to + 1% with respect to the results of bridge A1 (single-span integral steel composite option [Figure 6.47]), which are relative to the base case of 100%. The different constructions and the different maintenance strategies cause minor differences of the results. When integrating environmental external effects into the analyses, a different picture arises. The orange bars in Figure 6.57 indicate that bridge A1 causes the lowest external impacts compared to the other ones. Nevertheless, traffic controls lead to environmental external impacts of 60% compared to the bridge itself. The impacts generated by bridge A2-V (two-span steel composite [Figure 6.47]) are significantly higher, and the different maintenance scenarios analysed only lead to a slight variation of the results. Bridge A2-B (two-span reinforced concrete [Figure 6.47]) leads to the highest impacts generated by traffic controls.

If the bridge-related GWP is presented in absolute values, an impact potential of about 15 [kg CO_2-Eq.]/m^2 results. Compared to the findings in Table 4.9 (Chapter 4), this value is above average but still in the range of the results from other studies. Overall, in the impact category GWP, the same influence of external effects can be found as seen in the external cost calculation. Bridges, which minimise impacts on users and third parties, also achieve societal benefits.

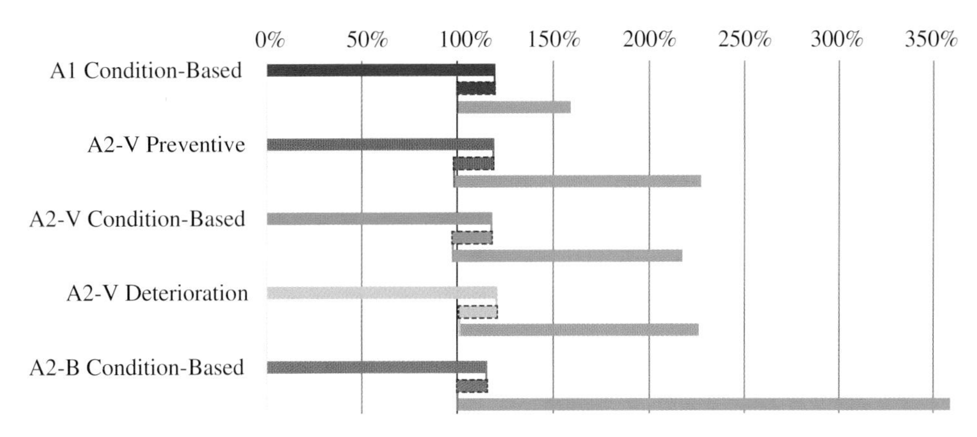

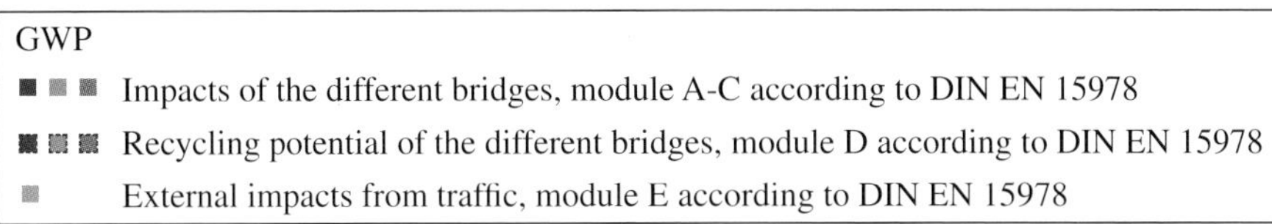

Figure 6.57 Relative LCA results for the impact category GWP with regard to the results of bridge A1 after settlement of recycling [44].

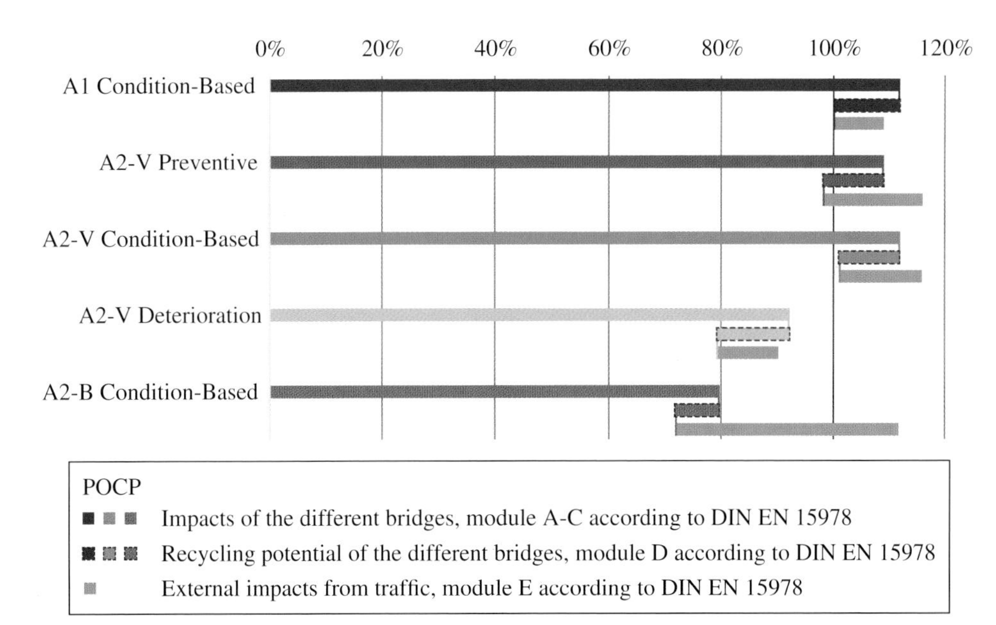

Figure 6.58 Relative LCA results for the impact POCP with regard to the results of bridge A1 after settlement of recycling [44].

Analysing the impact category POCP, the influence of the bridge structure on the results is much higher. In Figure 6.58, the results of bridge A1 are set to 100% again. The structure-related variation in results compared to the preventative and condition-based strategy of bridge A2-V is very small. However, the strategy of permitted deterioration of bridge A2-V and the results of bridge A2-B are about 20%–25% lower compared to the reference scenario. The reason can be found in the reduced number of measures for the renewal of corrosion protection. For the reinforced concrete bridge, no corrosion protection has to be applied. As a result, the impact category POCP is influenced much more by the design of the structure than impact category GWP.

Moreover, Figure 6.58 reveals that external environmental impacts resulting from traffic obstruction play a minor role compared to structure-related impacts. For steel composite bridges, external impacts account for 10%–20% compared to the impacts from the bridge itself. For bridge A2-B, a percentage of 40% arises. Overall, in this impact category the external effects should not be neglected, but they are not as important for the overall decision as in the GWP. In summary, no general conclusions can be made on the optimum solution for LCA because different impact categories lead to varying results. Therefore, it is always recommended to perform a complete LCA to model and capture all possible effects.

6.6.4 Additional indicators

All analyses presented in the previous section use models and established methods to create a common basis for the comparison of bridges. For development of a truly holistic assessment scheme, additional indicators have to be incorporated, especially adressing social aspects of sustainability. Some important factors, which are normally defined as social aspects, are regularly considered within external cost calculation, that is, time delays of users, additional fuel consumption and additional maintenance of cars. It must be noted that some aspects, such as air pollution, climate change or aspects of toxicology, are often assigned to the environmental as well as to the social dimension because a strict separation of these two issues of concern is not possible.

Social aspects, which are not captured within LCA or external cost calculation, are often designed as qualitative indicators, which use predefined assessment classes. The project specific attributes are assigned to the classes, and each class corresponds with a level of target achievement. The dissemination of qualitative indicators can be seen in Table 6.17, where most studies focus on a qualitative assessment. One crucial reason for the utilisation of qualitative indicators is the practical applicability. Practitioners and decision makers prefer easy-to-use tools that directly allow seeing benefits for the overall result when changing or adapting one aspect. On the other hand, predefined assessment classes do not allow a detailed comparison of different planning variants, and an adaption of bridge details and the impact of changed input parameters is not easily analysed.

Different indicator systems have been proposed to meet the special requirements for infrastructure projects; see Table 6.17. One example for an infrastructure-specific indicator is the assessment of the barrier effect. Constructions such as roads lead to a reduction of freedom of movement for residents, wildlife and the public. At the same time, the construction of infrastructures is often related to a strong interference with ecosystems and landscape. Therefore, public acceptance is a crucial issue and is considered in the aspects of public participation,

Table 6.17 Aspects proposed for the assessment of infrastructures in the social sustainability dimension; aspects considered in the external cost calculation or LCA are not listed.

Aspect	Impact on	IMPACT Study 2008 [37]	CEEQUAL System 2010 [33]	Fernández-Sánchez 2010 [34]	FOGIB Project 1997 [35]	Ugwu et al. 2007 [36]	Yao et al. 2011 [45]
Dynamic behaviour	User		q		q		
User safety	User			q	q	q	q
User comfort	User				q		
Accident costs	User and public	m	(q)		q	q, (m)	
Noise pollution	Public	m	q	q	q	q	q
Public participation	Public		q			q	
Human biodiversity access	Public	m	q	q			
Integration into the society	Public		q	q			
Barrier effect	Public	m	(q)	q	q	q, (m)	
Visual impact	Public		q	q	q	q	q
Respect for local customs	Public		q	q	q		
Job opportunities	Public		q			q	q
Historic and amenity values	Public		q			q	q
Short- and long-term health	Public					q	q

q = qualitative assessment; m = monetization (quantitative assessment);
(q) and (m) = aspect only considered partially or subordinated

integration into the society and visual impact. All the aspects presented can be integrated into an infrastructure assessment system by defining one or several indicators that measure the particular level of target achievement.

6.6.4.1 *Outlook*

Sustainability assessment has become increasingly important, and for bridges, several approaches are under development. Three established and powerful methods (LCC, external cost calculation and LCA) have been presented that are based on engineering models and generate results that can be represented on a relative scale. Such methods are well suited for a comparison of different planning variants and the assessment of the change of results when input parameters are varied. Additionally, they can be used to describe the complete life-cycle.

The application of the three methods for reference bridges shows that the construction phase dominates the results of these assessments. Impacts during service life and at the end of life are less important. Nevertheless, the comparison of three maintenance strategies leads to the result that different strategies can influence the LCCs by −5% to + 10%. The results of the external cost calculations show the same effects. External costs arising from additional travel time of the vehicles dominate the total external costs. When comparing the total external costs of the three bridge types, the integral steel composite bridge shows the best performance over the life-cycle and the reinforced concrete bridge the worst. The reason for this lies in the minimization of traffic obstruction, especially within the construction period. Overall, the external costs are 2.5–5 times higher compared to the LCCs. If external costs are neglected within the assessment of bridges, an overall dead weight loss can arise. The LCA shows the same effects in the impact category GWP, whereas for the POCP, effects resulting from traffic obstruction are much smaller. Here, bridge-related aspects like corrosion protection are much more important in terms of the overall result.

All engineering models can be used to analyse interdependencies and different bridge design variants in detail. For a practical application, normally additional indicators are incorporated that perform a qualitative assessment, for example, by means of a predefined classification scheme. For infrastructures and bridges in particular, different indicators have already been developed. However, the number of practically applied assessment schemes for infrastructure projects is still very small. A reason for this is the location dependency of the result. For example, a commuter highway will cause much higher impacts especially when calculating external effects than a highway in a rural area. As a result, an assessment of bridges should always take the very specific location-dependent boundary conditions into account. In the end, this could even lead to application-specific indicator sets for different kinds of bridges.

6.7 SUSTAINABILITY OF STEEL CONSTRUCTION FOR RENEWABLE ENERGY

Anne Bechtel, Peter Schaumann, Natalie Stranghöner and Jörn Berg

6.7.1 Offshore wind energy

The material steel represents 90% of the material mass used in offshore wind turbines (OWTs) and has the biggest effect to environmental aspects as pointed out by Wagner et al. [47]. Eighty per cent of the cumulated energy demand can be ascribed to the manufacturing and installation process of the steel structure. Altogether these facts present potentials and needs for an evaluation method to assess the sustainability of steel constructions for OWTs.

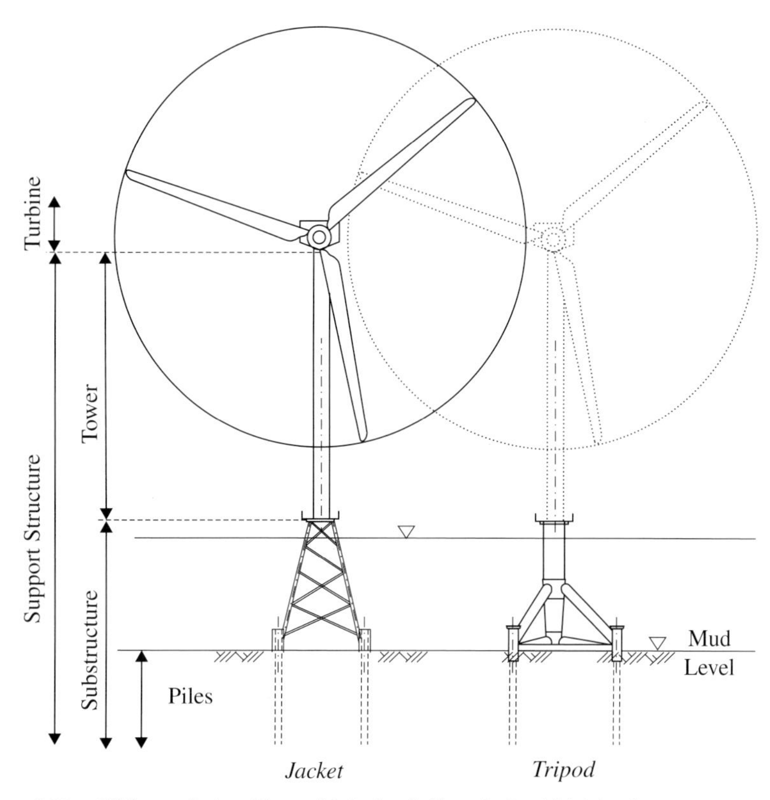

Figure 6.59 Offshore wind turbines with jacket (left) and tripod (right) substructure.

The support structure of an OWT consists of the tower and the substructure, whereas the substructure includes all structural components below the tower including the foundation. Depending on the water depth, turbine size and local conditions, different types of substructures have been developed. Even though the monopile is the common solution in Europe, for Germany, deeper water requires use of lattice structures like jacket or tripod (Figure 6.59). Detailed information on steel structures for OWTs can be found in [48].

Nowadays tower production is a highly automated process. Bending machines are used for the forming process of round plates, and submerged-arc welding is used to connect the steel tube segments. Even though relatively few employees are needed for the fabrication process, the quality control needs to be provided by highly trained employees. These affect the social and process quality regarding sustainability aspects.

Steel tube segments are taken to a location close to the sea, where they are assembled and manufactured. For final assembling of the tubes, large factory halls are required. Special lifting equipment is used to handle the heavy steel construction not only in the installation halls but also dockside to load the segments to the installation vessels. In addition, high logistic effort results from storage before shipping. Short weather-dependent time slots for installation on site influence the installation process and consequently the amount of stored components and steel

Table 6.18 Parameters used for the exemplary assessment of jacket and tripod (see [44]).

Substructure	Tripod	Jacket
Water depth	~30 m	~30 m
Pile length	~50 m	~30–45 m
Steel mass	~1300 t	~830 t
Corrosion protection	Anodes & coating	Anodes & coating
Pile recycling	Left in seabed	Left in seabed

structures on the dockside. Coordination of different technical crews for the final works at the steel structures, for example, welding processes at jackets, application of the corrosion protection is essential.

The installation process of OWTs includes impacts from but also to the environment. The piles are driven into the seabed by a hydraulic hammer producing noises, which influence the flora and fauna, especially whales in the German Exclusive Economic Zone. As shown in [46], one of the decisive components of OWTs regarding ecological sustainability criteria is the substructure. Depending on the substructure type and pile length, the substructure requires up to five times more steel than the tower. Hence, for a first comparison of results, the focus is made on jacket and tripod substructures (Figure 6.59), which have been analysed by use of the ecological indicators described in Chapter 4.13.

In addition to material masses, special aspects such as welds and corrosion protection were taken into account, reflecting a holistic view. Table 6.18 summarises the system parameters used for the ecological assessment of the substructures for an operational lifetime of 20 years. The material used for the primary structure is a steel grade S355. The corrosion protection for both substructures consists of a coating system in the splash zone and anodes underwater. Regarding the assessment, it was assumed that both substructures are coated by the same corrosion-protection system. Therefore, the systems differ only in the material masses caused by the different kind of structures.

The ecological assessment of the aforementioned substructures was analysed regarding the named life-cycle stages and the listed criteria in Chapter 4.13. To indicate the influence of different life-cycle stages, the environmental effects are analysed for each life-cycle stage A–D, shown in Figure 6.60. Comparing the selected common ecological impact indicators, it can be shown that the AP, the GWP, the water demand and the total primary energy demand reveal that stage A 'planning to construction' is the decisive life-cycle phase for jacket and tripod.

During construction of the support structures, the energy demand is relatively high due to manufacturing and installation processes. The fabrication and installation process causes the largest release of CO_2 emissions, which influence the GWP. Although during the operation stage cost and time-consuming monitoring intervals with offshore transportations are conducted, this stage has only minor impact to ecological factors. The primary energy demand is relatively small compared to stage A. During stage C, the removal of the construction is realised, whereas according to current assumptions, the driven piles are left in the seabed. This process of deconstructing the support structure has no significant influence

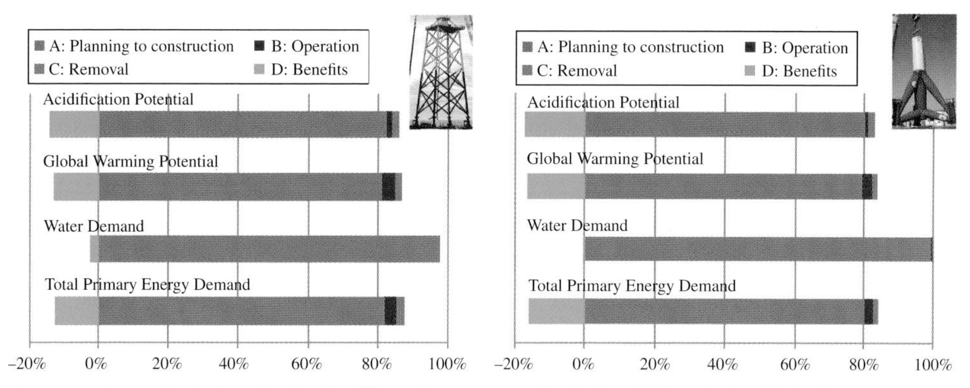

Figure 6.60 Ecological results for the life-cycle stages A–D for jacket (left) and tripod (right).

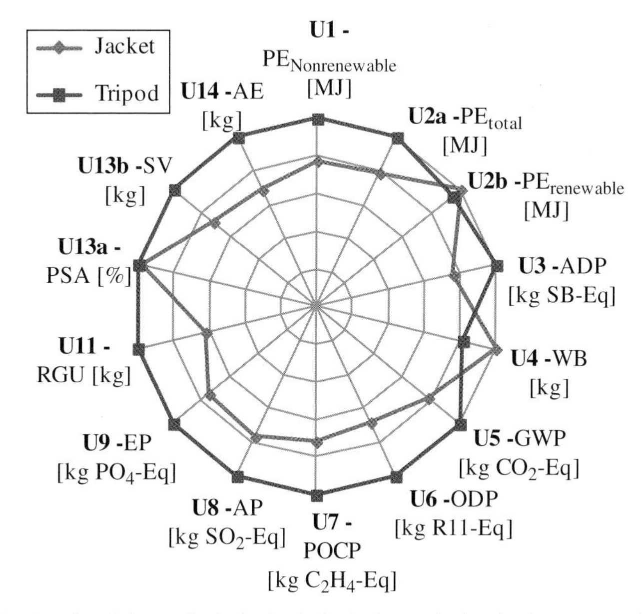

Figure 6.61 Results of the ecological criteria for jacket and tripod substructures depicted in a polar diagram (abbreviations of the criteria are given in Table 4.10 [Chapter 4.13]).

on the ecological indicators. Due to the almost full recyclability of steel, the 'benefits' stage has a negative output, impacting the holistic evaluation positively by reducing the total greenhouse emissions.

Additional environmental effects of using a tripod and jacket structure are shown in Figure 6.61 by a polar diagram. The impact indicator values for tripod and jacket are assigned to the diagram, whereas the diagram centre displays the value zero as a basis. The values are normalised to the maximum, which leads to the effect that the maximum values are on the outer ring of the polar diagram. The different results for tripod and jacket lead to different spanned areas that are normalised by the maximum value for each indicator.

In addition to the results presented in Figure 6.60, these spanned areas show that the adopted tripod has a worse ecological impact than the assumed jacket structure, which is mainly caused by the higher steel quantity. The relatively high manufacturing process for the jacket substructure leads to a high primary energy demand (U2b) than required for the manufacturing of the tripod. Furthermore, it was assumed that the jacket structure is equipped with considerably more anodes made of aluminium for corrosion protection. In the production of aluminium, a large amount of water is required, so the corresponding water demand (WB) value U4 is higher for the jacket than for the compared tripod structure; see Figure 6.61. By analysing and interpreting the data, it has to be considered that the reference structures jacket and tripod were reflected by assumptions with regard to the experience of the authors. Nevertheless, for other design and substructure concepts and boundary conditions, different results may be achieved.

In addition to the ecological category, exemplary calculations for the categories economy, sociology, technology and process can be found in [49].

Nevertheless, some decisive effects and parameters influencing the sustainability of OWTs can be mentioned. Economic effects are considered by the LCC calculation according to EN 15643-4 [46]. For OWT substructures, the material quantities and type, the transportation and the fabrication process are essential parameters influencing the structure´s sustainability. With regard to these investigations, the manufacturing and installation phase lead to a significant proportion of the LCCs.

Considering social criteria, for OWTs especially, occupational safety is an important parameter, as 'harsh' offshore conditions cause an additional high level of safety aspects that need to be considered. These environmental characteristics induce high requirements for the corrosion protection, which is one of the significant indicators for the technical criteria. Depending on the location of the steel components, active and passive corrosion protection systems may be used. In the totally submerged area, cathodic corrosion protection by anodes is provided, whereas in the splashing and areal zone, a layered coating system is applied. This coating system consists of polyurethane or epoxy resin, which is applied in three to five layers. Due to their chemical composition, these coatings have a significant impact on ecological criteria. As well as the preferable use of solvent-free coatings, special protective measures have to be adopted for the application. Considering technical aspects, a thorough surface preparation and uniform application are important for a high durability.

The process category encompasses procedural and planning implementations of the constructional processing. A decisive aspect is the registration and evaluation of transport quantity, goods, distances and type of transport carrier. The investigations have shown that especially the transport method, for example, ship or truck, and transport distances affect the sustainability of the jacket and tripod structures.

6.7.2 Digester for biogas power plants

Use of steel is an important factor in the design of biogas power plants. Plant components like stirrers, screw conveyors and pipelines are mainly manufactured of steel and stainless steel, respectively. In addition to these mechanical engineering parts, different tanks with varying dimensions and volumes are required. In general, one slurry tank, one or more main digester and one tank for digestate are part of a biogas power plant. These tanks can be designed with different types of construction and materials. The main digester plays an important role within a biogas power plant. Often, most of the digesters are manufactured as concrete structures, making them especially prone to corrosion attack by hydrogen sulphide. Steel structures provide promising alternatives. The advantages of steel structures like prefabrication, shortening of construction time or gas-tightness (using welded tanks) are mainly not utilized so far.

Steel digesters can be constructed both of stainless steel and coated carbon steel. Frequently used austenitic stainless steel grades are X5CrNi18-10 (1.4301) and the higher alloyed X6CrNiMoTi17-12-2 (1.4571) for the gas dome, as the highest anticorrosive requirements are needed for this zone. Steel grades S235J2 + N and S355J2 + N are mostly used for the type of construction made of coated carbon steel. Depending on the size of the biogas power plant, common plate thicknesses range from 5 to 12 mm for carbon steel tanks and 1.5 to 5 mm for stainless steel tanks. Different joining techniques can be applied during the installation. Steel plates can be assembled on site by welding, bolting or seaming. Adhesive bonding as an innovative joining technique of stainless steel plates is currently part of research investigations. The roof structures of the digesters mainly consist of membranes or steel roofs.

Depending on the size of the biogas power plant, which can range from approximately 75 kW_{el} up to 1 MW_{el} and higher, 10–75 tonnes of carbon steel and 10–20 tonnes of stainless steel are required for the fabrication of one digester, respectively in each case. The lifetime of biogas power plants is defined by 20 years due to rather low experience of using this construction and the guaranteed feed-in reward from the power generation.

To demonstrate the application of the sustainability assessment method, two different types of biogas digesters each with 2000 m^3 volume were investigated. The construction types mainly differ in the steel grade and the joining technique, which are, respectively welded and coated carbon steel, and bolted stainless steel (Figure 6.62). The system parameters are summarized in Table 6.19. In addition to the material and the joining technique, both digesters differ in their roof construction, which consists of a carbon steel roof (variant A) and a membrane roof made of coated fabric (variant B).

The following results of the sustainability assessment are only valid for these variants of construction and the assumptions in their evaluation and cannot be transferred to other types of construction in general.

The environmental assessment of the two variants of a steel biogas digester was carried out in accordance to the aforementioned assessment of the OWT substructures. The environmental effects are analyzed for each life cycle stage A–D and are presented in Figure 6.63. The most important life-cycle stage for most of the indicators is the production stage A1–3 resulting from relatively high energy input for the production of steel. Further significant input is accumulated within the construction stage A4–7, especially concerning the water demand for the digester made of carbon steel. Both operation stage B and removal stage C contribute lower input for

Figure 6.62 Biogas digesters of welded and coated carbon steel (left photo: © Schachtbau Nordhausen GmbH) and bolted, stainless steel.

Table 6.19 Parameters of the investigated systems of biogas digesters.

Parameter	Variant A	Variant B	
Material	Carbon steel	Stainless steel	
Steel grade	S235J2 + N	X5CrNi18-10	(1.4301)
	S355J2 + N	X6CrNi-MoTi17-2-2	(1.4571)
Anticorrosive system	Coating based on epoxy resin	–	
Joining technique	Welded	Bolted	
Diameter	13.9 m	20.1 m	
Height	14.0 m	6.3 m	
Volume	2000 m^3	2000 m^3	
Plate thickness	5–6 mm	1.5–2.5 mm	
Steel mass	35 t	8 t	
Lifetime	20 years	20 years	

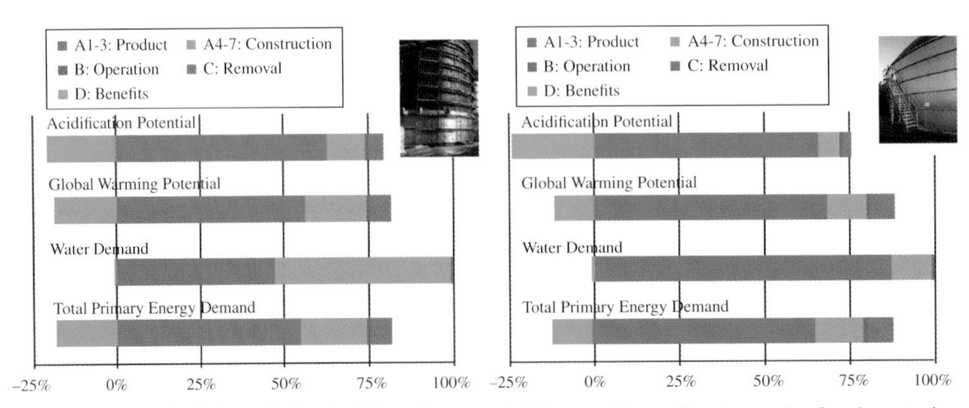

Figure 6.63 Ecological results for the life-cycle stages A–D for one biogas digester made of carbon steel (left) and stainless steel (right).

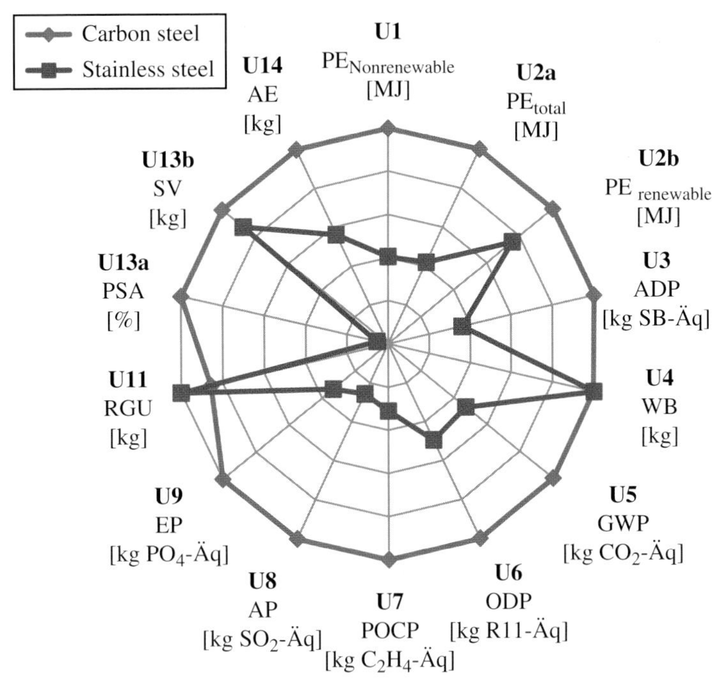

Figure 6.64 Results of the ecological criteria for the investigated variants of a biogas digester depicted in a polar diagram (abbreviations of the criteria are given in Table 4.10 [Chapter 4.13]).

the different indicators. The recyclability of both carbon steel and stainless steel results in considerable benefits within stage D.

The summation of the results for each indicator over the life-cycle stages of the structures is illustrated in a polar diagram, shown in Figure 6.64. With the exception of the criteria water demand (U4) and risk for human and environment (U11), the results for the variant made of stainless steel lead to considerably lower values than for the variant made of carbon steel. The results are correlated directly with the steel weights, which are approximately 4.5 times higher for the carbon steel design to that of stainless steel. For the criteria water demand (U4) and risk for human and environment (U11), the results for the stainless steel design exceed the results for the carbon steel design due to higher emissions of heavy metal as well as higher specific water demand during production of stainless steel. Explicit advantages can be seen for the stainless steel design concerning the indicator of steel obtained from the primary blast furnace route (U13a) compared to exclusive use of scrap for production of stainless steel.

The results of the LCA analysis cannot be compared to results of extended LCA analysis and EPDs due to limited input data and special assumptions. The approach for the sustainability assessment of the other categories economy, sociology, technology and process is equivalent to that of the category 'ecology' with representable results in polar diagrams [49]. Examples are given for the evaluation of the criteria of those categories applied to steel biogas digesters, as follows.

Welding of biogas digesters on site requires higher skills of employees in comparison to assembling of steel plates by bolting. Consequently, the type of joining technique influences both the criterion qualification index (S3) within the company as well as the installation costs (Ö1).

One of the decisive criteria in design is the corrosion protection system (T1). For steel biogas digesters, the corrosion protection is mainly provided by the use of stainless steel or coated and enamelled carbon steel. To quantify its sustainability effects, the evaluation of the criterion is correlated to the duration of protection according to EN ISO 12944-1, the number of inspections during its lifetime and the content of solvents of coated systems. Anticorrosive systems with multilayered coatings influence the criteria significantly due to their chemical composition. For this reason, the use of stainless steel has some benefits for the evaluation of this criterion compared to the use of solvent-based coatings.

In contrast to support structures of OWTs, the entire structure of steel biogas digesters can be removed and reused or recycled. This advantage effects the criterion 'removal', recycling friendliness and reusability (T5), which reflects the effort for the removal of the steel structure and the possibility to dismantle and reuse the steel components. The effort for dismantling of the structure and separation of the used materials is evaluated by qualitative indicators. Bolted steel digesters can be dismantled and rebuilt at a different site.

With the help of the sustainability rating system, different types of construction and construction variants can be compared. Finally, the rating system can be applied as an instrument to evaluate the advantages or disadvantages of one variant relative to another in terms of sustainability.

6.8 CONSIDERATION OF TRANSPORT AND CONSTRUCTION

Raban Siebers

6.8.1 Environmental impacts according to the origin of structural steel products

Additional environmental impacts arise for each tonne of steel that is transported. This depends on the transport distances from the steel mills to the workshop and construction site. Module A5 according to EN 15804 [5] includes these transports; see also Chapter 2. Structural steel that originates from Western Europe, Brazil or China has to travel the average distances to a construction site in Western Europe shown in Table 6.20.

To transport one tonne of a material over 1 kilometre (=1 tonne-kilometre or tkm), the environmental data shown in Table 6.21 can be found in national databases such as the ÖKOBAUDAT [1]. For simplification, packaging, such as containers, is not taken into consideration. For example, the environmental data for a single-storey building made of structural steel, 16.8 t of S355, can be calculated also including transport (see also Section 6.1). No consideration is

Table 6.20 Average distances travelled by steel products for a construction site in Western Europe.

Origin	Sea Freight km	Rail Freight km
Western Europe	–	500
Brazil	10,000	500
China	20,000	800

Table 6.21 Environmental data for selected methods of transport [1].

	GWP in kg CO_2/tkm	Total primary energy in MJ/tkm
Container ship	0.0156	0.190
Rail transport	0.0173	0.302
Road transport	0.0518	0.745

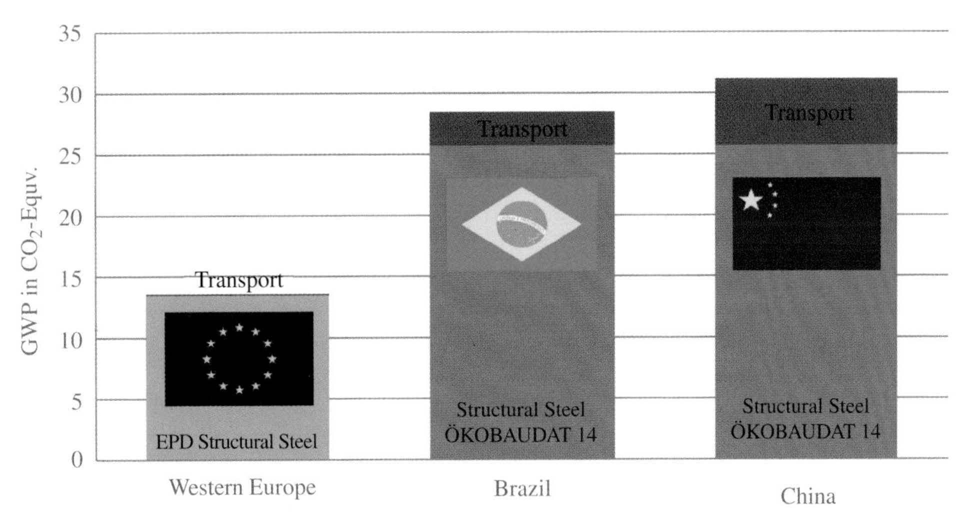

Figure 6.65 Global warming potential according to the origin of structural steel. 16.8 t of S355 for a typical single-storey building.

given to the foundations because ready-mixed concrete or precast concrete components are obtained elsewhere and the transport can thus be ignored.

Depending on the country of origin and the steel product, long transport distances lead up to an additional 30% environmental impacts (see Figures 6.65 and 6.66). This means the environmental data for transport also has to be taken into consideration in the LCA of a complete building. Where structural steel from Europe is used, the available EPDs from European producers should be applied. If the steel is supplied by a producer without a valid EPD, the average values for structural steel from national databases, for example, the ÖKOBAUDAT (see Chapter 3.10) have to be used. High strenth structural steel that is certified

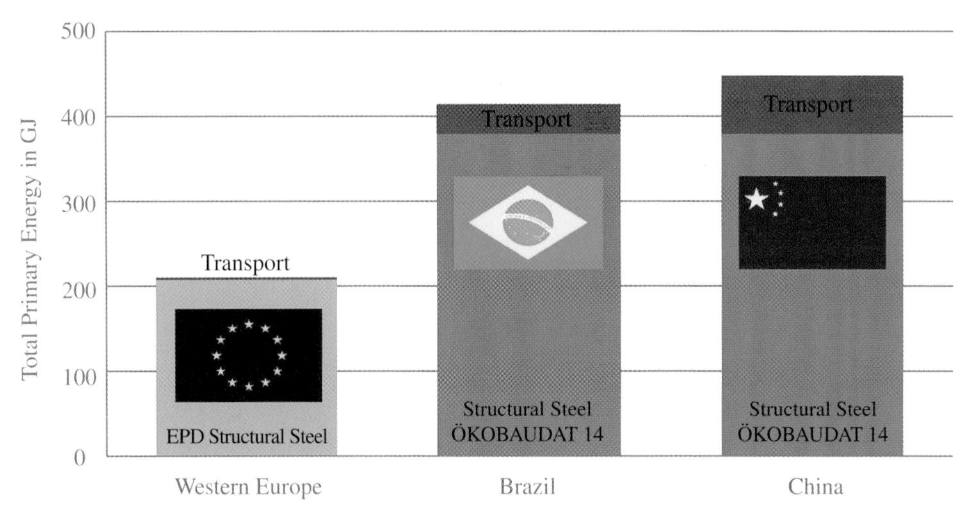

Figure 6.66 Total primary energy use according to the origin of structural steel. 16.8 t of S355 for a typical single-storey building.

with EPDs offers favourable environmental data and is readily available in Europe. When the additional environmental impacts are taken into consideration, the supposed economic benefit of imported steel from other regions is put into question. Structural steel from Europe in particular, which is recycled and returned to the industrial loop, is thus de facto also a local building material.

6.8.2 Comparison of expenses for transport and hoisting of large girders

The effects of different construction on the transport situation can be significant. Here the transportation of girders for a industrial single-storey building is considered. The building dimensions are 27 × 61 m with a girder span of 25 m. Two different girders are compared: precast prestressed concrete (Figure 6.67) and structural steel cellular beams (Figure 6.68). The design of the prestressed concrete beam is taken from a collection of examples for Eurocode 2 [50], [51]. The structural steel cellular beam is designed using the ACB+ program [52].

The masses of the considered components are compared in Figure 6.69 (see also [53]). For the nine cellular beams made of structural steel, only two heavy-duty transports are required with a payload up to 25 tonnes. An example for that kind of transportation is shown in Figure 6.71 The precast prestressed concrete girders have to be moved in nine individual heavy-duty transports up to 25 tonnes (see Figure 6.70). Heavy-duty lorries with higher payloads and thus fewer trips would be theoretically feasible, but they are not common and hence uneconomical due to the low availability of such special vehicles in Europe.

With the transport of members from the workshop to the construction site, additional environmental burden occurs. For the structural steel beams a

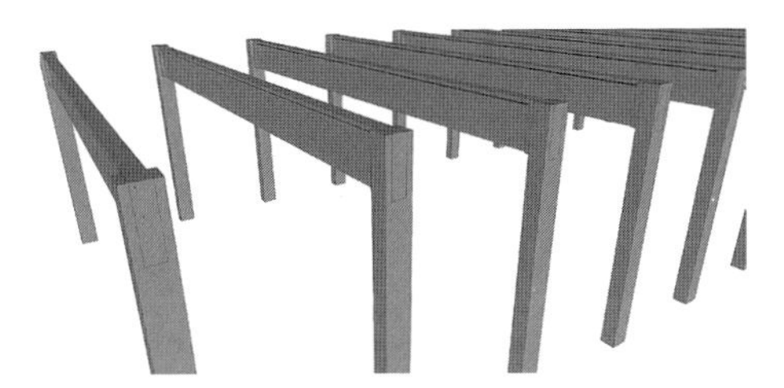

Figure 6.67 Typical single-storey building with prestressed concrete griders.

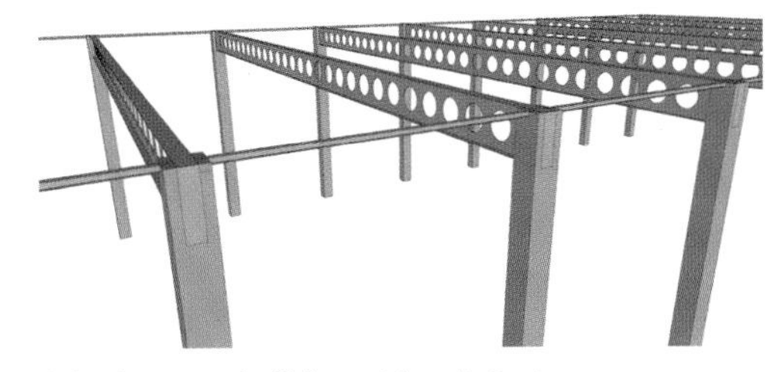

Figure 6.68 Typical single-storey building with cellular beams.

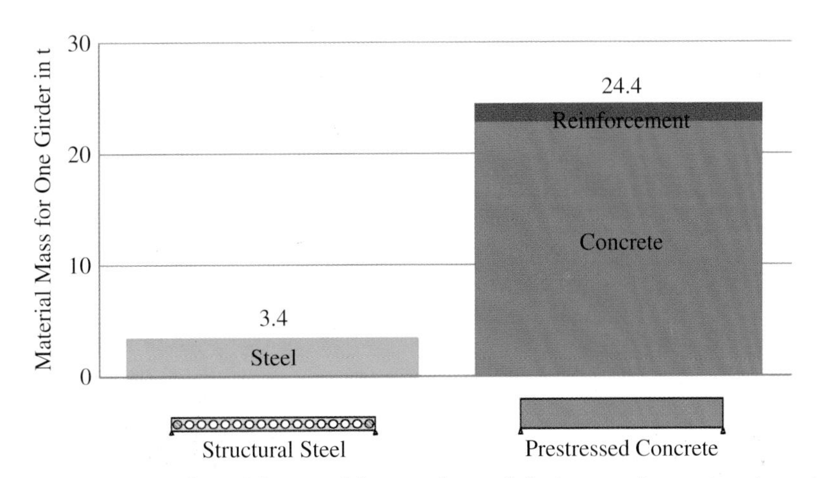

Figure 6.69 Comparison of the material masses for a cellular beam and precast prestressed concrete girder

transport distance of 500 km is assumed, and for the precast prestressed concrete girders a transport distance of 100 km. For the transport of 1 tonne of material over a distance of 1 km (=1 tkm), the relevant environmental data can be taken from national databases such as the ÖKOBAUDAT [1]; see Table 6.21. As an example, the nonrenewable primary energy for the transport is shown in Figure 6.72. Even with the assumed five times longer distance, less pollution by the transport of steel components is caused.

Considering the installation on site, the cost of the required lifting equipment also plays an important role. For the assembly of four cellular beams from one position a 60-tonne mobile crane can be used, whereas a 200-tonne mobile crane must hoist the prestressed concrete girders and just three beams can be fitted from

9 × Cellular Deams

2 × Heavy-duty Transport
4 or 5 Girders on One Lorry
3 -Axle Extendable Trailer
Payload up to 25t

9 × Precast Prestressed
Concrete Girders

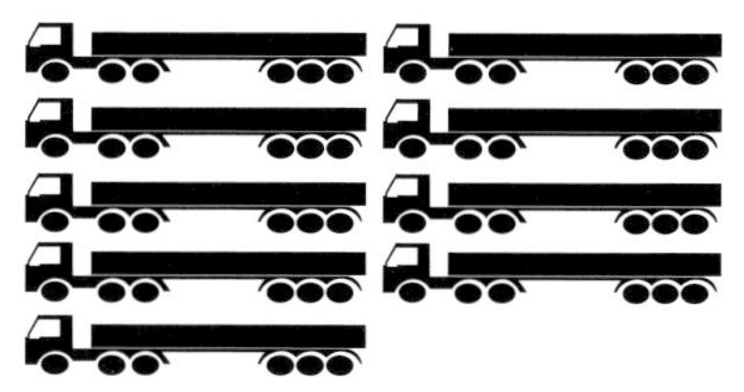

9 × Heavy-duty Transport
1 × Girder on One Lorry
3 × Axle Dolly
Payload up to 25t

Figure 6.70 Influence of the construction on the transportation to the construction site.

Figure 6.71 Transportation of several cellular beams – delivery length 29.8 m. © ArcelorMittal.

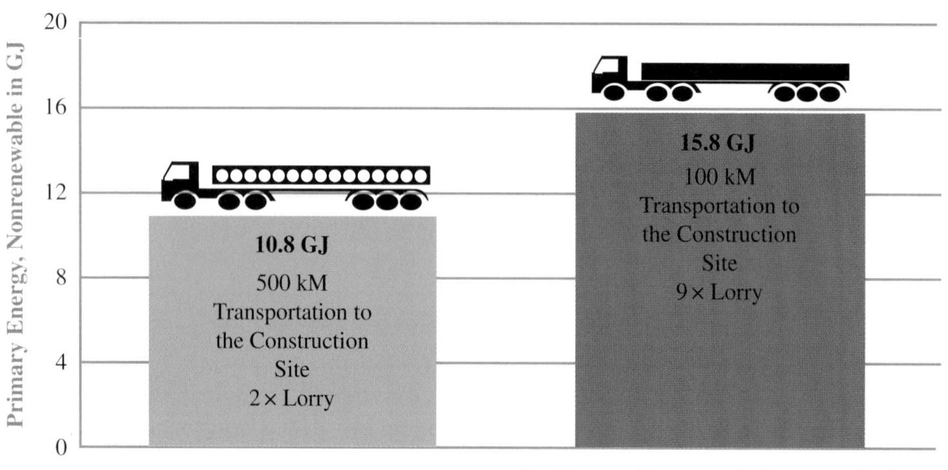

Figure 6.72 Primary energy, nonrenewable for the transportation of the nine required girders.

Table 6.22 Comparison of the installation expenses for cellular beams and prestressed concrete girders.

	Cellular beam	Prestressed concrete girder
Mobil crane	60 tonnes	200 tonnes
Costs including heavy-duty insurance	~75€/h	~250€/h
Costs for travel + assembly and disassembly	~200€	~2.000€

one position. In addition, the 200-tonne mobile crane needs longer set-up times in between the changes of positions. Alternatively, another type of mobile crane could be used, whereby the costs would be even higher. For the installation of both type of beams, the work of technicians and the costs for the provision of tools, working platforms and so forth can be considered the same. Table 6.22 shows the summary of costs for the installation of the two systems.

REFERENCES

[1] Bundesministeriums für Umwelt, Naturschutz, Bau und Reaktorsicherheit (BMUB). (2015) ÖKOBAUDAT 2015. Available at www.oekobaudat.de. Accessed 22 October 2015.

[2] EPD-IZB-2013411. (2013) *Environmental Product Declaration – Concrete C 20/25*. Berlin: Institute for Construction and Environment.

[3] EPD-IZB-2013431. (2013) *Environmental Product Declaration – Concrete C 30/37*. Berlin: Institute for Construction and Environment.

[4] EPD-BFS-20130094. (2013) *Environmental Product Declaration: Structural Steel: Open Rolled Profiles and Heavy Plates*. Berlin: Institute for Construction and Environment.

[5] EN 15804. (2011) *Sustainability of Construction Works – Environmental Product Declarations – Core Rules for the Product Category of Construction Products*. Brussels: European Committee for Standardization.

[6] EN 15978. (2011) *Sustainability of Construction Works – Assessment of Environmental Performance of Buildings – Calculation Method*. Brussels: European Committee for Standardization.

[7] Kuhnhenne, M., Döring, B. Pyschny, D. (2010) Ökobilanzierung von Typenhallen, RWTH Aachen, Aachen

[8] Siebers, R., Hubauer, A., Lange, J., Hauke, B. (2012) Eco efficiency of structural frames for low rise office buildings, In: *ECCS/Concepts and Methods for Steel Intensive Buildings*. Munich

[9] Mensinger, M., Stroetmann, R., Eisele, J., Feldmann, M., Lingnau, V., Zink, J., et al. (2015) Nachhaltigkeit von Stahl- und Verbundkonstruktionen bei Büro- und Verwaltungs-gebäuden, Abschlussbericht: AiF-Vorhaben-Nr. 373 ZBG, Düsseldorf.
[10] Stroetmann, R., Podgorski, C. (2014) Zur Nachhaltigkeit von Stahl- und Verbundkonstruktionen bei Büro- und Verwaltungsgebäuden – Tragkonstruktion en Teil 1. Stahlbau 83(4), pp. 245–256.
[11] Stroetmann R., et al. (2014) Ganzheitliche Planung nachhaltiger Bürogebäude in Stahl- und Stahlverbundbauweise. Stahlbau 83(7).
[12] EPD-IFBS-2013211. (2013) *Environmental Product Declaration – Trapezoidal Steel Profile*. Berlin: Institute for Construction and Environment.
[13] EPD Gypsum Products. (2009) *Environmental Product Declaration – Gypsum Fire Protection Boards*. Available at www.gips.de. Accessed 30 October 2015.
[14] M-EPD-SFA-000003. (2013) *Environmental Product Declaration – Steel/Stainless Steel Facades*. Rosenheim: ift Rosenheim GmbH.
[15] EPD-DRW-2012131. (2012) *Environmental Product Declaration – Mineral Wool High Density*. Berlin: Institute for Construction and Environment.
[16] EPD-FPX-2010111. (2010) *Environmental Product Declaration – XPS, Extruded Polystyrene Foam*. Berlin: Institute for Construction and Environment.
[17] DIN V 18599. (2011) Energetische Bewertung von Gebäuden – Berechnung des Nutz-, End- und Primärenergiebedarfs für Heizung, Kühlung, Lüftung, Trinkwarmwasser und Beleuchtung – Deutsches Institut für Normung e.V. – Berlin
[18] Stroetmann, R. (2011) High strength steel for improvement of sustainability. In: EUROSTEEL 2011, 6th European Conference on Steel and Composite Structures, Proceedings Volume C, pp. 1959–1964. Budapest.
[19] Stroetmann, R. (2011) Sustainable design of steel structures with high-strength steels. Dillinger Colloquium on Constructional Steelwork on 1 December 2011, Dillingen.
[20] Stroetmann, R., Deepe, P., Tröger, A. (2012) Tragwerke aus höherfesten Stählen – Planung, Berechnung und Ausführung, Düsseldorf: bauforumstahl e.V.
[21] Stroetmann, R., Deepe, P., Rasche, Chr., Kuhlmann, U. (2012) Bemessung von Tragwerken aus höherfesten Stählen bis S700 nach EN 1993-1-12. Stahlbau 81(4), pp. 332–342.
[22] Stroetmann, R., Podgorski, C. (2014) Zur Nachhaltigkeit von Stahl- und Verbundkonstruktionen bei Büro- und Verwaltungsgebäuden – Tragkonstruktion en Teil 2. Stahlbau 83(9), pp. 599–607.
[23] Lindner, J. (1987) Stabilisierung von Trägern durch Trapezbleche. Stahlbau 56(1), pp. 9–15.
[24] Friemann, H., Stroetmann, R. (1998) Zum Nachweis ausgesteifter biegedrillknick-gefährdeter Träger. Stahlbau 67(12), pp. 936–955.
[25] EN 1993-1-1. (2010) *Eurocode 3: Design of Steel Structures – Part 1-1: General Rules and Rules for Buildings*. Brussels: European Committee for Standardization.
[26] Ritter, F., Huang, L. (2012) An early design tool for sustainable steel and steel composite structures under the use of a genetic algorithm. Proceedings des 25 Forum Bauinformatik.
[27] SketchUp software. Available at http://www.sketchup.com/. Accessed 23 October 2015.
[28] SOD, Sustainable Office Designer plugin for SketchUp. Available at http://www.metallbau.bgu.tum.de/forschung/ab-gebiete/190-sod. Accessed 21 October 2015.
[29] Goldberg, D.E. (1989) *Genetic Algorithms in Search, Optimization and Machine Learning*. Reading, MA: Addison-Wesley.
[30] EN 1994-1-1. (2010) *Eurocode 4: Design of Composite Steel and Concrete Structures. Part 1-1: General Rules and Rules for Buildings*. Brussels: European Committee for Standardization.

[31] German Sustainable Building Council - DGNB e.V. (2012) *Neubau Büro- und Verwaltungsgebäude – DGNB Handbuch für nachhaltiges Bauen Version 2012*. Stuttgart.

[32] Mensinger, M., et al. (2011) Nachhaltige Bürogebäude mit Stahl. Stahlbau 80(10), pp. 740–749.

[33] CEEQUAL. (2010) *CEEQUAL: Scheme Description and Assessment Process Handbook, Version 4.1 for Projects*. Surrey.

[34] Fernández-Sánchez, G., Rodríguez-López, F. (2010) A methodology to identify sustainability indicators in construction project management – application to infrastructure projects in Spain. Ecological Indicators 10, pp. 1193–1201.

[35] Fogib. (1997) *Ingenieurbauten - Wege zu einer ganzheitlichen Bewertung, Band 1–3*. Universität Stuttgart. Final report of the DFG research group civil engineering structures. Stuttgart.

[36] Ugwu, O., Haupt, T. (2007) Key performance indicators and assessment methods for infrastructure sustainability – a South African construction industry perspective. Building and Environment 42, pp. 665–680.

[37] Maibach, M., et al. (2008) *Handbook on Estimation of External Costs in the Transport Sector. Internalisation Measures and Policies for All External Cost of Transport (IMPACT)*. CE Delft - Solutions for environment, economy and technology. Publication number: 07.4288.52. Delft.

[38] Zinke, T. (2016) Nachhaltigkeit von Infrastrukturbauwerken – Ganzheitliche Bewertung von Autobahnbrücken unter besonderer Berücksichtigung externer Effekte. Karlsruhe: KIT Scientific Publishing, Dissertation.

[39] BMVBS. (2007) *Ri-Wi-Brü: Richtlinie zur Durchführung von Wirtschaftlichkeitsuntersuchungen im Rahmen von Instandsetzungs-/Erneuerungsmaßnahmen bei Straßenbrücken*. Teil der Richtlinien für die Erhaltung von Ingenieurbauwerken. Ausgabe 2007. Berlin: Bundesministerium für Verkehr, Bau- und Stadtentwicklung.

[40] Zinke, T., Ummenhofer, T., Mensinger, M., Pfaffinger, M. (2013) The role of traffic in sustainability assessment of road bridges. In: Petzek, E., Bancila, R., eds., *The Eighth International Conference 'Bridges in Danube Basin'*. New Trends in Bridge Engineering and Efficient Solutions for Large and Medium Span Bridges, Proceedings, Timisoara, Romania, 4–5 October 2013. pp. 255–266. Heidelberg: Springer.

[41] Ressel, W. (1994) *Untersuchungen zum Verkehrsablauf im Bereich der Leistungsfähigkeit an Baustellen auf Autobahnen*. München-Neubiberg: Univ. d. Bundeswehr.

[42] Zinke, T., Schmidt-Thrö, G., Ummenhofer, T. (2012) Entwicklung und Verwendung von externen Kosten für die Nachhaltigkeitsbewertung von Verkehrsinfrastruktur. Beton- und Stahlbetonbau 107, pp. 524–532.

[43] EN ISO 14040. (2009) *Environmental Management – LCA – Principles and Framework*. Brussels: European Committee for Standardization.

[44] NaBrü. (2014) *Ganzheitliche Bewertung von Stahl- und Verbundbrücken nach Kriterien der Nachhaltigkeit*. Kuhlmann, U., Maier, P., Zinke, T., Ummenhofer,, T., Schneider, S., Beck, T., et al., eds. Research project no. IGF 353 ZN. Stuttgart.

[45] Yao, H., Shen, L., Tan, Y., Hao, J. (2011) Simulating the impacts of policy scenarios on the sustainability performance of infrastructure projects. Automation in Construction, pp. 1060–1069. doi:10.1016/j.autcon.2011.04.007.

[46] EN 15643-4. (2012). *Sustainability of Construction Works – Assessment of Building – Part 4: Framework for the Assessment of Economic Performance*. Brussels: European Committee for Standardization.

[47] Wagner, H.-J., et al. (2010). LCA of the offshore wind farm alpha ventus. Münster: LIT-press (in German).

[48] Schaumann, P., Böker, C., Bechtel, A., Lochte-Holtgreven, S. (2011). *Support Structures of Wind Energy Converters*. CISM Course and Lectures, Vol. 531 Environmental Wind

Engineering and Design of Wind Energy Structures, pp. 191–253. Udine, Italy: Springer.

[49] Schaumann, P., et al. (2014) *Final Report – Sustainability of Steel Structures of Renewable Power Plants (NaStafEE)*. Düsseldorf: FOSTA.

[50] EN 1992-1-1. (2011) *Eurocode 2: Design of Concrete Structures – General Rules and Rules for Buildings*. Brussels: European Committee for Standardization.

[51] Deutscher Beton- und Bautechnik Verein e.V. (2011) Beispiele zur Bemessung nach Eurocode 2 Band 1: Hochbau, Beispiel 8 Spannbetonbinder. Berlin: Ernst.

[52] ACB+ v3.07 Software ArcelorMittal Lochstegträger mit kreisförmigen Öffnungen. Available at www.sections.arcelormittal.com/de/downloadcenter/entwurfssoftware/lochstegtraeger.html. Accessed 28 October 2015.

[53] Hauke, B., Siebers, R., May, M., Girkes, H., Olitzscher, F., Weigel, W. (2015) Lochstegträger als wirtschaftliche Dachbinder, Bauingenieur Jahresausgabe 2015/2016, pp. 114–119. Düsseldorf: Springer.

Index

Sustainable Steel Buildings: A Practical Guide for Structures and Envelopes, First Edition.
Edited by Bernhard Hauke, Markus Kuhnhenne, Mark Lawson and Milan Veljkovic.